Food Science Text Series

The **Food Science Text Series** provides faculty with the leading teaching tools. The Editorial Board has outlined the most appropriate and complete content for each food science course in a typical food science program and has identified textbooks of the highest quality, written by the leading food science educators.

B. Pam Ismail • S. Suzanne Nielsen

Editors

Nielsen's
Food Analysis
Laboratory Manual

Fourth Edition

 Springer

Editors
B. Pam Ismail
Food Science and Nutrition
University of Minnesota
St. Paul, MN, USA

S. Suzanne Nielsen
Department of Food Science
Purdue University
West Lafayette, IN, USA

ISSN 1572-0330   ISSN 2214-7799   (electronic)
Food Science Text Series
ISBN 978-3-031-44969-7   ISBN 978-3-031-44970-3   (eBook)
https://doi.org/10.1007/978-3-031-44970-3

For this fourth edition of the Food Analysis Laboratory Manual, Springer has added "Nielsen's" at the beginning of the title. This is meant to recognize Dr. Suzanne Nielsen's efforts to edit this book over the past 20 years, starting with publication of the first edition in 2003. The fourth edition of the laboratory manual will be her last. We are grateful for her years of dedication and tenacity which made the textbook and lab manual the gold standard. Most importantly, we admire and applaud her dedication to students. Tentative plans are in place for editors of the future edition of the laboratory manual. Thank you, Dr. Nielsen.

Preface and Acknowledgments

This laboratory manual, 4th edition, was written to accompany the textbook, *Nielsen's Food Analysis*, 6th edition. The laboratory exercises are tied closely to 24 of the 35 chapters in the textbook. Compared to the 3rd edition of this laboratory manual, the 4th edition contains three new laboratory exercises: "Food Forensics" and "Mid Infrared Spectroscopy," which are linked to the textbook chapters on these topics, and "Preparation of Solutions and Buffers," which is linked to the existing chapter in the laboratory manual with background information on reagents and buffers (Chap. 2). In addition to these three new chapters, the laboratory exercises on rheological measurements and color analysis have been greatly revised. The chapters related to mineral analysis (Chap. 11) and proximate analysis (moisture, ash, fat, and protein, to then calculate total carbohydrate) (Chaps. 13, 14, 15, and 16) have been modified so the same food products are suggested for these analyses. Other chapters have been updated and corrected as appropriate.

As with previous editions of this laboratory manual, instructors using these laboratory exercises should note the following:

1. Use of Introductory Chapters: The four chapters in this section of the book include basic information that compliments both the textbook chapters and the laboratory exercises. These chapters are recommended for students prior to starting food analysis laboratory exercises. The introductory chapters can provide the background to help students make chemical solutions, do the relevant calculations in the laboratory exercises, and do the data analysis. Three of the introductory chapters include example problems and their solutions within the chapters, plus additional practice problems (with answers at the end of the laboratory manual).

2. Content of Laboratory Exercises: Most of the laboratory exercises include the following sections: background, reading assignment, objective, principle of method, chemicals (with CAS number and hazards), reagents, precautions and waste disposal, supplies, equipment, procedure, data and calculations, questions, and resource materials.

3. Order of Laboratory Exercises: The order of laboratory exercises is fairly consistent with the order of chapters in the textbook, *Nielsen's Food Analysis*, 6th edition. However, each laboratory exercise stands alone, so they can be covered in any order.

4. Customizing Laboratory Procedures: It is recognized that the time and equipment available for teaching food analysis laboratory sessions vary considerably between schools, as do student numbers and their level in school. Therefore, instructors may need to modify the laboratory procedures to fit their needs and situation (e.g., number of samples analyzed, replicates). Some experiments include numerous parts/methods, and it is not assumed that an instructor uses all parts of the experiment as written. It may be logical to have students work in pairs or a group of 3–5 (depending on the complexity and steps of the experiment) to make things go faster. Also, it may be logical to have some students do one part of the experiment/one type of sample and other students do another part of the experiment/type of sample.

5. Use of Chemicals: The information on hazards and precautions in use of the chemicals for each experiment is not comprehensive but should make students and

laboratory assistants aware of major concerns in handling and disposing of the chemicals.

6. Reagent Preparation: It is recommended in the text of the experiments that a laboratory assistant prepare many of the reagents, because of the time limitations for students in a laboratory session. The lists of supplies and equipment for experiments do not necessarily include those needed by the laboratory assistant to prepare reagents for the laboratory session. Note that Chap. 2 ("Reagents and Buffers") and Chap. 7 ("Preparation of Solutions and Buffers") will be helpful to students and laboratory assistants in preparing solutions needed for various laboratory exercises.

7. Data and Calculations: The laboratory exercises provide details on recording data and doing calculations. In requesting laboratory reports from students, instructors will need to specify if they require just sample calculations or all calculations.

Even though this is the 4th edition of this laboratory manual, there are sure to be inadvertent omissions and mistakes. We will very much appreciate receiving suggestions for revisions from instructors, including input from lab assistants and students.

We maintain a website with additional teaching materials related to both *Nielsen's Food Analysis* textbook and laboratory manual. Instructors are welcome to contact Dr. B. Ismail for access to this website. To compliment the laboratory manual, the website contains more detailed versions of select introductory chapters and Excel sheets related to numerous laboratory exercises.

Both Drs. Nielsen and Ismail extend special thanks to the persons who developed the new laboratory experiments for this edition, and to the authors of chapters with significant updating and expansion. This will be the last edition of S. Nielsen's involvement in the production of this laboratory manual. She thanks all current and past chapter authors for their contributions, and her past graduate students for their help with initial development of the first edition of the book. Lastly, she thanks her former post doc, Dr. B. Ismail, for co-editing this edition with her, and for all the input she provided on previous editions. Likewise, B. Ismail thanks Dr. S. Nielsen for the opportunity to work on this edition with her; it has been quite an enriching journey.

St. Paul, MN, USA   B. Pam Ismail
West Lafayette,   S. Suzanne Nielsen
IN, USA

Contents

Contributors

Young-Hee Cho Department of Food Science, Purdue University, West Lafayette, IN, USA

Jinping Dong Core Research & Development, Cargill Research and Development Center, Cargill, Inc., Plymouth, MN, USA

M. Monica Giusti Department of Food Science and Technology, The Ohio State University, Columbus, OH, USA

B. Pam Ismail Department of Food Science and Nutrition, University of Minnesota, St. Paul, MN, USA

Helen S. Joyner Premier Nutrition, Emeryville, CA, USA

Gonzalo Miyagusuku-Cruzado Department of Food Science and Technology, The Ohio State University, Columbus, OH, USA

Andrew P. Neilson Department of Food, Bioprocessing and Nutrition Sciences, North Carolina State University, Raleigh, NC, USA

S. Suzanne Nielsen Department of Food Science, Purdue University, West Lafayette, IN, USA

Shreya Madhav Nuguri Department of Food Science and Technology, The Ohio State University, Columbus, OH, USA

Qinchun Rao Department of Health, Nutrition, and Food Sciences, Florida State University, Tallahassee, FL, USA

Luis Rodriguez-Saona Department of Food Science and Technology, The Ohio State University, Columbus, OH, USA

Ann M. Roland Owl Software, Columbia, MO, USA

Denise M. Smith School of Food Science, Washington State University, Pullman, WA, USA

Catrin Tyl Department of Chemistry, Biotechnology and Food Science, Norwegian University of Life Sciences, As, Norway

Danielle M. Voss Department of Food Science and Technology, The Ohio State University, Columbus, OH, USA

Robert E. Ward Department of Nutrition, Dietetics and Food Sciences, Utah State University, Logan, UT, USA

Siyu Yao Department of Nutrition and Food Hygiene, School of Public Health, Southeast University, Nanjing, China

Laboratory Standard Operating Procedures

1

Andrew P. Neilson

Contents

A. P. Neilson (✉)
Department of Food, Bioprocessing and Nutrition Sciences, North
Carolina State University, Raleigh, NC, USA
e-mail: aneilso@ncsu.edu

© The Author(s), under exclusive license to Springer Nature Switzerland AG 2024
B. P. Ismail, S. S. Nielsen (eds.), *Nielsen's Food Analysis Laboratory Manual*, Food Science Text Series,
https://doi.org/10.1007/978-3-031-44970-3_1

1.1 INTRODUCTION

This chapter is designed to cover essential "standard operating procedures" (SOPs), or best practices, for a general food analysis laboratory. The topics covered in this chapter include balances, mechanical pipettes, glassware, reagents, precision and accuracy, data handling, data reporting, and safety. These procedures apply to all the laboratory experiments in this manual, and therefore a thorough review of general procedures will be invaluable for successful completion of these laboratory exercises as well as development of sound laboratory skills.

This manual covers many of the basic skills and information that are necessary for one to be a good analytical food chemist. Much of this material is the type that one typically "picks up" from experience over time working in a laboratory. Nothing can replace actual lab experience as a learning tool, but hopefully this manual will help students learn proper lab techniques early rather than having to correct improper habits later. When one reads this manual, your reaction may be "is all of this attention to detail necessary?" Admittedly, the answer is "not always." This brings to mind an old Irish proverb that "the best person for a job is the one that knows what to ignore." There is much truth to this proverb, but a necessary corollary is that one must know what they are ignoring. The decision to use something other than the "best" technique must be conscious decision and not one made from ignorance. This decision must be based not only upon knowledge of the analytical method being used but also on how the resulting data will be used. Much of the information in this manual has been obtained from an excellent publication by the US Environmental Protection Agency entitled *Handbook for Analytical Quality Control in Water and Wastewater Laboratories*.

1.2 PRECISION AND ACCURACY

To understand many of the concepts in this chapter, rigorous definition of the terms "precision" and "accuracy" is required here. These terms are often used incorrectly, or are used interchangeably (also incorrect). Precision refers to the **reproducibility** of replicate observations, typically measured as **standard deviation** (SD), **standard error** (SE), or **coefficient of variation** (CV). Refer to Chap. 4 in this laboratory manual and Smith (2024) for a more complete discussion of precision and accuracy. The smaller these values are, the more reproducible or precise the measurement is. Precision is determined not on reference standards, but by the use of actual food samples, which cover a range of concentrations and a variety of interfering materials usually encountered by the analyst. Obviously, such data should not be collected until the analyst is familiar with the method and has obtained a reproducible standard curve (a mathematical relationship between the analyte concentration and the analytical response). There are a number of different methods available for the determination of precision. One method follows:

1. Three separate concentration levels should be studied, including a low concentration near the sensitivity level of the method, an intermediate concentration, and a concentration near the upper limit of application of the method.
2. Seven replicate determinations should be made at each of the concentrations tested.
3. To allow for changes in instrument conditions, the precision study should cover at least 2 hours of normal laboratory operation.
4. To permit the maximum interferences in sequential operation, it is suggested that the samples be run in the following order: high, low, and intermediate. This series is then repeated seven times to obtain the desired replication.
5. The precision statement should include a range of standard deviations over the tested range of concentration. Thus, three standard deviations will be obtained over a range of three concentrations.

Accuracy refers to the degree (absolute or relative) of difference between observed and "actual" or "true" values. The "actual" value is often difficult to ascertain or unknown. It may be the value obtained by a standard reference method (the accepted manner of performing a measurement). Another means of evaluating accuracy is by the addition of a known amount of the material being analyzed for the food sample and then calculation of **% recovery**. This latter approach entails the following steps:

1. Known amounts of the particular constituent are added to actual samples at concentrations for which the precision of the method is satisfactory. It is suggested that amounts be added to the low-concentration sample, sufficient to double that concentration, and that an amount be added to

the intermediate concentration, sufficient to bring the final concentration in the sample to approximately 75% of the upper limit of application of the method.

2. Seven replicate determinations at each concentration are made.
3. Accuracy is reported as the percent recovery at the final concentration of the spiked sample. Percent recovery at each concentration is the mean of the seven replicate results.

A fast, less rigorous means to evaluate precision and accuracy is to analyze a food sample and replicate a spiked food sample, and then calculate the recovery of the amount spiked. An example is shown in Table 1.1.

The accuracy can then be measured by calculating the % of the spike (0.75 g/L) detected by comparing the measured values from the unspiked and spiked samples:

$$\text{accuracy} \approx \% \text{ recovery} =$$
$$\frac{\text{measured spiked sample}}{\text{measured sample} + \text{amount of spike}} \times 100\% \quad (1.1)$$

$$\text{accuracy} \approx \% \text{ recovery} =$$
$$\frac{2.0955 \text{g} / \text{L}}{1.2955 \text{g} / \text{L} + 0.75 \text{g} / \text{L}} \times 100\% = 102.44\%$$

The method measured the spike to within 2.44%. By adding 0.75 g/L Ca to a sample that was measured to have 1.2955 g/L Ca, a perfectly accurate method would result in a spiked sample concentration of:

1.2955 g/L + 0.75 g/L = 2.0455 g/L.

The method actually measured the spiked sample at 2.0955 g/L, which is 2.44% greater than it should be. Therefore, the accuracy is estimated at ~2.44% relative error.

Table 1.1 Measured calcium content (g/L) of milk and spiked milk

Replicate	Milk	Milk + 0.75 g Ca/L
1	1.29	2.15
2	1.40	2.12
3	1.33	2.20
4	1.24	2.27
5	1.23	2.07
6	1.40	2.10
7	1.24	2.20
8	1.27	2.07
9	1.24	1.74
10	1.28	2.01
11	1.33	2.12
Mean	1.2955	2.0955
SD	0.062	0.138
%CV	4.8	6.6

1.3   BALANCES

1.3.1   Types of Balances

Balances quantify mass, or amount of matter (commonly referred to as "weight," although weight is technically a function of mass and the force of gravity). This is a key measurement for analytical chemistry, particularly quantitative methods. Two general types of balances are used in most laboratories. These are **top loading balances** and **analytical balances**. Top loading balances usually are sensitive to 0.1–0.001 g, depending on the specific model in use (this means that they can measure differences in the mass of a sample to within 0.1–0.001 g). In, general, as the **capacity** (largest mass that can be measured) increases, the sensitivity decreases. In other words, balances that can measure larger masses generally measure differences in those masses to fewer decimal places. Analytical balances are usually sensitive to 0.001–0.00001 g, depending on the specific model. It should be remembered, however, that **sensitivity** (ability to detect small differences in mass) is not necessarily equal to **accuracy** (the degree to which the balances correctly report the actual mass). The fact that a balance can be read to 0.01 mg does not necessarily mean it is accurate to 0.01 mg. What this means is that the balance can distinguish between masses that differ by 0.01 mg, but may not accurately measure those masses to within 0.01 mg of the actual masses (because the last digit is often rounded). The accuracy of a balance is independent of its sensitivity.

1.3.2   Choice of Balance

Which type of balance to use depends on "how much accuracy" is needed in a given measurement. One way to determine this is by calculating how much **relative (%) error** would be introduced by a given type of balance. For instance, if 0.1 g of a reagent was needed, weighing it on a top loading balance accurate to within only ±0.02 g of the actual mass would introduce approximately 20% error:

$$\% \text{ error in measured mass}$$
$$= \frac{\text{absolute error in measured mass}}{\text{measured mass}} \times 100\% \quad (1.2)$$

$$\% \text{ error in measured mass} = \frac{0.02 \text{g}}{0.1 \text{g}} \times 100\% = 20\%$$

This would clearly be unacceptable in most situations. Therefore, a more accurate balance would be needed. However, the same balance (with accuracy to within

±0.02 g) would probably be acceptable for weighing out 100 g of reagent, as the error would be approximately 0.02%:

$$\% \text{ error in measured mass} = \frac{0.02\text{g}}{100\text{g}} \times 100\% = 0.02\%$$

The decision on "how much accuracy" is needed can only be answered when one knows the function of the reagent in the analytical method. This is one reason why it is necessary to understand the principles involved in an analytical method, and not to simply approach an analytical method in a cookbook fashion. Therefore, a general guideline regarding which balance to use is hard to define.

Another situation in which care must be exercised in determining what type of balance to use is when a difference in masses is to be calculated. For instance, a dried crucible to be used in a total ash determination may weigh 20.05 g on a top loading balance, crucible plus sample = 25.05 g, and the ashed crucible 20.25 g. It may appear that the use of the top loading balance with its accuracy of ±0.02 g would introduce approximately 0.1% error, which would often be acceptable. Actually, since a difference in weight (0.20 g) is being determined, the error would be approximately 10% and thus unacceptable. In this case, an analytical balance is definitely required because sensitivity is required in addition to accuracy.

1.3.3 Use of Top Loading Balances

These instructions are generalized but apply to use of most models of top loading balances:

1. Level the balance using the bubble level and the adjustable feet (leveling is required so that the balance performs correctly).
2. Either *zero* the balance (so the balance reads 0 with nothing on the pan) or **tare** the balance so that the balance reads 0 with a container that will hold the sample (empty beaker, weighing boat, etc.) on the weighing pan. The tare function is conveniently used for "subtracting" the weight of the beaker or weighing boat into which the sample is added.
3. Weigh the sample.

1.3.4 Use of Analytical Balances

It is always wise to consult the specific instruction manual for an analytical balance before using it. Speed and accuracy are both dependent on one being familiar with the operation of an analytical balance. If it has been a while since you have used a specific type of analytical balance, it may be helpful to "practice" before actually weighing a sample by weighing a spatula or other convenient article. The following general rules apply to most analytical balances and should be followed to ensure that accurate results are obtained and that the balance is not damaged by improper use:

1. Analytical balances are expensive precision instruments; treat them as such.
2. Make sure that the balance is level and is on a sturdy table or bench free of vibrations.
3. Once these conditions are met, the same procedure specified above for top loading balances is used to weigh the sample on an analytical balance.
4. Always leave the balance clean.

1.3.5 Additional Information

Other points to be aware of regarding the use of balances are the following:

1. Many analyses (moisture, ash, etc.) require weighing of the final dried or ashed sample with the vessel in which the sample was dried or ashed. The mass of the vessel must be known so that it can be subtracted from the final mass to get the mass of the dried sample or ash. Therefore, make sure to obtain the mass of the vessel before the analysis. This can be done by either weighing the vessel before taring the balance and then adding the sample or obtaining the mass of the vessel and then the mass of the vessel plus the sample.
2. The accumulation of moisture from the air or fingerprints on the surface of a vessel will add a small mass to the sample. This can introduce errors in mass that affect analytical results, particularly when using analytical balances. Therefore, beakers, weigh boats, and other weighing vessels should be handled with tongs or with gloved hands. For precise measurements (moisture, ash, and other measurements), weighing vessels should be pre-dried and stored in a desiccator before use, and then stored in a desiccator after drying, ashing, etc., prior to weighing the cooled sample.
3. Air currents or leaning on the bench can cause appreciable error in analytical balances. It is best to take the reading after closing the side doors of an analytical balance.
4. Most balances in modern laboratories are electric balances. Older lever-type balances are no longer in wide use, but they are extremely reliable.

1.4 MECHANICAL PIPETTES

Mechanical pipettes (i.e., **automatic pipettors**) are standard equipment in most analytical laboratories. This is due to their convenience, precision, and acceptable accuracy *when used properly, and when calibrated*. Although these pipettes are often viewed as being easier to use than conventional glass volumetric pipettes, this does not mean that the necessary accuracy and precision can be obtained without attention to proper pipetting technique. Just the opposite is the case; if mechanical pipettes are used incorrectly, this will usually cause greater error than the misuse of glass volumetric pipettes. Improper pipetting technique is widespread and contributes significant error to analytical measurements. The proper use of glass volumetric pipettes is discussed in the section on glassware. The *pipetman* mechanical pipette (Gilson Inc.) is an example of a continuously adjustable design. The proper use of this type of pipette, as recommended by the manufacturer, will be described here. Other brands of mechanical pipettes are available, and although their specific instructions should be followed, their proper operation is usually very similar to that described here.

1.4.1 Operation

1. Set the desired volume on the digital micrometer/volumeter. For improved precision, always approach the desired volume by dialing downward from a larger volume setting. Make sure not to wind it up beyond its maximum capacity; this will break it beyond repair.
2. Attach a disposable tip to the shaft of the pipette and press on firmly with a slight twisting motion to ensure a positive, airtight seal.
3. Depress the **plunger** to the **first positive stop**. This part of the stroke is the calibrated volume displayed. Going past the first positive stop will cause inaccurate measurement.
4. Holding the mechanical pipette vertically, immerse the disposable tip into sample liquid to a depth indicated (Table 1.2), specific to the maximum volume of the pipette (P-20, 100, 200, 500, 1000, and 5000, which correspond to maximum volumes of 20, 100, 200, 500, 1000, and 5000 µL, respectively).

Table 1.2 Appropriate pipette depth for automatic pipettors

Pipette	Depth (mm)
P-20D, P-100D, P-200D	1–2
P-500D, P-1000D	2–4
P-5000D	3–6

Table 1.3 Appropriate dispense wait time for automatic pipettors

Pipette	Time (s)
P-20D, P-100D, P-200D	1
P-500D, P-1000D	1–2
P-5000D	2–3

5. Allow the plunger to *slowly* return to the "up" position. *Never permit it to snap up* (this will suck liquid up into the pipette mechanism, causing inaccurate measurement and damaging the pipette).
6. Wait 1–2 s to ensure that full volume of sample is drawn into tip. If the solution is viscous such as glycerol, you need to allow more time.
7. Withdraw tip from sample liquid. Should any liquid remain on outside of tip, wipe carefully with a lint-free cloth, taking care not to touch the tip opening.
8. To dispense sample, place tip end against side wall of vessel and depress plunger *slowly* past the first stop until the **second stop** (fully depressed position) is reached.
9. Wait (Table 1.3).
10. With plunger fully depressed, withdraw mechanical pipette from the vessel carefully with tip sliding along wall of vessel.
11. Allow plunger to return to top position.
12. Discard tip by depressing **tip-ejector button** smartly.
13. A fresh tip should be used for the next measurement if:
    (a) A different solution or volume is to be pipetted.
    (b) A significant residue exists in the tip (not to be confused with the visible "film" left by some viscous or organic solutions).

1.4.2 Pre-Rinsing

Pipetting very viscous solutions or organic solvents will result in a significant film being retained on the inside wall of the tip. This will result in an error that will be larger than the tolerance specified if the tip is only filled once. Since this film remains relatively constant in successive pipettings with the same tip, accuracy may be improved by filling the tip, dispensing the volume into a waste container, refilling the tip a second time, and using this quantity as the sample. This procedure is recommended in all pipetting operations when critical reproducibility is required, whether or not tips are reused (same solution) or changed (different solutions/different volumes). Note that the "nonwettability" of the polypropylene tip is not absolute and that pre-rinsing will improve the precision and accuracy when pipetting any solution.

1.4.3 Pipetting Solutions of Varying Density or Viscosity

Compensation for solutions of varying viscosity or density is possible with any adjustable pipette by setting the digital micrometer slightly higher or lower than the required volume. The amount of compensation is determined empirically by comparing the amount actually dispensed to the amount selected (usually using a balance and the density of the liquid). Also, when dispensing viscous liquids, it will help to *wait* 1 s longer at the first stop before depressing to the second stop.

1.4.4 Performance Specifications

The manufacturer of *pipetman* mechanical pipettes provides the information in Table 1.4, on the precision and accuracy of their mechanical pipettes.

1.4.5 Selecting the Correct Pipette

Although automatic pipettes can dispense a wide range of volumes, you may often have to choose the "best" pipette with the most accuracy/precision from among several choices. For example, a P5000 (i.e., 5 mL) automatic pipettor could theoretically pipette anywhere between 0 and 5 mL. However, there are several limitations that dictate which pipettes to use. The first is a practical limitation: mechanical pipettes are limited by the graduations (the increments) of the pipette. The P5000 and P1000 are typically adjustable in increments of 0.01 mL (10 μL). Therefore, these pipettes cannot dispense volumes of <10 μL, nor can they dispense volumes with more precision that 10 μL (such as 4007 μL, or 4.007 mL). However, just because these pipettes can technically be adjusted to 10 μL does not mean

Table 1.4 Accuracy and precision of PIPETMAN mechanical pipettes

Model	Accuracy[a]	Reproducibility[a] (standard deviation)
P-2OD	<0.1 μL @ 1–10 μL	<0.04 μL @ 2 μL
	<1% @ 10–20 μL	<0.05 μL @ 10 μL
P-200D	<0.5 μL @ 20–60 μL	<0.15 μL @ 25 μL
	<0.8% @ 60–200 μL	<0.25 μL @ 100 μL
		<0.3 μL @ 200 μL
P-1000D	<3 μL @ 100–375 μL	<0.6 μL @ 250 μL
	<0.8% @ 375–1000 μL	<1.0 μL @ 500 μL
		<1.3 μL @ 1000 μL
P-5000D	<12 μL @ 0.5–2 mL	<3 μL @ 1.0 mL
	<0.6% @ 2.0–5.0 mL	<5 μL @ 2.5 mL
		<8 μL @ 5.0 mL

[a]Aqueous solutions, tips prerinsed once

Table 1.5 Recommended volume ranges for mechanical pipettors

Maximum volume	Lowest recommended volume
5 mL (5000 μL)	1 mL (1000 μL)
1 mL (1000 μL)	0.1–0.2 mL (100–200 μL)
0.2 mL (200 μL)	0.02–0.04 mL (20–40 μL)
0.1 mL (100 μL)	0.01–0.02 mL (10–20 μL)
0.05 mL (50 μL)	0.005–0.01 mL (5–10 μL)
0.02 mL (20 μL)	0.002–0.004 mL (2–4 μL)
0.01 mL (10 μL)	0.001–0.002 mL (1–2 μL)

that they should be used to measure volumes anywhere near this small. Most pipettes are labeled with a working range that lists the minimum and maximum volume, but this is not the range for ideal performance. Mechanical pipettes should be operated from 100% down to 10–20% of their maximum capacity (Table 1.5). Below 10–20% of their maximum capacity, performance (accuracy and precision) suffers. A good way of thinking of this is to use the largest pipette capable of dispensing the volume in a single aliquot.

Mechanical pipettes are invaluable pieces of laboratory equipment. If properly treated and maintained, they can last for decades. However, improper use can destroy them in seconds. Mechanical pipettes should be calibrated, lubricated, and maintained at least yearly by a knowledgeable pipette technician. Weighing dispensed water repeatedly to assess both accuracy and precision is a good check to see if the pipette needs calibration.

1.5 GLASSWARE

1.5.1 Types of Glassware/Plasticware

Glass is the most widely used material for construction of laboratory vessels. There are many grades and types of glassware to choose from, ranging from student grade to others possessing specific properties such as resistance to thermal shock or alkali, low boron content, and super strength. The most common type is a highly resistant borosilicate glass, such as that manufactured by Corning Glass Works under the name "Pyrex," or by Kimble Glass Co. as "Kimax." Brown/amber actinic glassware is available, which blocks UV and IR light to protect light-sensitive solutions and samples. The use of vessels, containers, and other apparatus made of Teflon, polyethylene, polystyrene, and polypropylene is common. Teflon stopcock plugs have practically replaced glass plugs in burets, separatory funnels, etc., because lubrication to avoid sticking (called "freezing") is not required. Polypropylene, a methylpentene polymer, is available as laboratory bottles, graduated cylinders, beakers, and even volumetric flasks. It is crystal clear, shatter-proof, autoclavable, chemically resistant, but relatively expensive as compared to glass. Teflon (polytetrafluoroethylene, PTFE) vessels are available, although they are very expensive. Finally, most glassware has a polar surface. Glassware can be treated to derivatize the surface (typically, tetramethylsilane, or *tms*) to make it nonpolar, which is required for some assays. However, acid washing will remove this nonpolar layer.

1.5.2 Choosing Glassware/Plasticware

Some points to consider in choosing glassware and/or plasticware are the following:

1. Generally, special types of glass are not required to perform most analyses.
2. Reagents and standard solutions should be stored in borosilicate or polyethylene bottles.
3. Certain dilute metal solutions may plate out on glass container walls over long periods of storage. Thus, dilute metal standard solutions should be prepared fresh at the time of analysis.
4. Strong mineral acids (such as sulfuric acid) and organic solvents will readily attack polyethylene; these are best stored in glass or a resistant plastic.
5. Borosilicate glassware is not completely inert, particularly to alkalis; therefore, standard solutions of silica, boron, and the alkali metals (such as NaOH) are usually stored in polyethylene bottles.
6. Certain solvents dissolve some plastics, including plastics used for pipette tips, serological pipettes, etc. This is especially true for acetone and chloroform. When using solvents, check the compatibility with the plastics you are using. Plastics dissolved in solvents can cause various problems, including binding/precipitating the analyte of interest, interfering with the assay, clogging instruments, etc.
7. Ground-glass stoppers require care. Avoid using bases with any ground glass because the base can cause them to "freeze" (i.e., get stuck). Glassware with ground-glass connections (burets, volumetric flasks, separatory funnels, etc.) are very expensive and should be handled with extreme care.

For additional information, the reader is referred to the catalogs of the various glass and plastic manufacturers. These catalogs contain a wealth of information as to specific properties, uses, sizes, etc.

1.5.3 Volumetric Glassware

Accurately calibrated glassware for accurate and precise measurements of volume has become known as **volumetric glassware**. This group includes **volumetric flasks**, **volumetric pipettes**, and accurately **calibrated burets**. Proper use of volumetric glassware is essential for quality analytical work, particularly quantitative analysis. Less accurate types of glassware, including **graduated cylinders**, **serological pipettes**, and **measuring pipettes**, also have specific uses in the analytical laboratory when exact volumes are unnecessary. Volumetric flasks are to be used in preparing standard solutions, but not for storing reagents. The precision of an analytical method depends in part upon the accuracy with which volumes of solutions can be measured, due to the inherent parameters of the measurement instrument. For example, a 10 mL volumetric flask will typically be more precise (i.e., have smaller variations between repeated measurements) than a 1000 mL volumetric flask, because the neck on which the "fill to" line is located is narrower, and therefore smaller errors in liquid height above or below the neck result in smaller volume differences compared to the same errors in liquid height for the larger flask. However, accuracy and precision are often independent of each other for measurements on similar orders of magnitude. In other words, it is possible to have precise results that are relatively inaccurate and vice versa. There are certain sources of error, which must be carefully considered. The volumetric apparatus must be read correctly; the bottom of the **meniscus** should be tangent to the calibration mark. There are other sources of error, however, such as changes in temperature, which result in changes in the actual capacity of glass apparatus and in the volume of the solutions. The volume capacity of an ordinary 100 mL glass flask increases by 0.025 mL for each 1 ° rise in temperature, but if made of borosilicate glass, the increase is much less. One thousand mL of water (and of most solutions that are ≤0.1 *N*) increases in volume by approximately 0.20 mL per 1 °C increase at room temperature. Thus, solutions must be measured at the temperature at which the apparatus was calibrated. This temperature (usually 20 °C) will be indicated on all volumetric ware. There may also be errors of calibration of the adjustable measurement apparatus (e.g., measuring pipettes), that is, the volume marked on the apparatus may not be the true volume. Such errors can be eliminated only by recalibrating the apparatus (if possible) or by replacing it.

A volumetric apparatus is calibrated either "**to contain**" or "**to deliver**" a definite volume of liquid. This will be indicated on the apparatus with the letters "**TC**" (to contain) or "**TD**" (to deliver). Volumetric flasks are calibrated to contain (TC) a given volume, which means that the flask contains the specified volume ± a defined tolerance (error). The certified TC volume only applies to the volume contained by the flask. It does not take into account the volume of solution that will stick to the walls of the flask if the liquid is poured out. Therefore, for example, a TC 250 mL volumetric flask will hold 250 mL ± a defined tolerance; if the liquid is poured out, slightly less than 250 mL will be dispensed due to solution retained on the walls of the flask (this is the opposite of "to deliver," or TD, glassware discussed below). They are available in various shapes and sizes ranging from 1 to 2000 mL capacity. Graduated cylinders, on the other hand, can be either TC or TD. For accurate work, the difference may be important. TC glassware is used to prepare solutions with specified accuracy and precision; TD glassware is used to transfer liquids with specified accuracy and precision.

Volumetric pipettes are typically calibrated to deliver a fixed volume (TD). The usual capacities are 1–100 mL, although micro-volumetric pipettes are also available. The proper technique for using volumetric pipettes is as follows (this technique is for TD pipettes, which are much more common than TC pipettes):

1. Draw the liquid to be delivered into the pipette above the line on the pipette. Always use a pipette bulb or pipette aid to draw the liquid into the pipette. Never pipette by mouth.
2. Remove the bulb (when using the pipette aid, or bulbs with pressure release valves, you can deliver without having to remove it) and replace it with your index finger.
3. Withdraw the pipette from the liquid and wipe off the tip with tissue paper. Touch the tip of the pipette against the wall of the container from which the liquid was withdrawn (or a spare beaker). Slowly release the pressure of your finger (or turn the scroll wheel to dispense) on the top of the pipette and allow the liquid level in the pipette to drop so that the bottom of the meniscus is even with the line on the pipette.
4. Move the pipette to the beaker or flask into which you wish to deliver the liquid. Do not wipe off the tip of the pipette at this time. Allow the pipette tip to touch the side of the beaker or flask. Holding the pipette in a vertical position, allow the liquid to drain from the pipette.
5. Allow the tip of the pipette to remain in contact with the side of the beaker or flask for several seconds. Remove the pipette. There will be a small amount of liquid remaining in the tip of the pipette. Do not blow out this liquid with the bulb, as TD pipettes are calibrated to account for this liquid that remains.

Note that some volumetric pipettes have calibration markings for both TC and TD measurements. Make sure to be aware which marking refers to which measurement (for transfers, use the TD marking). The TC marking will be closer to the dispensing end of the pipette (TC does not need to account for the volume retained on the glass surface, whereas TD does account for this).

Measuring and serological pipettes should also be held in a vertical position for dispensing liquids; however, the tip of the pipette is only touched to the wet surface of the receiving vessel *after* the outflow has ceased. Some pipettes are designed to have the small amount of liquid remaining in the tip blown out and added to the receiving container; such pipettes have a frosted band near the top. If there is no frosted band near the top of the pipette, do not blow out any remaining liquid.

1.5.4 Using Volumetric Glassware to Perform Dilutions and Concentrations

Typically, dilutions are performed by adding a liquid (water or a solvent) to a sample or solution. Concentrations may be performed by a variety of methods, including rotary evaporation, shaking vacuum evaporation, vacuum centrifugation, boiling, oven drying, drying under N_2 gas, or freeze drying.

For bringing samples or solutions up to a known volume (typically used when precise/accurate total final volumes are needed for preparing precise/accurate concentrations), the "gold standard" providing maximal accuracy and precision is a **Class A** glass volumetric flask (Fig. 1.1a). During manufacture, glassware to be certified as Class A is calibrated and tested to comply with tolerance specifications established by the American Society for Testing and Materials (ASTM, West Conshohocken, PA). These specifications are the standard for laboratory glassware. Class A glassware has the tightest tolerances and therefore the best accuracy and precision. These flasks are rated TC. Therefore, volumetric flasks are used to bring samples and solutions up to a defined volume. They are not used to quantitatively deliver or transfer samples because the delivery volume is not known. Other types of glassware (non-Class A flasks, graduated cylinders, Erlenmeyer flasks, round-bottomed flasks, beakers, bottles, etc., Fig. 1.1b) are less accurate and less precise. They should not be used for quantitative volume dilutions or concentrations if Class A volumetric flasks are available.

For transferring a known volume of a liquid sample for a dilution or concentration, the "gold standard" providing maximal accuracy and precision is a Class A glass volumetric pipette (Fig. 1.2a). These pipettes are rated "to deliver" (TD), which means that the pipette will deliver the specified volume ± a defined tolerance (error). The certified TD volume takes into account the volume of solution that will stick to the walls of the pipette as well as the volume of the drop of solution that typically remains in the tip of the pipette after delivery (again, you should not attempt to get this drop out, since it is already accounted for). Therefore, for example, a TD 5 mL pipette will hold slightly more than 5 mL but will deliver (dispense) 5 mL ± a defined tolerance (the opposite of TC glassware). It is important to note that volumetric pipettes are used only to deliver a known amount of solution. Typically, they should not be used to determine the

Fig. 1.1 Class A volumetric flask (**a**) and other types of non-Class A volume measuring glassware; graduated cylinder (**b**), Erlenmeyer flask (**c**), beaker (**d**), and bottle (**e**)

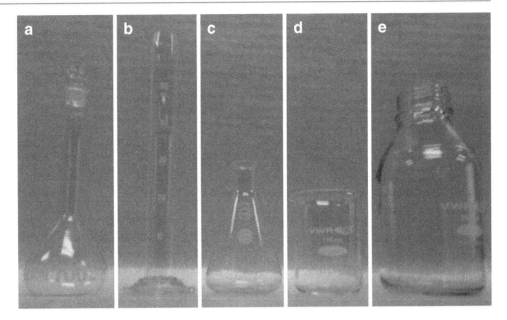

Fig. 1.2 Class A volumetric pipette (**a**) and non-volumetric pipettes: adjustable pipettors (**b**), reed pipettor (**c**), and serological pipettes (**d**)

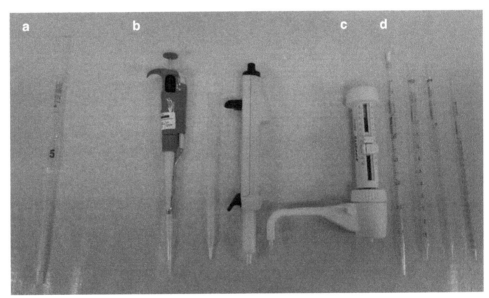

final volume of the solution unless the liquids dispensed are the only components of the final solution. For example, if a sample is dried down and then liquid from a volumetric pipette is used to resolubilize the solutes, it is unknown if the solutes significantly affect the volume of the resulting solution, unless the final volume is measured, which may be difficult to do. Although the effect is usually negligible, it is best to use volumetric glassware to assure that the final volume of the resulting solution is known (the dried solutes could be dissolved in a few mL of solvent and then transferred to a volumetric flask for final dilution). However, in some cases it may be acceptable to add several solutions together using TD volumetric pipettes and then add the individual volumes together to calculate the final volume. However, using a single volumetric flask to dilute to a final

volume is still the favored approach, as using one measurement for the final volume reduces the uncertainty. (The errors, or tolerances, of the amounts added are also added together; therefore, using fewer pieces of glassware lowers the uncertainty of the measurement even if the tolerances of the glassware are the same.) For example, suppose you need to measure out 50 mL of solution. You have access to a 50 mL volumetric flask and a 25 mL volumetric pipette, both of which have tolerances of ±0.06 mL. If you obtain 50 mL by filling the volumetric flask, the measured volume is 50 mL ± 0.06 mL (or somewhere between 49.94 and 50.06 mL). If you pipette 25 mL twice into a beaker, the tolerance of each measurement is 25 mL ± 0.06 mL, and the tolerance of the combined volume is the sum of the means and the errors:

$$(25mL \pm 0.06mL) + (25mL \pm 0.06mL) = 50mL \pm 0.12mL$$
$$= 49.88 \text{ to } 50.12mL$$

This additive property of tolerances, or errors, compounds further as more measurements are combined; conversely, when the solution is brought to volume using a volumetric flask, only a single tolerance factors into the error of the measurement.

Other types of pipettes (non-Class A volumetric glass pipettes, adjustable pipettors, automatic pipettors, reed pipettors, serological pipettes, etc., Fig. 1.2b) and other glassware (graduate cylinders, etc.) are less accurate and less precise. They should not be used for quantitative volume transfers. Pipettes are available (but rare) that are marked with lines for both TC and TD. For these pipets, the TD line would represent the volume delivered when the drop at the tip is dispensed and TC when the drop remains in the pipette.

Information typically printed on the side of the pipette or flask includes the class of the pipette or flask, whether the glassware is TD or TC, the TC or TD volume, and the defined tolerance (error) (Fig. 1.3). Note that the specifications are typically valid at a specified temperature, typically 20 °C. Although it is rare that scientists equilibrate solutions to exactly 20 °C before volume measurement, this temperature is assumed to be approximate room temperature. Be aware that the greater the deviation from room temperature, the greater the error in volume measurement. The specific gravity (density) of water at 4, 20, 60, and 80 °C relative to 4 °C is 1.000, 0.998, 0.983, and 0.972. This means that a given mass of water has lower density (greater volume for given mass) at temperatures above 20 °C. This is sometimes seen when a volumetric flask is brought exactly to volume at room temperature and then is placed in an ultrasonic bath to help dissolve the chemicals, warming the solution. A solution that was exactly at the volume marker at room temperature will be above the volume when the solution is warmer. To minimize this error, volumes should be measured at room temperature.

Volumetric glassware (flasks and pipettes) should be used for quantitative volume measurements during dilutions and concentrations whenever possible to maximize the accuracy and precision of the procedure. For both volumetric flasks and pipettes, the level of the liquid providing the defined volume is indicated by a line (usually white or red) etched or printed on the neck of the glassware. To achieve the TD or TC volume, the bottom of the meniscus of the liquid should be at the line as shown in Fig. 1.4.

For a volumetric flask, the proper technique for achieving the correct volume is to carefully pour the liquid into the flask using a funnel until the meniscus is close to the marking line, and then add additional liquid dropwise (with a manual pipette or Pasteur pipette) until the bottom (*not* the top or middle) of the meniscus is at the line with your eye level to the line. (If you do not look straight at the line, so that your eye and the line are at the same level, a phenomenon known as "parallax" can occur, making it appear that the bottom of the meniscus is at the line when in fact it is not, resulting in errors in volume measurement.) If the level of the liquid is too high, the liquid can be removed using a clean pipette (or the liquid poured out and start again). However, be aware that this cannot be done when preparing a reagent for which the solutes were accurately measured into the flask and you are adding liquid to make up to volume. In this case, you must start over. For this reason, the best practice is to add liquid slowly, and use a pipette to add liquid dropwise when approaching the desired volume.

For a volumetric pipette, the proper technique for achieving the correct volume is to draw liquid into the pipette until the meniscus is above the line, and then withdraw the pipette from the liquid and dispense the excess liquid from the pipette until the bottom of the meniscus is at the line. It is critical that the pipette be withdrawn from the solution for this step. If the level of the liquid goes below the line, additional liquid is drawn up, and the process is repeated. Proper volumetric measurements require practice and should be repeated until they are performed correctly. Improper volu-

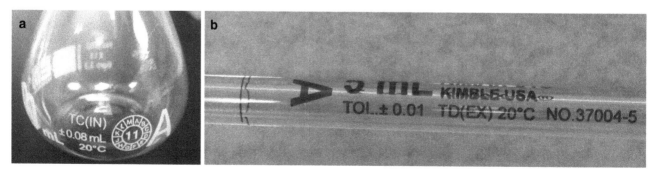

Fig. 1.3 Image of the label on a Class A volumetric flask pipette (**a**) and Class A volumetric pipette (**b**)

metric measurements can result in significant error being introduced into the measurement.

Typical tolerances for lab glassware are presented in Tables 1.6 and 1.7. References for *astm* specifications are found at http://www.astm.org/.

A comparison of Tables 1.6 and 1.7 reveals some important points. First, even for Class A glassware, the tolerances for volumetric transfer pipettes (pipettes with a single TD measurement) are much tighter than for graduated measuring pipettes (pipettes with graduations that can be used to measure a wide range of volumes) of the same volume. Second, even for Class A glassware, the tolerances for volu-

Table 1.7 Volume tolerances of non-Class A glassware required by ASTM specifications

| Volume (mL) | Tolerance (± mL) | | | |
	Buret	Volumetric (transfer) pipette	Volumetric flask	Graduated cylinder
0.5		0.012		
1		0.012		
2		0.012		
3		0.02		
4		0.02		
5		0.02		0.10
10	0.04	0.04	0.04	0.20
25	0.06	0.06	0.06	0.34
50	0.10	0.10	0.24	0.50
100	0.20	0.16	0.40	1.00
250			0.60	2.00
500				4.00
1000				6.00

metric transfer pipettes and volumetric flasks are much tighter than for graduated cylinders of the same volume. Therefore, volumetric transfer pipettes and volumetric flasks are preferred for dilutions and concentrations. For example, a 1000 mL Class A volumetric flask has a tolerance of ±0.015 mL (the actual TC volume is somewhere between 999.985 and 1000.015 mL), while a 1000 mL graduated cylinder has a tolerance of ±3.00 mL (the actual TC volume is somewhere between 997 and 1003 mL). This is a 200-fold larger potential error in the measurement of 1000 mL! Finally, tolerances for non-Class A glassware are much broader than for Class A, and thus Class A should be used if available. Proper use of the glassware with the tightest tolerances available will significantly improve the quality (accuracy/precision) of analytical data.

Fig. 1.4 Image of a liquid meniscus at the line for a Class A volumetric flask

1.5.5 Conventions and Terminology

To follow the analytical procedures described in this manual and perform calculations correctly, common terminology and conventions (a convention is a standard or generally accepted way of doing or naming something) must be understood. A common phrase in dilutions and concentrations is "**diluted to**" or "**diluted to a final volume of**." This means that the sample or solution is placed in a volumetric flask, and the final volume is adjusted to the specified value. In contrast, the phrase "**diluted with**" means that the specified amount is added to the sample or solution. In this latter case, the final mass/volume must be calculated by adding the sample mass/volume and the amount of liquid added. For example, suppose you take a 1.7 mL initial volume and either (1) dilute to 5 mL with methanol or (2) dilute with 5 mL methanol. In the first case, this means that the sample (1.7 mL) is

Table 1.6 Volume tolerances of Class A glassware required by ASTM specifications

| Volume (mL) | Tolerance (± mL) | | | | |
	Buret	Volumetric (transfer) pipette	Measuring (graduated) pipettes	Volumetric flask	Graduated cylinder
0.5		0.006			
1		0.006		0.010	
2		0.006	0.01	0.015	
3		0.01	0.02	0.015	
4		0.01	0.03	0.020	
5		0.01	0.05	0.020	0.05
10	0.02	0.02	0.08	0.020	0.10
25	0.03	0.03	0.10	0.030	0.17
50	0.05	0.05		0.050	0.25
100	0.10	0.08		0.080	0.50
250				0.012	1.00
500				0.013	2.00
1000				0.015	3.00

placed in a volumetric flask and methanol (~3.3 mL) is added so that the final volume is 5 mL total. In the second case, the sample (1.7 mL) is combined with 5 mL methanol, and the final volume is 6.7 mL. As you can see, these are very different values. This will always be the case except when one of the volumes is much larger than the other. For example, if you were working with a 10 μL sample, diluting it "to 1 L" or "with 1 L" would result in final volumes of 1 L and 1.00001 L, respectively. It is important to understand the differences between these two conventions to perform procedures correctly and interpret data accurately.

Another common term in dilutions/concentrations is the term "**fold**" or "**X**." This refers to the ratio of the final and initial concentrations (or volumes and masses) of the sample or solution during each step. An "X-fold dilution" means that the concentration of a sample decreases (and typically the volume increases) by a given factor. For example, if 5 mL of an 18.9% NaCl solution is diluted 10-fold (or 10X) with water, water is added so that the final volume is 50 mL (10-fold or 10X greater than 5 mL) and the final concentration is 1.89% NaCl (10-fold or 10X less than 18.9%). Conversely, an "X-fold concentration" means that the concentration of a sample increases (and typically the volume decreases) by the stated factor. For example, if 90 mL of a 0.31 ppm salt solution is concentrated 10-fold (10X), the volume is decreased to 9 mL (either by reducing to 9 mL or drying completely and reconstituting to 9 mL, 10-fold or 10X lower than 90 mL), and the final concentration is 3.1 ppm salt (10-fold or 10X more than 0.31 ppm). Although 10-fold or 10X was used for these examples, any value can be used. In microbiology, values of 10X, 100X, 1000X, etc., are commonly used due to the log scale used in that field. However, less standard dilutions of any value are routinely used in analytical chemistry.

The last terminology system for dilutions and concentrations involves **ratios**. This system is somewhat ambiguous and is not used in the *Food Analysis* textbook or lab manual. This system refers to dilutions as "X:Y," where X and Y are the masses or volumes of the initial and final solutions/samples. For example, it may be stated that "the solution was diluted 1:8." This system is ambiguous for the following reasons:

1. The first and last numbers typically refer to the initial and final samples, respectively (therefore, a 1:8 dilution would mean 1 part initial sample and 8 parts final sample). However, there is no standard convention. Therefore, an "X:Y" dilution could be interpreted either way.
2. There is no standard convention as to whether this system describes the "diluted to" or "diluted with" (as described above) approach. Therefore, diluting a sample 1:5 could be interpreted as either (1) diluting 1 mL sample with

4 mL for a final volume of 5 mL ("diluted to") or (2) diluting 1 mL sample with 5 mL for a final volume of 6 mL ("diluted with").

Because of these ambiguities, the ratio system is discouraged in favor of the "X-fold" terminology. However, ratio dilutions still appear in some literature. If possible, it is recommended that you investigate to clarify what is meant by this terminology.

Another factor to consider is that liquid volumes are often not strictly additive. For example, exactly 500 mL 95% v/v ethanol aq. added to 500 mL distilled water will not equal exactly 1000 mL final volume; in fact, the new volume will be closer to 970 mL. Where did the missing 30 mL go? Polar molecules such as water undergo different three-dimensional intermolecular bonding and thus spatial arrangements in a pure solution versus in a mixture with other solute or chemicals such as ethanol. The difference in bonding causes an apparent contraction in this case. As well, addition of solute to an exact volume of water will change the volume after dissolved. To account for this effect, volumetric glassware is used to bring mixed solutions up to a final volume after initial mixing. When two liquids are mixed, the first liquid is volumetrically transferred into a volumetric flask, and then the second liquid is added to volume, with intermittent swirling or vortexing to mix the liquids as they are being combined. For mixing solids into solvents, the solids are first placed in a volumetric flask, dissolved in a partial volume, and then brought to exact volume with additional solvent.

1.5.6 Burets

Burets are used to deliver definite volumes, particularly for titrations. The more common types are usually of 25 or 50 mL capacity, graduated to tenths of a milliliter, and are provided with stopcocks. For precise analytical methods in microchemistry, microburets are also used. Microburets generally are of 5 or 10 mL capacity, graduated in hundredths of a milliliter division. General rules in regard to the manipulation of a buret are as follows:

1. Do not attempt to dry a buret that has been cleaned for use, but rather rinse it two or three times with a small volume of the solution with which it is to be filled.
2. Do not allow alkaline solutions to stand in a buret, because the glass will be chemically eroded, and the stopcock, unless made of Teflon, will tend to freeze.
3. A 50 mL buret should not be emptied faster than 0.7 mL per second; otherwise, too much liquid will adhere to the walls; as the solution drains down, the meniscus will gradually rise, giving a high false reading.

It should be emphasized that improper use of and/or reading of burets can result in serious calculation errors. Buret use will be addressed in more detail in titration laboratory exercises.

1.5.7 Cleaning of Glass and Porcelain

In the case of all apparatus for delivering liquids, the glass must be absolutely clean so that the film of liquid never breaks at any point. Careful attention must be paid to this fact or the specified amount of solution will not be delivered. The method of cleaning should be adapted to both the substances that are to be removed and the determination to be performed. Water-soluble substances are simply washed out with hot or cold water, and the vessel is finally rinsed with successive small amounts of distilled water. Other substances more difficult to remove, such as lipid residues or burned material, may require the use of a detergent, organic solvent, nitric acid, or aqua regia (25% v/v conc. HNO_3 in conc. HCl). In all cases, it is good practice to rinse a vessel with distilled water as soon as possible after use. Material allowed to dry on glassware is much more difficult to remove.

1.6 REAGENTS

Chemical reagents, solvents, and gases are available in a variety of **grades of purity**, including technical grade, analytical reagent grade, and various "ultrapure" grades. The purity of these materials required in analytical chemistry varies with the type of analysis. The parameter being measured and the sensitivity and specificity of the detection system are important factors in determining the purity of the reagents required. **Technical grade** is useful for making cleaning solutions, such as the nitric acid and alcoholic potassium hydroxide solutions mentioned previously. For many analyses, **analytical reagent grade** is satisfactory. Other analyses, for example, trace organic, HPLC (high performance liquid chromatography) and LC-MS (liquid chromatography-mass spectrometry), frequently require special "ultrapure" reagents and solvents. In methods for which the purity of reagents is not specified, it is intended that analytical reagent grade be used. Reagents of lesser purity than that specified by the method should not be used.

There is some confusion as to the definition of the terms **analytical reagent grade**, **reagent grade**, and **ACS analytical reagent grade**. A review of the literature and chemical supply catalogs indicates that the three terms are synonymous. National Formulary (NF), US Pharmaceutical (USP), and Food Chemicals Codex (FCC) are grades of chemicals certified for use as food ingredients. It is important that only NF, USP, or FCC grades be used as food additives if the product is intended for consumption by humans, rather than for chemical analysis.

1.6.1 Acids

The concentration of common commercially available acids is given in Table 1.8.

1.6.2 Distilled Water

Distilled or **demineralized water** is used in the laboratory for dilution, preparation of reagent solutions, and final rinsing of washed glassware. Ordinary distilled water is not actually very "pure" (100% water with no contaminants at any level). It may be contaminated by dissolved gases and by materials leached from the container in which it has been stored. Volatile organics distilled over from the original source feed water may be present, and nonvolatile impurities may occasionally be carried over by the steam, in the form of a spray. The concentration of these contaminants is usually quite small, and distilled water is used for many analyses without further purification. There are a variety of methods for purifying water, such as distillation, filtration, and ion exchange. Distillation employs boiling of water and condensation of the resulting steam, to eliminate nonvolatile impurities (such as minerals). Ion exchange employs cartridges packed with ionic residues (typically negatively charged) to remove charged contaminants (typically positively charged minerals) when water is passed through the cartridge. Finally, filtration and reverse osmosis remove insoluble particulate matter above a specific size.

Table 1.8 Concentration of common commercial strength acids

Acid	Molecular weight (g/mol)	Concentration (M)	Specific gravity
Acetic acid, glacial	60.05	17.4	1.05
Formic acid	46.02	23.4	1.20
Hydriodic acid	127.9	7.57	1.70
Hydrochloric acid	36.5	11.6	1.18
Hydrofluoric acid	20.01	32.1	1.167
Hypophosphorous acid	66.0	9.47	1.25
Lactic acid	90.1	11.3	1.2
Nitric acid	63.02	15.99	1.42
Perchloric acid	100.5	11.65	1.67
Phosphoric acid	98.0	14.7	1.70
Sulfuric acid	98.0	18.0	1.84
Sulfurous acid	82.1	0.74	1.02

1.6.3 Water Purity

Water purity has been defined in many different ways, but one generally accepted definition states that high purity water is water that has been distilled and/or deionized so that it will have a specific resistance of 500,000 Ω (2.0 $\mu\Omega/cm$ conductivity) or greater. This definition is satisfactory as a base to work from, but for more critical requirements, the breakdown shown in Table 1.9 has been suggested to express degrees of purity.

Distilled water is usually produced in a steam-heated metal still. The feed water is (or should be) softened prior to distillation to remove calcium and magnesium to prevent scale (Ca or Mg carbonate) formation. Several companies produce ion-exchange systems that use resin-packed cartridges for producing "distilled water." The lifespan of an ion-exchange cartridge is very much a function of the mineral content of the feed water. Thus, the lifespan of the cartridge is greatly extended by using distilled or reverse osmosis-treated water as the incoming stream. This procedure can also be used for preparing ultrapure water, especially if a low flow rate is used and the ion-exchange cartridge is of "research" grade.

1.6.4 Carbon-Dioxide-Free Water

Carbon dioxide (CO_2) dissolved in water can interfere with many chemical measurements. Thus, CO_2-free water may need to be produced. CO_2-free water may be prepared by boiling distilled water for 15 min and cooling to room temperature. As an alternative, distilled water may be vigorously aerated (sparged or bubbled) with a stream of inert gas (e.g., N_2 or He_2) for a period sufficient to achieve CO_2 removal. The final pH of the water should lie between 6.2 and 7.2. It is not advisable to store CO_2-free water for extended periods. To ensure that CO_2-free water remains that way, an **ascarite trap** should be fitted to the container such that air entering the container (as boiled water cools) is CO_2-free. Ascarite is silica coated with NaOH, and it removes CO_2 by the following reaction:

$$2NaOH + CO_2 \rightarrow Na_2CO_3 + H_2O$$

Table 1.9 Classification of water purity

Degree of purity	Maximum conductivity ($\mu\Omega/cm$)	Approximate concentration of electrolytes (mg/L)
Pure	10	2–5
Very pure	1	0.2–0.5
Ultrapure	0.1	0.01–0.02
Theoretically pure	0.055	0.00

Ascarite should be sealed from air except when water is being removed from the container.

1.6.5 Preparing Solutions and Reagents

The accurate and reproducible preparation of laboratory reagents is essential to good laboratory practice. Liquid reagents are prepared using volumetric glassware (pipettes and flasks) as appropriate.

The following procedures are used to prepare solutions from solid reagents (such as sodium hydroxide):

1. Determine the amount of solid reagent needed.
2. Fill the TC volumetric flask ~¼–½ full with the solvent.
3. Add the solid reagent using a funnel (it is best to predissolve solids in a beaker with a small amount of liquid, and then add this to the flask; rinse the smaller beaker thoroughly and also put the rinses into flask).
4. Swirl to mix until essentially dissolved.
5. Fill the flask to volume with the solvent.
6. Cap and invert the flask ~10–20 times to completely mix the solution.

(*Note:* that it is not appropriate to simply combine the solid reagent with the final volume and assume that the final volume does not change. This is particularly true for high % concentrations. For example, 1 L of a 10% aqueous NaOH solution is correctly made by filling a 1 L flask with ~25–500 mL water, adding 100 g NaOH, mixing until dissolved, and diluting to 1 L. It would be incorrect to simply combine 100 g NaOH with 1 L water, as the dissolved solid will take up some volume in solution. (Note that solid NaOH is difficult to dissolve, requires a stir bar, and is exothermic, releasing heat upon dissolution; therefore, do not handle the glass with bare hands.) Additionally, if a stir bar is used, make sure to remove this after the solution is dissolved (rinsing the bar off and adding the rinse to the solution) but BEFORE diluting to volume. Note that sonication is preferred to using a stir bar in a volumetric flask.)

The following similar procedures are used to prepare reagents from two or more liquids:

1. Determine the total volume of the final reagent.
2. Obtain a TC volumetric flask (if possible) equal to the final volume.
3. Use TD volumetric glassware to add the correct amount of the liquids with the smallest volumes.
4. Dilute to volume with the liquid with the largest volume, gently swirling during addition.
5. Cap and invert the flask ~10–20 times to completely mix the solution.

Note that a TC volumetric flask should be used whenever possible to bring the solution to final volume. For example, the correct way to prepare 1 L of a 5% ethanol in water solution is to use a 50 mL TD pipette to dispense 50 mL ethanol into a 1 L TC flask and then fill the flask to volume with water. It would be incorrect to simply combine 50 mL ethanol and 950 mL water, since complex physical properties govern the volume of a mixture of liquids, and it cannot be assumed that two liquids of different densities and polarities will combine to form a volume equal to the sum of their individual volumes. If the final volume is not a commonly available TC flask size, then use TD glassware to deliver all reagents.

The use of graduated cylinders and beakers should be avoided for measuring volumes for reagent preparation.

1.7 DATA HANDLING AND REPORTING

1.7.1 Significant Figures

The term **significant figure** is used rather loosely to describe some judgment of the number of reportable digits in a result. Often the judgment is not soundly based and meaningful digits are lost or meaningless digits are accepted. Proper use of significant figures gives an indication of the reliability of the analytical method used. Thus, reported values should contain only significant figs. A value is made up of significant figures when it contains all digits known to be true and one last digit in doubt. For example, if a value is reported at 18.8 mg/L, the "18" must be a firm value, while the "0.8" is somewhat uncertain and the actual value may be between "0.7" or "0.9." The number zero may or may not be a significant figure:

1. Final zeros after a decimal point are always significant figures. For example, 9.8 grams to the nearest mg is reported as 9.800 g.
2. Zeros before a decimal point with other preceding digits are significant. With no preceding digit, a zero before the decimal point is not significant.
3. If there are no digits preceding a decimal point, the zeros after the decimal point but preceding other digits are not significant. These zeros only indicate the position of the decimal point.
4. Final zeros in a whole number may or may not be significant. In a conductivity measurement of 1000 $\mu\Omega$/cm, there is no implication that the conductivity is 1000 ± 1 $\mu\Omega$/cm. Rather, the zeros only indicate the magnitude of the number.

A good measure of the significance of one or more zeros before or after another digit is to determine whether the zeros can be dropped by expressing the number in exponential form. If they can, the zeros are not significant. For example, no zeros can be dropped when expressing a weight of 100.08 g in exponential form; therefore, the zeros are significant. However, a weight of 0.0008 g can be expressed in exponential form as $8 \times 10{-}4$ g, and the zeros are not significant. Significant figures reflect the limits of the particular method of analysis. If more significant figures are needed, selection of another method will be required to produce an increase in significant figures.

Once the number of significant figures is established for a type of analysis, data resulting from such analyses are reduced according to set rules for rounding off.

1.7.2 Rounding Off Numbers

Rounding off numbers is a necessary operation in all analytical areas. However, it is often applied in chemical calculations incorrectly by blind rule or prematurely and, in these instances, can seriously affect the final results. Rounding off should normally be applied only as follows:

1. If the figure following those to be retained is less than 5, the figure is dropped, and the retained figures are kept unchanged. As an example, 11.443 is rounded off to 11.44.
2. If the figure following those to be retained is greater than 5, the figure is dropped, and the last retained figure is raised by 1. As an example, 11.446 is rounded off to 11.45.
3. When the figure following those to be retained is 5 and there are no figures other than zeros beyond the 5, the figure is dropped, and the last place figure retained is increased by 1 if it is an odd number, or it is kept unchanged if an even number. As an example, 11.435 is rounded off to 11.44, while 11.425 is rounded off to 11.42.

1.7.3 Rounding Off Single Arithmetic Operations

Addition: When adding a series of numbers, the sum should be rounded off to the same numbers of decimal places as the addend with the smallest number of places. However, the operation is completed with all decimals places intact and rounding off is done afterward. As an example:

$$11.1 + 11.12 + 11.13 = 33.35$$

The sum is rounded off to 33.4

Multiplication: When two numbers of unequal digits are to be multiplied, all digits are carried through the operation, then the product is rounded off to the number of significant digits of the less accurate number.

Division: When two numbers of unequal digits are to be divided, the division is carried out on the two numbers using all digits. Then the quotient is rounded off to the lower number of significant digits between the two values.

Powers and roots: When a number contains n significant digits, its root can be relied on for n digits, but its power can rarely be relied on for n digits.

1.7.4   Rounding Off the Results of a Series of Arithmetic Operations

The rules for rounding off are reasonable for simple calculations. However, when dealing with two nearly equal numbers, there is a danger of loss of all significance when applied to a series of computations that rely on a relatively small difference in two values. Examples are calculation of variance and standard deviation. The recommended procedure is to carry several extra figures through the calculation and then to round off the final answer to the proper number of significant figures. This operation is simplified by using the memory function on calculators, which for most calculators is a large number, often 10 or more, digits.

1.8   BASIC LABORATORY SAFETY

1.8.1   Safety Data Sheets

Safety Data Sheets (SDSs), formerly called Material Safety Data Sheets (MSDSs), are informational packets that are "intended to provide workers and emergency personnel with procedures for handling or working with that substance in a safe manner, and includes information such as physical data (melting point, boiling point, flash point, etc.), toxicity, health effects, first aid, reactivity, storage, disposal, protective equipment, and spill-handling procedures" (http://en.wikipedia.org/wiki/Material_safety_data_sheet#United_States).

SDSs are available for all reagents, chemicals, solvents, gases, etc., used in your laboratory. You can consult these documents if you have questions regarding how to safely handle a material, the potential risks of the material, how to properly clean up a spill, etc. They should be available to you in a centralized location (typically, a binder) the lab. If not available, you may request these from your instructor or find them online. Generally, the following information is available on a MSDS or SDS in a 16-section format:

1. Identification of the substance/mixture
2. Hazard identification
3. Composition/information on ingredients
4. First aid measures

5. Firefighting measures
6. Accidental release measures
7. Handling and storage
8. Exposure controls/personal protection
9. Physical and chemical properties
10. Stability and reactivity
11. Toxicological information
12. Ecological information
13. Disposal considerations
14. Transport information
15. Regulatory information
16. Other information

1.8.2   Hazardous Chemicals

Food analysis laboratories, like any chemical laboratory, often contain hazardous compounds, including the following:

1. Acids (hydrochloric acid, sulfuric acid, etc.)
2. Bases (e.g., sodium hydroxide)
3. Corrosives and oxidizers (sulfuric acid, nitric acid, perchloric acid, etc.)
4. Flammables (organic solvents such as hexane, ether, alcohols).

These materials can cause harm if not used properly. Follow all safety warnings in written laboratory instructions, on labels and from instructors and teaching assistants.

1.8.3   Personal Protective Equipment and Safety Equipment

It is important to understand the location and use of lab safety equipment. The purpose of this is threefold:

1. To prevent accidents and/or injuries in the lab.
2. To quickly and effectively respond to any accident and/or injury in the lab.
3. To be able to perform laboratory procedures without excessive worrying about lab hazards.

Your laboratory instructor should provide instruction regarding basic laboratory safety equipment. You should be aware of these general rules and the existence of this equipment.

Proper clothing is required to work in any chemical laboratory. The following standards and rules regarding dress are generally applicable, although standards may vary between laboratories:

1. Close-toed shoes (no flip-flops, sandals, or other "open" footwear).

2. Long pants (dresses, skirts, and shorts may be allowed in some laboratories).
3. No excessively loose clothing or accessories.
4. Long hair should be pulled back from the face into a ponytail or otherwise restrained.

You should be able to obtain and wear the following personal protective equipment (PPE) and understand their proper use:

1. Safety glasses, goggles, and face shields
2. Lab coat or apron
3. Shoe covers
4. Latex or acetonitrile gloves
5. Puncture-resistant gloves
6. Heat-resistant gloves

You should be aware of the locations of the following safety equipment items, and their proper use:

1. First aid kit
2. Bodily fluids cleanup kit
3. Acid, base, and solvent spill kits
4. Fire extinguisher and fire blanket
5. Safety shower and eyewash station
6. Solid, liquid, chlorinated, and biohazard waste disposal containers, if applicable
7. Sharps and broken glass disposal containers, if applicable

1.8.4 Eating, Drinking, Etc.

Your hands may become contaminated with substances used in the lab simply by touching lab benches, glassware, etc. This may happen even without your knowledge. Even if you are not handling hazardous substances, previous lab occupants may not have cleaned benches and glassware, leaving behind hazardous substances that you are unaware of. To avoid spreading potentially harmful substances from your hands to your face, eyes, nose, and mouth (where they may irritate sensitive or be introduced to circulation by mucus membranes, ingestion, or inhalation), the following activities are prohibited in chemical laboratories: eating, drinking, smoking, chewing tobacco or snuff, and applying cosmetics (including lip balm, etc.). The following should not even be brought into a chemical laboratory: food, water, beverages, tobacco, and cosmetics. Some unconscious activities (touch-ing your face and eyes, etc.) are difficult to avoid. However, wearing gloves in the laboratory may minimize these actions.

1.8.5 Miscellaneous Information

The following general rules and guidelines apply to working in the laboratory:

1. When combining acid and water, always add the acid to water (instead of adding water to the acid). When acid dissolves in water, heat is released. This can cause splattering of the solution. By adding the acid to water, the heat is dissipated, and splattering is reduced or eliminated.
2. Be aware that dissolving sodium hydroxide in water generates heat. Making high concentrations of aqueous sodium hydroxide can lead to very hot solutions that can burn bare hands. Allow these solutions to cool, or handle with heat-resistant gloves.
3. Broken glass and other sharps (razor blades, scalpel blades, needles, etc.) should be disposed of in puncture-resistant sharps containers.
4. Do not pour waste or chemicals down the drain. This practice can damage the building's plumbing and harm the environment. Dispose of liquid, solid, chlorinated, radioactive, and biohazard wastes into the appropriate containers provided by the lab instructor. If you are unsure how to properly dispose of waste, ask your instructor or teaching assistant.
5. Handle volatile, noxious, or corrosive compounds in the fume hood with appropriate PPE.

Acknowledgments The author of this chapter thanks Dennis A. Lonergan (The Vista Institute, Eden Prairie, MN, USA) and S. Suzanne Nielsen (Department of Food Science, Purdue University, West Lafayette, IN, USA), who developed the original content of this chapter.

RESOURCE MATERIALS

Anonymous (2023) Pipetman® Product Guide. Pipetman® Classic User's Guide. Gilson, Inc., Middleton, WI. https://www.gilson.com/pub/media/docs/Product_Guide_2023.pdf.   https://www.gilson.com/pub/media/docs/PIPETMANCLASSIC_UG_LT801120-F.pdf

Smith JS (2024) Evaluation of analytical data, Ch. 4, In: Ismail, BP, Nielsen, SS (eds) Nielsen's food analysis, 6th edn. Springer, New York

Reagents and Buffers

2

Catrin Tyl and B. Pam Ismail

Contents

2.1 REAGENTS OF SPECIFIED CONCENTRATIONS

Virtually every analytical method involving wet chemistry starts with preparing reagent solutions. This usually involves dissolving solids in a liquid or diluting from stock solutions. The concentration of analytes in solution can be expressed in weight (kg, g, or lower submultiples) or in the amount of substance (mol), per a unit volume (interchangeably L or dm^3, mL or cm^3, and lower submultiples). Preparing reagents of correct concentrations is crucial for the validity and reproducibility of any analytical method. Example A1 and A2 is a sample calculation to prepare a calcium chloride reagent of a particular concentration.

Example A1

How much calcium chloride do you need to weigh out to get 2 L of a 4 mM solution?

Solution The molarity (M) equals the number of moles (n) in the volume (v) of 1 L:

$$M\left(\frac{mol}{L}\right) = \frac{n(mol)}{v[L]} \qquad (2.1)$$

The desired molarity is 4 mM = 0.004 M; the desired volume is 2 L. Rearrange Eq. 2.1 so that:

$$n(mol) = M\left(\frac{mol}{L}\right) \times v(L) \qquad (2.2)$$

The mass (m) of 1 mole $CaCl_2$ (110.98 g/mol) is specified by the molecular weight (MW) as defined through Eq. 2.3. The mass to be weighed is calculated by rearranging into Eq. 2.4:

$$MW\left(\frac{g}{mol}\right) = \frac{m(g)}{n(mol)} \qquad (2.3)$$

$$m(g) = n(mol) \times MW\left(\frac{g}{mol}\right) \qquad (2.4)$$

C. Tyl (✉)
Department of Chemistry, Biotechnology and Food Science, Norwegian University of Life Sciences, As, Norway
e-mail: catrin.tyl@nmbu.no

B. P. Ismail
Department of Food Science and Nutrition, University of Minnesota, St. Paul, MN, USA
e-mail: bismailm@umn.edu

Now substitute Eq. 2.2 into Eq. 2.4. Some of the units that can be canceled out are indicated as strike-throughs:

$$m(g) = M\left(\frac{mol}{L}\right) \times v(L) \times MW\left(\frac{g}{mol}\right) \quad (2.5)$$

$$m(g) = 0.004\left(\frac{mol}{L}\right) \times 2(L) \times 110.98\left(\frac{g}{mol}\right)$$
$$= 0.888[g] \quad (2.6)$$

Example A2

How much would you need to weigh out if you were using the dihydrate (i.e., $CaCl_2 \cdot 2H_2O$) to prepare 2 L of a 4 mM solution of calcium chloride?

Solution During crystallization of salts, water may be incorporated into the crystal lattice. Examples include phosphate salts, calcium chloride, and certain sugars. The names of these compounds are amended by the number of bound water molecules. For example, $Na_2HPO_4 \cdot 7H_2O$ is called sodium monophosphate heptahydrate, and $CaCl_2 \cdot 2H_2O$ is called calcium chloride dihydrate. The water is tightly bound and is not visible (the dry reagents do not look clumped). Many commercially available salts are sold as hydrates that can be used analogously to their dry counterparts after adjusting for their increased molecular weight, which includes the bound water molecules. The molecular weight of $CaCl_2$ changes from 110.98 g/mol to 147.01 g/mol to account for two molecules of water at about 18 g/mol. Hence, Eq. 2.6 needs to be modified to:

$$m(g) = 0.004\left(\frac{mol}{L}\right) \times 2(L) \times 147.01\left(\frac{g}{mol}\right)$$
$$= 1.176[g] \quad (2.7)$$

Thus, 1.176 g of $CaCl_2$ would be weighed out and the volume made up to 2 L in a volumetric flask.

Other commonly used ways to express concentrations are listed in Table 2.1. For instance, for very low concentrations as encountered in residue analysis, parts per million (e.g., µg/mL or mg/L) and parts per billion (e.g., µg/L) are preferred units. Concentrated acids and bases

are often labeled in percent mass by mass or percent mass per volume. For instance, a 28% wt/wt solution of ammonia in water contains 280 g ammonia per 1000 g of solution. On the other hand, 32% wt/vol NaOH solution contains 320 g of NaOH per L. For dilute solutions in water, the density is approximately 1 kg/L at room temperature (the density of water is exactly 1 kg/L only at 5 °C), and thus, wt/wt and wt/vol are almost equal. However, concentrated solutions or solutions in organic solvents can deviate substantially in their density. Therefore, for concentrated reagents, the correct amount of reagent needed for dilute solutions is found by accounting for the density, as illustrated in Example A3.

Example A3

Prepare 500 mL of a 6 M sulfuric acid solution from concentrated sulfuric acid. The molecular weight is 98.08 g/mol, and the manufacturer states that it is 98% wt/wt and has a density (d) of 1.84 g/mL.

Solution The density specifies the mass per volume of the concentrated H_2SO_4:

$$d = \frac{m}{v}\left(\frac{g}{mL}\right) \quad (2.8)$$

The molarity of the concentrated sulfuric acid is determined by rearranging Eqs. 2.3 and 2.8 to express the unknown number of moles (n) in terms of known quantities (MW and d). However, because molarity is specified in moles per L, the density needs to be multiplied by 1000 to obtain the g per L (the unit of density is g per mL or kg per L).

$$\text{Rearrange Eq.2.3}: n(mol) = \frac{m(g)}{MW\left(\frac{g}{mol}\right)} \quad (2.9)$$

$$\text{Rearrange Eq.2.8}: m(g) = d\left(\frac{g}{L}\right) \times 1000 \times v(L) \quad (2.10)$$

Substitute Eq.2.10 into Eq.2.9:

$$n(mol) = \frac{d \times 1000\left(\frac{g}{L}\right) \times v(L)}{MW\left(\frac{mol}{g}\right)} \quad (2.11)$$

Substitute Eq. 2.11 into Eq. 2.1. The 98% wt/wt (see Table 2.1) can be treated like a proportionality factor: every g of solution contains 0.98 g H_2SO_4. To account for this, multiply Eq. 2.1 with the % wt:

$$M\left(\frac{mol}{L}\right) = \frac{d \times \cancel{v} \times 1000}{MW} \times \frac{1}{\cancel{v}} \times \%w\left(\frac{g}{g}\right) = \quad (2.12)$$

$$M\left(\frac{mol}{L}\right) = \frac{d \times 1000\left(\frac{g}{L}\right)}{MW\left(\frac{g}{mol}\right)} \times \%w\left(\frac{g}{g}\right) \quad (2.13)$$

$$M\left(\frac{mol}{L}\right) = \frac{1.84\left(\frac{g}{L}\right) \times 1000 \times 0.98\left(\frac{g}{g}\right)}{98.08\left(\frac{g}{mol}\right)}$$

$$= 18.39\left(\frac{mol}{L}\right) \quad (2.14)$$

The necessary volume to supply the desired amount of moles for 500 mL of a 6 M solution is found by using Eq. 2.15:

$$v_1(L) \times M_1\left(\frac{mol}{L}\right) = v_2(L) \times M_2\left(\frac{mol}{L}\right) \quad (2.15)$$

$$v\,stock\,solution\,(L) = \frac{v\,of\,diluted\,solution\,(L) \times M\,of\,diluted\,solution}{M\,of\,stock\,solution} \quad (2.16)$$

$$v(L) = \frac{0.5(L) \times 6\,\cancel{mol/L}}{18.39\,\cancel{mol/L}} \quad (2.17)$$

$$= 0.163\ L\ or\ 163\ mL$$

Hence, to obtain 500 mL, 163 mL concentrated sulfuric acid would be combined with 337 mL of water (500–163 mL). The dissolution of concentrated sulfuric acid in water is an exothermic process, which may cause splattering, and glassware can get very hot (most plastic containers are not suited for this purpose!). The recommended procedure would be to add some water to a 500 mL volumetric glass flask, e.g., ca. 250 mL, then add the concentrated acid, allow the mixture to cool down, mix, and bring up to volume with water.

Table 2.1 Concentration expression terms

Unit	Symbol	Definition	Relationship
Molarity	M	Number of moles of solute per liter of solution	$M = \dfrac{mol}{liter}$
Normality	N	Number of equivalents of solute per liter of solution	$N = \dfrac{equivalents}{liter}$
Percent by weight (parts per hundred)	wt %	Ratio of weight of solute to weight of solute plus weight of solvent × 100	$wt\% = \dfrac{wt\,solute \times 100}{total\,wt}$
	wt/vol %	Ratio of weight of solute to total volume × 100	$wt/vol\% = \dfrac{wt\,solute \times 100}{total\,volume}$
Percent by volume	vol %	Ratio of volume of solute to total volume × 100	$vol\% = \dfrac{vol\,of\,solute \times 100}{total\,volume}$
Parts per million	ppm	Ratio of solute (wt or vol) to total weight or volume × 1,000,000	$ppm = \dfrac{mg\,solute}{kg\,solution}$ $= \dfrac{\mu g\,solute}{g\,solution}$ $= \dfrac{mg\,solute}{L\,solution}$ $= \dfrac{\mu g\,solute}{mL\,solution}$
Parts per billion	ppb	Ratio of solute (wt or vol) to total weight or volume × 1,000,000,000	$ppb = \dfrac{\mu g\,solute}{kg\,solution}$ $= \dfrac{ng\,solute}{g\,solution}$ $= \dfrac{\mu g\,solute}{L\,solution}$ $= \dfrac{ng\,solute}{mL\,solution}$

2.2   USE OF TITRATION TO DETERMINE CONCENTRATION OF ANALYTES

A wide range of standard methods in food analysis, such as the iodine value, peroxide value, and titratable acidity, involve the following concept:

- A **reagent** of **known** concentration (i.e., the titrant) is titrated into a solution of **analyte** with **unknown** concentration. The used up volume of the reagent solution is measured.
- The ensuing reaction converts the reagent and analyte into products.
- When all of the analyte is used up, there is a measurable change in the system, for example, in color or pH.
- The concentration of the titrant is known; thus, the amount of converted reactant can be calculated. The stoichiometry of the reaction allows for the calculation of the concentration of the analyte, for example, in the case of iodine values, the absorbed grams of iodine per 100 g of sample; in the case of peroxide value, milliequivalents of peroxide per kg of sample; or in the case of titratable acidity, % wt/vol acidity.

There are two principal types of reactions for which this concept is in widespread use: acid-base and redox reactions. For both reaction types, the concept of normality (N) plays a role, which signifies the number of equivalents of solute per L of solution. The number of equivalents corresponds to the number of transferred H^+ for acid-base reactions and transferred electrons for redox reactions. Normality equals the product of the molarity with the number of equivalents (typically 1–3 for common acids and bases with low molecular weight), i.e., it is equal to or higher than the molarity. For instance, a 0.1 M sulfuric acid solution would be 0.2 N, because two H^+ are donated per molecule H_2SO_4. On the other hand, a 0.1 M NaOH solution would still be 0.1 N, as indicated by Eq. 2.18:

$$\text{Normality} = \text{Molarity} \times \text{number of equivalents} \quad (2.18)$$

Substituting N for M, reaction equivalence can be expressed through a modified version of Eq. 2.15 (which is the same as Ref. [1], Sect. 22.2.2, Eq. 2.1):

$$v \text{ of titrant} \times N \text{ of titrant} =$$
$$v \text{ of analyte solution} \times N \text{ of analyte solution} \quad (2.19)$$

Some analyses (such as in titratable acidity) require the use of **equivalent weights** instead of molecular weights. These can be obtained by dividing the molecular weight by the number of equivalents transferred over the course of the reaction.

$$\text{Equivalent weight (g)} = \frac{\text{molecular weight}\left(\dfrac{g}{mol}\right)}{\text{number of equivalents}\left(mol\right)} \quad (2.20)$$

Using equivalent weights can facilitate calculations, because it accounts for the number of reactive groups of an analyte. For H_2SO_4, the equivalent weight would be $\dfrac{98.08}{2} = 49.04$ g/mol, whereas for NaOH it would be equal to the molecular weight since there is only one OH group. To illustrate the concept, the reaction of acetic acid with sodium hydroxide in aqueous solution is stated below:

$$CH_3COOH + NaOH \rightarrow CH_3COO^- + H_2O + Na^+$$

This reaction can form the basis for quantifying acetic acid contents in vinegar (for which it is the major acid):

Example B1
100 mL of vinegar is titrated with a solution of NaOH that is exactly 1 M. (Chap. 22, Sect. 22.2, in this laboratory manual describes how to standardize titrants.) If 18 mL of NaOH are used up, what is the corresponding acetic acid concentration in vinegar?

Solution The reaction equation shows that both NaOH and CH_3COOH have an equivalence number of 1, because they each have only one reactive group. Thus, their normality and molarity are equal. Eq. 2.19 can be used to solve Example B1, and the resulting N will, in this case, equal the M:

$$18(\text{mL}) \times 1\left(\frac{mol}{L}\right) = 100(\text{mL}) \times$$
$$\text{Normality of acetic acid solution}\left(\frac{mol}{L}\right) \quad (2.21)$$

$$M \text{ and } N \text{ of acetic acid in vinegar} \left(\frac{mol}{L}\right) =$$
$$\frac{18(mL)}{100(mL)} \times 1\left(\frac{mol}{L}\right) = 0.18\left(\frac{mol}{L}\right) \quad (2.22)$$

Sometimes the stoichiometry of a reaction is different, such as when NaOH reacts with malic acid, the main acid found in apples and other fruits.

COOH — CHOH + 2 NaOH — CH₂ — COOH → COO⁻ — CHOH + 2 Na⁺ + 2 H₂O — CH₂ — COO⁻

For a practical application of how normality is used to calculate results of titration experiments, see Chap. 22 in this laboratory manual.

2.3 BUFFERS

A **buffer** is an aqueous solution containing **comparable molar amounts** of either a **weak acid and its corresponding base** or a **weak base and its corresponding acid**. A buffer is used to keep a pH constant. In food analysis, buffers are commonly used in methods that utilize enzymes, but they also arise whenever weak acids or bases are titrated. To explain how a weak acid or base and its charged counterpart manage to maintain a certain pH, the "comparable molar amounts" part of the definition is key: For a buffer to be effective, its components must be present in a certain molar ratio. This section is intended to provide guidance on how to solve calculation and preparation problems relating to buffers. While it is important for food scientists to master calculations of buffers, the initial focus will be on developing an understanding of the chemistry. Figures 2.1 and 2.2 show an exemplary buffer system, and how the introduction of strong acid would affect it.

Only **weak acids** and **bases** can form buffers. The distinction of strong versus weak acids/bases is made based on how much of either H_3O^+ or OH^- is generated, respectively. Most

Example B2

Assume that 100 mL of apple juice are titrated with 1 M NaOH, and the volume used is 36 mL. What is the molarity of malic acid in the apple juice?

Solution Malic acid contains two carboxylic groups, and thus, for every mole of malic acid, 2 moles of NaOH are needed to fully ionize it. Therefore, the normality of malic acid is 2 times the molarity. Again, use Eq. 2.19 to solve Example B2:

$$36\,(\text{mL}) \times 1\left(\frac{\text{mol}}{\text{L}}\right) = 100\,(\text{mL}) \times 2 \times$$

$$\text{M of malic acid solution}\left(\frac{\text{mol}}{\text{L}}\right) \tag{2.23}$$

M of malic acid in apple juice $\left(\dfrac{\text{mol}}{\text{L}}\right)$

$$= \frac{36\,(\cancel{\text{mL}})}{100\,(\cancel{\text{mL}})} \times 1\left(\frac{\text{mol}}{\text{L}}\right) \times \frac{1}{2} = 0.18\left(\frac{\text{mol}}{\text{L}}\right) \tag{2.24}$$

This is the same value as obtained above for acetic acid. However, as a direct consequence of malic acid having two carboxyl groups instead of one, twice the amount of NaOH was needed.

Tables, calculators, and other tools for calculating molarities, normalities, and % acidity are available in print and online literature. However, it is important for a scientist to know the stoichiometry of the reactions involved, and the reactivity of reaction partners to correctly interpret these tables.

The concept of normality also applies to redox reactions; only electrons instead of protons are transferred. For instance, potassium dichromate, $K_2Cr_2O_7$, can supply six electrons, and therefore, the normality of a solution would be six times its molarity. While the use of the term normality is not encouraged by the IUPAC, the concept is ubiquitous in food analysis because it can simplify and speed up calculations.

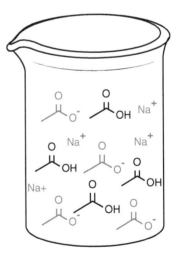

Fig. 2.1 A buffer solution composed of the weak acetic acid, CH_3COOH, and a salt of its corresponding base, sodium acetate, $CH_3COO^-Na^+$. In aqueous solution, the sodium acetate dissociates into CH_3COO^- (acetate ions) and Na^+ (sodium ions), and for this reason, these ions are drawn spatially separated. The Na^+ ions do not participate in buffering actions and can be ignored for future considerations. Note that the ratio of acetic acid and acetate is equal in our example. Typical buffer systems have concentrations between 1 and 100 mM

acids found in foods are weak acids; hence, once the equilibrium has been reached, only trace amounts have dissociated, and the vast majority of the acid is in its initial, undissociated state. For the purpose of this discussion, we will refer to the buffer components as **acid AH, undissociated state**, and corresponding **base A⁻, dissociated state**. Equilibrium concentrations can be measured and are published in the form of **dissociation constants, K_a**, or their **negative logarithms to base 10, pK_a** values. These values can be found on websites of reagent manufacturers as well as numerous other websites and textbooks. Table 2.2 lists the pK_a values of some common acids either present in foods or often used to prepare buffers, together with the acid's molecular and equivalent weights. However, reported literature values for K_a/pK_a can be different for the same compound. For instance, they range between 6.71 and 7.21 for $H_2PO_4^-$. Dissociation constants depend on the ionic strength of the system, which is influenced by the concentrations of all ions in the system, even if they do not buffer. In addition, the pH in a buffer system is, strictly speaking, determined by activities, not by concentrations (see Refs. [1, 2], Sect. 22.3.2.1). However, for concentrations <0.1 M, activities are approximately equal to concentrations, especially for monovalent ions. For $H_2PO_4^-$, the value 7.21 is better suited for very dilute systems. For buffers intended for media or cell culture, the system typically contains several salt components, and 6.8 would be a commonly used value. If literature specifies several pK_a values, try finding information on the ionic strength where these values were obtained and calculate/estimate the ionic strength of the solution where the buffer is to be used. However, even for the same ionic strength, there may be different published values, depending on the analytical method. When preparing a buffer (see notes below), *always test* and, if necessary, *adjust the pH*, even for commercially available dry buffer mixes that only need to be dissolved. The values listed in Table 2.2 are for temperatures of 25 °C and ionic strengths of 0 M, as well as 0.1 M, if available.

Maximum buffering capacity always occurs around the pK_a value of the acid component. At the pK_a, the ratio of acid/corresponding base is 1:1 (not, as often erroneously

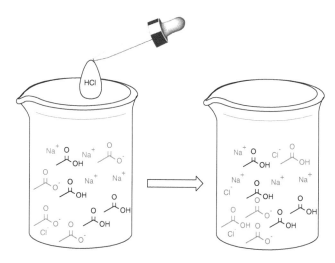

Fig. 2.2 Changes induced by addition of the strong acid HCl, to the buffer from Fig. 2.1. HCl can be considered as completely dissociated into Cl⁻ and H⁺. The H⁺ combines with CH_3COO^- instead of H_2O, because CH_3COO^- is the stronger base. Thus, instead of H_3O^+, additional CH_3COOH is formed. This alters the ratio between CH_3COOH and CH_3COO^-, resulting in a different pH as illustrated in problem C2. The Cl⁻ ions are merely counterions to balance charges but do not participate in buffer reactions and can thus be ignored

Table 2.2 Properties of common food acids

	Formula	pK_{a1}	pK_{a2}	pK_{a3}	Molecular weight	Equivalent weight
Acetic acid	CH_3COOH	4.75[a]	–	–	60.06	60.06
Carbonic acid	H_2CO_3	6.4[a] 6.1[b]	10.3[a] 9.9[b]	–	62.03	31.02
Citric acid	$HOOCCH_2C(COOH)$ $OHCH_2COOH$	3.13[a] 2.90[b]	4.76[a] 4.35[b]	6.40[a] 5.70[b]	192.12	64.04
Formic acid	$HCOOH$	3.75[a]	–	–	46.03	46.03
Lactic acid	$CH_3CH(OH)COOH$	3.86	–	–	90.08	90.08
Malic acid	$HOOCCH_2CHOHCOOH$	3.5[a] 3.24[b]	5.05[a] 4.68[b]	–	134.09	67.05
Oxalic acid	$HOOCCOOH$	1.25[a] 1.2[b]	4.27[a] 3.80[b]	–	90.94	45.02
Phosphoric acid	H_3PO_4	2.15[a] 1.92[b]	7.20[a] 6.71[b]	12.38[a] 11.52[b]	98.00	32.67
Potassium acid phthalate	$HOOCC_6H_4COO^-K^+$	5.41[a]	–	–	204.22	204.22
Tartaric acid[c]	$HOOCCH(OH)CH(OH)COOH$	3.03	4.45	–	150.09	75.05

Data from Refs. [3, 4]
[a]At ionic strength of 0 M
[b]At ionic strength of 0.1 M
[c]Dissociation constants depend on the stereoisomer (D/L vs. meso-form). Values given for the naturally occurring R,R enantiomer (L-form)

assumed, 100:0). Therefore, a certain acid/corresponding base pair is suitable for buffering a pH range of $pK_a \pm 1$. The **molarity** of a buffer refers to the sum of concentration for acid and corresponding base. The resulting pH of a buffer is governed by their concentration ratio, as described through the **Henderson-Hasselbalch equation**:

$$pH = pK_a + \log\frac{\left[A^-\right]}{\left[AH\right]} \qquad (2.25)$$

Weak bases such as ammonia, NH_3, can also form buffers with their corresponding base, in this case NH_4^+. The Henderson-Hasselbalch equation would actually not change, since NH_4^+ would serve as the acid, denoted BH, and the base NH_3 is denoted as B. The acid component is always the form with more H^+ to donate. However, you may find the alternative equation:

$$pOH = pK_b + \log\frac{\left[BH^+\right]}{\left[B\right]} \qquad (2.26)$$

The pH would then be calculated as $14 - pOH$. The term pOH refers to the concentration of OH^-, which increases when bases are present in the system (see Ref. [1], Sect. 22.3, for details). However, using Eq. 2.25 and BH^+ instead of AH as well as B in place of A^- gives the same result, as the pK_b is related to the pK_a through $14 - pK_b = pK_a$.

Example C1, C2 and C3 are several examples of buffer preparation using the Henderson-Hasselbalch equation:

Example C1

What is the pH of a buffer obtained by mixing 36 mL of a 0.2 M Na_2HPO_4 solution and 14 mL of a 0.2 M NaH_2PO_4 solution, after adding water to bringing the volume to 100 mL to obtain a 0.1 M buffer?

Solution The Henderson-Hasselbalch equation requires knowledge of the acidic component's pK_a value and the concentrations of both buffer components. Rearrange Eq. 2.15 so that:

$$M \text{ in buffer}\left(\frac{mol}{L}\right) =$$

$$\frac{M \text{ of stock solutions}\left(\frac{mol}{L}\right) \times v \text{ of stock solutions}(L)}{v \text{ of buffer}(L)}$$

$$(2.27)$$

$$M \text{ of } Na_2HPO_4\left(\frac{mol}{L}\right) = \frac{0.2\left(\frac{mol}{L}\right) \times 0.036(L)}{0.1(L)}$$

$$= 0.072\left(\frac{mol}{L}\right) \qquad (2.28)$$

$$M \text{ of } Na_2HPO_4\left(\frac{mol}{L}\right) = \frac{0.2\left(\frac{mol}{L}\right) \times 0.014(L)}{0.1(L)}$$

$$= 0.028\left(\frac{mol}{L}\right) \qquad (2.29)$$

The other necessary step to solve Example C1 is finding the correct pK_a.

$$H_3PO_4 \underset{k_{-1}}{\overset{k_1}{\rightleftharpoons}} H_2PO_4^- + H^+ \underset{k_{-2}}{\overset{k_2}{\rightleftharpoons}} H_2PO_4^- +$$

$$H^+ \underset{k_{-3}}{\overset{k_3}{\rightleftharpoons}} PO_4^{3-} + H^+$$

In this buffer, $H_2PO_4^-$ acts as the acid, as it donates H^+ more strongly than HPO_4^{2-}. The acid component in a buffer is always the one with more acidic H^+ attached. Hence, the relevant pK_a value is pk_2. As listed in Table 2.2, this value is 6.71.

The solution to Example C1 is now only a matter of inserting values into Eq. 2.25:

$$pH = 6.71 + \log\frac{[0.072]}{[0.028]} \qquad (2.30)$$

$$pH = 6.71 + 0.41 = 7.12 \qquad (2.31)$$

Example C2

How would the pH of the buffer in example C1 change upon addition of 1 mL of 2 M HCl? (*Note:* You may ignore the slight change in volume caused by the HCl addition.)

Solution Calculate the moles of HCl supplied using Eq. 2.1. HCl converts HPO_4^{2-} into $H_2PO_4^-$, because HPO_4^{2-} is a stronger base than $H_2PO_4^-$, as apparent by its higher pK_a value. This changes the ratio of [AH]:[A$^-$]. The new ratio needs to be inserted into Eq. 2.24. To calculate the new ratio, account for the volume of the buffer, 0.1 L, and calculate the amount of HPO_4^{2-} and $H_2PO_4^-$ present:

$$n \text{ of buffer components} (\text{mol}) = M \times v \quad (2.2)$$

$$n \text{ of Na2HPO4} = 0.072 \times 0.1 = 0.0072 (\text{mol}) \quad (2.32)$$

$$n \text{ of NaH2PO4} = 0.028 \times 0.1 = 0.0028 (\text{mol}) \quad (2.33)$$

$$n \text{ of HCl} (\text{mol}) = M \times v = 2 \times 0.001 = 0.002 (\text{mol}) \quad (2.34)$$

n of Na_2HPO_4 after HCl addition (mol):

$$0.0072 - 0.002 = 0.0052 \quad (2.35)$$

n of NaH_2PO_4 after HCl addition (mol):

$$0.0028 + 0.002 = 0.0048 \quad (2.36)$$

pH of buffer after HCl addition and conversion of n into M through Eq. 2.1:

$$pH = 6.71 + \log \frac{[0.052]}{[0.048]} = 6.74 \quad (2.37)$$

(*Note:* The ratio of A$^-$ to AH stays the same whether concentrations or amounts are inserted, since units are canceled, so technically n does not need to be converted into M to obtain the correct result.)

(5) and the pK_a (4.76), the values are inserted into Eq. 2.25:

$$\text{Molarity of buffer} \left(\frac{\text{mol}}{\text{L}}\right) = [AH] + [A^-] \quad (2.38)$$

$$0.1 = [A^-] + [AH] \quad (2.39)$$

$$[A^-] = 0.1 - [AH] \quad (2.40)$$

$$5 = 4.76 + \log \frac{0.1 - [AH]}{[AH]} \quad (2.41)$$

$$0.24 = \log \frac{0.1 - [AH]}{[AH]} \quad (2.42)$$

$$1.7378 = \frac{0.1 - [AH]}{[AH]} \quad (2.43)$$

$$1.7378 \times [AH] = 0.1 - [AH] \quad (2.44)$$

$$1.7378 \times [AH] + [AH] = 0.1 \quad (2.45)$$

$$[AH] \times (1.7378 + 1) = 0.1 \quad (2.46)$$

$$[AH] = \frac{0.1}{1.7378 + 1} \quad (2.47)$$

$$[AH] = 0.0365 \left(\frac{\text{mol}}{\text{L}}\right) \quad (2.48)$$

$$[A^-] = 0.1 - 0.0365 = 0.0635 \left(\frac{\text{mol}}{\text{L}}\right) \quad (2.49)$$

Example C3

Prepare 250 mL of 0.1 M acetate buffer with pH 5. The pK_a of acetic acid is 4.76 (see Table 2.2). The molecular weights of acetic acid and sodium acetate are 60.06 and 82.03 g/mol, respectively.

Solution This example matches the tasks at hand in a lab better than Example C1. One needs a buffer to work at a certain pH, looks up the pK_a value, and decides on the molarity and volume needed. The molarity of a buffer equals [A$^-$] + [AH]. To solve Example C3, one of those concentrations needs to be expressed in terms of the other, so that the equation only contains one unknown quantity. For this example, it will be [A$^-$], but the results would be the same if [AH] had been chosen. Together with the target pH

There are three ways to prepare such a buffer:

1. Prepare 0.1 M acetic acid and 0.1 M sodium acetate solutions. For our example, 1 L will be prepared. Sodium acetate is a solid and can be weighed out directly using Eq. 2.5. For acetic acid, it is easier to pipet the necessary amount. Rearrange Eq. 2.8 to calculate the volume of concentrated acetic acid (density = 1.05) needed to prepare a volume of 1 L. Then express the mass through Eq. 2.5.

$$m \text{ of sodium acetate} (\text{g}) = M \times v \times MW \quad (2.5)$$

$$m \text{ of sodium acetate} (\text{g}) = 0.1 \left(\frac{\text{mol}}{\text{L}}\right) \times 1 (\text{L}) \times 82 \left(\frac{\text{g}}{\text{mol}}\right) = 8.2 (\text{g}) \quad (2.50)$$

$$v \text{ of acetic acid}(\text{mL}) = \frac{m(\text{g})}{d\left(\dfrac{\text{g}}{\text{mL}}\right)} \quad (2.51)$$

$$v \text{ of acetic acid}(\text{mL}) =$$
$$\frac{M\left(\dfrac{\text{mol}}{\text{L}}\right) \times v(\text{L}) \times MW\left(\dfrac{\text{g}}{\text{mol}}\right)}{d\left(\dfrac{\text{g}}{\text{mL}}\right)} \quad (2.52)$$

$$v \text{ of acetic acid}(\text{mL}) =$$
$$\frac{60.06\left(\dfrac{\text{g}}{\text{mol}}\right) \times 0.1\left(\dfrac{\text{mol}}{\text{L}}\right) \times 1(\text{L})}{1.05\left(\dfrac{\text{g}}{\text{mL}}\right)} = 5.72(\text{mL}) \quad (2.53)$$

Dissolving each of these amounts of sodium acetate and acetic acid in 1 L of water gives two 1 L stock solutions with a concentration of 0.1 M. To calculate how to mix the stock solutions, use their concentrations in the buffer, that is, 0.0365 as obtained from Eq. 2.48 and Eq. 2.49: 0.0365 $\left(\dfrac{\text{mol}}{\text{L}}\right)$ for acetic acid and 0.0635 $\left(\dfrac{\text{mol}}{\text{L}}\right)$ for sodium acetate and Eq. 2.15:

$$v \text{ of acetic acid stock solution}(\text{L}) =$$
$$\frac{M \text{ of acetic acid in buffer}\left(\dfrac{\text{mol}}{\text{L}}\right) \times v \text{ of buffer}(\text{L})}{M \text{ of stock solution}\left(\dfrac{\text{mol}}{\text{L}}\right)} \quad (2.54)$$

$$v \text{ of acetic acid stock solution}(\text{L}) =$$
$$\frac{0.0365 \times 0.25}{0.1} = 0.091(\text{L}) \quad (2.55)$$

$$v \text{ of sodium acetate stock solution}(\text{L}) =$$
$$\frac{0.0635 \times 0.25}{0.1} = 0.159(\text{L}) \quad (2.56)$$

Combining the 0.091 L of acetic acid stock and 0.159 L of sodium acetate stock solution gives 0.25 L of buffer with the correct pH and molarity.

2. Directly dissolve appropriate amounts of both components in the same container. Eq. 2.48 and Eq. 2.49 yield the molarities of acetic acid and sodium acetate in a buffer, that is, the moles per 1 L. Just like for approach 1 above, use Eq. 2.5 and Eq. 2.52 to calculate the m for sodium acetate and v for acetic acid, but this time use the buffer volume of 0.25 [L] to insert for v:

For sodium acetate:

$$m(\text{g}) = M\left(\dfrac{\text{mol}}{\text{L}}\right) \times v(\text{L}) \times MW\left(\dfrac{\text{g}}{\text{mol}}\right) \quad (2.57)$$
$$= 0.0635 \times 0.25(\text{L}) \times 82.03 = 1.3 \text{ g}$$

For acetic acid:

$$v(\text{mL}) = \frac{M\left(\dfrac{\text{mol}}{\text{L}}\right) \times v(\text{L}) \times MW\left(\dfrac{\text{g}}{\text{mol}}\right)}{d\left(\dfrac{\text{g}}{\text{mL}}\right)} \quad (2.58)$$
$$= \frac{0.0365 \times 0.25 \times 60.02}{1.05} = 0.52 \text{ mL}$$

Dissolve both reagents in the same glassware in 200 mL water, adjust the pH if necessary, and bring up to 250 mL after transferring into a volumetric flask.

(*Note:* It does not matter in how much water you initially dissolve these compounds, but it should be >50% of the total volume. Up to a degree, buffers are independent of dilution; however, you want to ensure complete solubilization and leave some room to potentially adjust the pH.)

3. Pipet the amount of acetic acid necessary for obtaining 250 mL of a 0.1 M acetic acid solution, but dissolve in <250 mL, e.g., 200 mL. Then add concentrated NaOH solution drop-wise until pH 5 is reached, and make up the volume to 250 mL. The amount of acetic acid is found analogously to Eq. 2.52:

$$v \text{ acetic acid}(\text{mL}) = \frac{M \times v \times MW}{d} \quad (2.59)$$
$$= \frac{0.1 \times 0.25 \times 60.02}{1.05} = 1.43(\text{mL})$$

You will find all three approaches described above if you search for buffer recipes online and in published methods. For instance, AOAC Method 991.43 for total dietary fiber involves approach 3. It requires dissolving 19.52 g of 2-(N-morpholino)ethanesulfonic acid (MES) and 12.2 g 2-amino-2-hydroxymethyl-propane-1,3-diol (Tris) in 1.7 L water, adjusting the pH to 8.2 with 6 M NaOH, and then making up the volume to 2 L.

The most common approach for preparing buffers is approach 1. It has the advantage that once stock solutions are prepared, they can be mixed in different ratios to obtain a range of pH values, depending on the experiment. One disadvantage is that if the pH needs to be adjusted, then either some acid or some base needs to be added, which slightly alters the volume and thus the concentrations. This problem can be solved by preparing stock solutions of higher concentrations and adding water to the correct volume, like in Example C1, for which 36 mL and 14 mL of 0.2 M stock solutions were combined and brought to a volume of 100 mL to give a 0.1 M buffer. This way, one also corrects for potential volume contraction effects that may occur when mixing solutions. Another potential disadvantage of stock solutions is that they are often not stable for a long time (see Notes below).

2.4   NOTES ON BUFFERS

When choosing an appropriate buffer system, the most important selection criterion is the pK_a of the acid component. However, depending on the system, additional factors may need to be considered, as detailed below (listed in no particular order of importance):

1. The buffer components need to be well soluble in water. Some compounds require addition of acids or bases to fully dissolve.
2. Buffer recipes may include salts that do not participate in the buffering process, such as sodium chloride for phosphate-buffered saline. However, the addition of such salts changes the ionic strength and affects the acid's pK_a. Therefore, combine all buffer components before adjusting the pH.
3. If a buffer is to be used at a temperature other than room temperature, heat or cool it to this intended temperature before adjusting the pH. Some buffer systems are more affected than others, but it is always advisable to check. For instance, 2-[4-(2-hydroxyethyl)piperazin-1-yl]ethanesulfonic acid [HEPES] is a widely used buffer component for cell culture experiments. At 20 °C, its pK_a is 7.55, but its change in pH from 20 to 37 °C is −0.014 $\Delta pH/°C$ (3). Thus, the pK_a at 37 °C would be:

$$pKa\ at\ 37°C = pH\ at\ 20°C - \Delta pH \times (T1 - T2) \qquad (2.60)$$

$$7.55 - 0.014 \times (37 - 20) = 7.31 \qquad (2.61)$$

4. Do not bring the buffer up to volume before adjusting the pH. Use relatively concentrated acids and bases for this purpose, so that the volumes needed for pH adjustment are small.
5. Ensure that buffer components do not interact with the test system. This is especially important when performing experiments on living systems, such as cell cultures, but even in vitro systems are affected, particularly when enzymes are used. For instance, phosphate buffers tend to precipitate with calcium salts or affect enzyme functionality. For this reason, a range of zwitterionic buffers with sulfonic acid and amine groups has been developed for use at physiologically relevant pH values.
6. Appropriate ranges for pH and molarities of buffer systems that can be described through the Henderson-Hasselbalch equation are roughly 3–11 and 0.001–0.1 M, respectively.
7. The calculations and theoretical background described in this chapter apply to aqueous systems. Consult appropriate literature if you wish to prepare a buffer in an organic solvent or water/organic solvent mixtures.
8. If you store non-autoclaved buffer or salt solutions, be aware that over time microbial growth or precipitation may occur. Visually inspect the solutions before use, and discard if cloudy or discolored. If autoclaving is an option, check if the buffer components are suitable for this process.
9. When developing a standard operating procedure for a method that includes a buffer, include all relevant details (e.g., reagent purity) and calculations. This allows tracing mishaps back to an inappropriate buffer.
10. There are numerous online tools that can help with preparing buffer recipes. They can save time and potentially allow you to verify calculations. However, they are not a substitute for knowing the theory and details about the studied system.

2.5   PRACTICE PROBLEMS

(Note: Answers to problems are in the last section of the laboratory manual.)

1. (a) How would you prepare 500 mL of 0.1 M NaH_2PO_4 starting with the solid salt?
   (b) When you look for NaH_2PO_4 in your lab, you find a jar with $NaH_2PO_4 \cdot 2\ H_2O$. Can you use this chemical instead, and if so, how much do you need to weigh out?
2. How many g of dry NaOH pellets (molecular weight: 40 g/mol) would you weigh out for 150 mL of 10% wt/vol sodium hydroxide?
3. What is the normality of a 40% wt/vol sodium hydroxide solution?
4. How many mL of 10 N NaOH would be required to neutralize 200 mL of 2 M H_2SO_4?
5. How would you prepare 250 mL of 2 N HCl starting with concentrated HCl? The supplier states that its concentration is 37% wt/wt, and the density is 1.2.
6. How would you prepare 1 L of 0.04 M acetic acid starting with concentrated acetic acid (density = 1.05)? The manufacturer states that the concentrated acetic acid is >99.8%, so you may assume that it is 100% pure.
7. Is a 1% wt/vol acetic acid solution the same as a 0.1 M solution? Show calculations.
8. Is a 10% wt/vol NaOH solution the same as a 1 N solution? Show calculations.
9. What would the molarity and normality of a solution of 0.2 g potassium dichromate (molecular weight = 294.185)

in 100 mL water be? It acts as an oxidizer that can transfer 6 electrons per reaction.

10. How would you make 100 mL of a 0.1 N KHP solution? (*Note:* for this practice problem, you only need to calculate the amount to weigh out; not explain how to standardize it. Chapter 22 of this laboratory manual provides further information about standardization of acids and bases.)

11. You want to prepare a standard for atomic absorption spectroscopy measurements containing 1000 ppm Ca. How much $CaCl_2$ do you need to weigh out for 1000 mL stock solution? The atomic mass unit of Ca is 40.078, and the molecular weight of $CaCl_2$ is 110.98 g/mol.

12. Outline how you would prepare 250 mL of 0.1 M acetate buffer at pH 5.5 for enzymatic glucose analysis.

13. Complexometric determination of calcium requires an ammonium buffer with 16.9 g ammonium chloride (molecular weight, 53.49) in 143 mL concentrated ammonium hydroxide solution (28% wt/wt, density = 0.88; molecular weight = 17; pK_b = 4.74). It is also to contain 1.179 g of $Na_2EDTA\cdot2H_2O$ (molecular weight = 372.24 g/mol) and 780 mg $MgSO_4\cdot7H_2O$ (molecular weight = 246.47 g/mol). After combining all reagents, the volume is brought up to 250 mL. What are the molarities for EDTA and $MgSO_4$, and what pH does this buffer have?

14. Your lab uses 0.2 M stock solutions of NaH_2PO_4 and Na_2HPO_4 for buffer preparation.
    (a) Calculate the amounts weighed out for preparing 0.5 L of fresh 0.2 M stock solutions of $NaH_2PO_4\cdot H_2O$ (molecular weight = 138) and $Na_2HPO_4\cdot7\ H_2O$ (molecular weight = 268).
    (b) You want to make 200 mL of a 0.1 M buffer with pH 6.2 from these stock solutions. How many mL of each stock solution do you have to take?
    (c) How would the pH change if 1 mL of a 6 M NaOH solution was added (*note:* you may ignore the volume change)?

15. Tris (2-amino-2-hydroxymethyl-propane-1,3-diol) is an amino compound suitable for the preparation of buffers in physiological pH ranges such as for the dietary fiber assay; however, its pK_a is highly affected by temperature. The reported pKa at 25 °C is 8.06 [2]. Assuming a decline in pK_a of approximately 0.023 ΔpH/°C, what pH would a Tris buffer with a molar acid/base ratio of 4:1 have at 60 °C versus at 25 °C?

16. Ammonium formate buffers are useful for LC-MS experiments. How would you prepare 1 L of a 0.01 M buffer of pH 3.5 with formic acid (pK_a = 3.75, 98% wt/wt, density = 1.2) and ammonium formate (molecular weight 63.06 g/mol)? (*Note:* You may ignore the contribution of ammonium ions to the pH and focus on the ratio of formic acid/formate anion. You may treat the formic acid as pure for the calculation, i.e., ignore the % wt/wt.)

REFERENCES

1. Tyl C (2024) pH and titratable acidity. Ch. 22. In: Ismail BP, Nielsen SS (eds) Nielsen's food analysis, 6th edn. Springer, New York

2. Harakany AA, Abdel Halim FM, Barakat AO (1984) Dissociation constants and related thermodynamic quantities of the protonated acid form of tris-(hydroxymethyl)-aminomethane in mixtures of 2-methoxyethanol and water at different temperatures. J. Electroanalytical Chemistry and Interfacial Electrochemistry 162:285–305

3. Albert A, Serjeant EP (1984) The determination of ionization constants. A laboratory manual, 3rd edn. Chapman and Hall, New York

4. Harris D, Lucy CA (2019) Quantitative Chemical Analysis, 10th edn W.H. Freeman and Company, New York

Dilutions and Concentrations

3

Andrew P. Neilson

Contents

3.1 INTRODUCTION

This chapter builds upon the information presented in the previous Chap. 1. This chapter covers the following topics:

1. Reasons for performing dilutions and concentrations in food analysis

2. Basic calculations and strategies for calculating the final concentration, given the initial concentration and a known dilution/concentration scheme (and vice versa)

3. Strategies for designing and performing dilutions to obtain standard curves

4. Example and practice problems

This information will be used repeatedly in a food analysis course (e.g., homework, laboratories, and exams). More importantly, the principles described in this chapter are essential for virtually all laboratory or bench work in food science, including quality assurance/quality control, analysis for food labeling, and product formulation. Understanding

A. P. Neilson (✉)
Department of Food, Bioprocessing and Nutrition Sciences, North Carolina State University, Raleigh, NC, USA
e-mail: aneilso@ncsu.edu

© The Author(s), under exclusive license to Springer Nature Switzerland AG 2024
B. P. Ismail, S. S. Nielsen (eds.), *Nielsen's Food Analysis Laboratory Manual*, Food Science Text Series,
https://doi.org/10.1007/978-3-031-44970-3_3

how to perform dilutions and concentrations, and calculate the resulting solution concentrations, is an essential skill for both quantitative and qualitative analysis.

3.2   REASONS FOR DILUTIONS AND CONCENTRATIONS

There are various reasons why dilutions and concentrations are performed during food analysis, including:

1. Dilution:
   (a) To reduce high analyte concentrations in the sample down to levels within the operating range or optimal range of a method/instrument
   (b) To reduce high analyte concentrations in the sample down to levels within the linear region of a method/instrument or to levels within a defined standard curve (see *Practice Problems 5 and 7*)
   (c) To dilute the background matrix to levels that do not interfere with the analysis
   (d) Reagent addition, which dilutes the sample by increasing the volume
   (e) Solvent extraction, which dilutes the sample by using large volumes of solvent to favor transfer of the analyte from sample to solvent (see *Practice Problem 1*)
2. Concentration:
   (a) To increase low analyte concentrations in the sample up to levels within the operating range or optimal range of a method/instrument
   (b) To increase low analyte concentrations in the sample up to levels in the linear region of a method or to within a standard curve (see *Practice Problem 1*)
   (c) Post-extraction evaporation of solvent (see *Practice Problem 1*)

3.3   USING VOLUMETRIC GLASSWARE TO PERFORM DILUTIONS AND CONCENTRATIONS

The proper use of laboratory glassware was covered in Chap. 1. However, a few key points are sufficiently important for accurate calculations regarding dilutions and concentrations to mention again:

1. Class A glass volumetric flasks are used for bringing samples or solutions up to a known "**to contain**" (TC) volume.
2. Class A glass volumetric pipettes are used for transferring (delivering) a known "**to deliver**" (TD) volume of sample.

3. Other types of pipettes (non-Class A volumetric glass pipettes, adjustable pipettors, automatic pipettors, reed pipettors, graduated measuring pipettes) and other glassware (graduated cylinders, beakers, etc.) are less accurate and less precise and should not be used for quantitative volume measurements.
4. Dilutions can be performed as "**diluted to**" or "**diluted to a final volume of**" or as "**diluted with**." These are very different terms (see Chap. 1 for definitions). It is critical to understand the differences between these types of dilutions to perform lab procedures and calculations correctly.

3.4   CALCULATIONS FOR DILUTIONS AND CONCENTRATIONS

3.4.1   Introduction

Calculations involving dilutions or concentrations are critical for quantitative measurements as well as preparing reagents for many qualitative analyses. Many analytical methods (such as spectrophotometric assays, chromatography, titrations, protein determinations, etc.) are performed on diluted or concentrated samples, and the analytical result must be converted to the concentration in the undiluted or unconcentrated sample for labeling or research purposes. Instances for which these calculations are used include, but are not limited to:

1. Preparation of standard curves (see *Practice Problems 3 and 4*)
2. Determining the necessary dilution required to obtain a sample concentration within a specified range (see *Practice Problems 5 and 7*)
3. Determining the necessary range of dilutions for preparing a range of concentrations for a standard curve from a stock standard solution (see *Practice Problems 5 and 7*)
4. For converting an analytical result obtained from a diluted or concentrated sample to the undiluted or concentrated food (see *Practice Problems 1 and 6*)

3.4.2   Expressing Concentration

Recall that the concentration of an analyte in a sample or solution is defined as follows:

$$C = \frac{X}{m} \quad \text{or} \quad C = \frac{X}{V} \tag{3.1}$$

C = concentration
X = amount of analyte (g, mol, etc.)
V = sample volume
m = sample mass

Note that concentrations also can be expressed in terms of **percentages** (%), parts per million (ppm), etc. (see Chap. 2, Table 2.1). Percentage is a particular problem as it can refer to ratios expressed as weight per weight, weight per volume, or volume per volume. Simply saying 5% ethanol is unclear as it could represent 5% w/w, w/v, or v/v. Therefore, % are typically expressed with the accompanying notation (w/w, w/v, v/v, etc.) to clarify the meaning of the % value. Rearranging Eq. 3.1, the amount of analyte can be expressed as:

$$X = Cm \quad \text{or} \quad X = CV \tag{3.2}$$

For each step in a dilution or concentration, a portion (mass or volume) of a sample or solution is either diluted with additional liquid or reduced in volume. The mass, volume, and concentration of the sample changes, but the amount (mol, g, etc.) of the analyte present in the amount of the sample that is diluted or concentrated does not change. Therefore, the following is true:

$$X_1 = X_2 \tag{3.3}$$

X_1 = the amount of analyte in the sample before the dilution/concentration step
X_2 = the amount of analyte in the sample after the dilution/concentration step

---

Example A1
Suppose that a stock solution of thiamine is made by dissolving 168 mg thiamine to a volume of 150 mL, and then 0.25 mL of that stock solution is added to a volumetric flask and water is added so that the final volume is 200 mL. The concentration of thiamine in the stock solution is:

$$C = \frac{X}{V} = \frac{168 \text{ mg thiamine}}{150 \text{ mL solution}} = \frac{1.12 \text{ mg thiamine}}{\text{mL solution}}$$

The amount of thiamine in the 0.25 mL that is diluted is:

$$X = CV = \left(0.25 \text{ mL solution}\right)\left(\frac{1.12 \text{ mg thiamine}}{\text{mL solution}}\right)$$

$$= 0.28 \text{ mg thiamine}$$

Note that we are only concerned with the amount of analyte in the portion of the sample that is diluted/concentrated (0.25 mL in this example) and *not* the amount

---

of analyte in the whole initial sample (150 mL in this case). When the 0.25 mL of the stock solution (containing a total of 0.28 mg thiamine) is diluted to a final volume of 200 mL, the total amount of thiamine (0.28 mg) does not change because the added water does not contain any thiamine. Therefore:

$$X_1 = X_2 = 0.28 \text{ mg thiamine}$$

However, the concentration of thiamine changes, because now the 0.28 mg thiamine is present in 200 mL instead of 0.25 mL:

$$C = \frac{X}{V} = \frac{0.28 \text{ mg thiamine}}{200 \text{ mL solution}} = \frac{0.0014 \text{ mg thiamine}}{\text{mL solution}}$$

Therefore, the thiamine solution has been diluted, as the final concentration (0.0014 mg/mL) is less than the initial concentration (1.68 mg/mL).

It is critical to understand the fundamental principle that "concentration" and "amount" are very distinct concepts. **Concentration** is a ratio of the amount of analyte to the amount of sample. The concentration (mg/mL, M, N, %, ppm, etc.) of an analyte present in a sample is not dependent on the amount (g, mL, etc.) of sample.

Example A2
If a pesticide is present at 2.8 parts per billion (ppb) (note that 1 ppb = 1 µg/kg) in a sample of applesauce, the pesticide is uniformly distributed throughout the applesauce, and therefore, the concentration is 2.8 ppb regardless of whether 1 µg, 1 mg, 1 g, 1 kg, 1 mL, or 1 L of applesauce is analyzed. However, the total amount of the analyte present does depend on the size of the sample. We can easily see that 0.063 kg and 2.2 kg of applesauce contain the same concentration, but vastly different total amounts, of the pesticide:

$$X = Cm \tag{3.4}$$

$$0.063 \text{ kg applesauce}\left(\frac{2.8 \text{ } \mu g \text{ pesticide}}{\text{kg applesauce}}\right)$$

$$= 0.176 \text{ } \mu g \text{ pesticide}$$

$$2.2 \text{ kg applesauce}\left(\frac{2.8 \text{ } \mu g \text{ pesticide}}{\text{kg applesauce}}\right)$$

$$= 6.16 \text{ } \mu g \text{ pesticide}$$

Once we know the concentration of a sample or solution, we can take any part, or all, of the sample. This changes the amount of the analyte we have, but not the concentration. Then, diluting or concentration the aliquot chosen changes the concentration, but not the amount, of the analyte. These are two critical concepts that must be implicitly understood to master the concepts of dilution and concentrations. Since the total amount of analyte does not change during each dilution or concentration step, the following equations can be derived:

$$X_i = X_f \tag{3.5}$$

and $X = Cm$ or $X = CV$, therefore:

$$C_i V_i = C_f V_f \quad \text{or} \quad C_i m_i = C_f m_f \tag{3.6}$$

i = initial (before dilution or concentration)
f = final (after dilution or concentration).

This equation is the critical equation for solving all dilution and concentration problems.

3.4.3 Forward Calculations

The above relationship between the initial and final concentration can be used to determine the starting and/or final concentration if one concentration and the masses or volumes used in the dilution are known. If the starting concentration is known, the final concentration can be calculated (a **"forward" calculation**, working forward from initial to final):

$$C_f = \frac{C_i V_i}{V_f} \quad \text{or} \quad C_f = \frac{C_i m_i}{m_f} \tag{3.7}$$

Similarly, if the final concentration is known, the starting concentration can be calculated (a **"back" calculation**, working back from final to initial). From this relationship, it becomes evident that the final concentration can be expressed as the initial concentration multiplied by the ratio of the initial and final concentrations (or vice versa for the initial concentration):

$$C_f = \frac{C_i V_i}{V_f} = C_i \left(\frac{V_i}{V_f} \right) \quad \text{or} \quad C_f = \frac{C_i m_i}{m_f} = C_i \left(\frac{m_i}{m_f} \right) \tag{3.8}$$

This is intuitive, as the ratio of the final to the initial volumes, masses, or concentrations is referred to as the "fold" of the dilution. The ratio of the starting and final masses or volumes for each step is referred to as the **dilution factor** (DF) for that step (see *Practice Problem 2*).

$$DF = \frac{V_i}{V_f} \tag{3.9}$$

The final concentration is the product of the initial concentration and the DF (the DF is a "multiplier" that can be used to convert the initial concentration to the final concentration):

$$C_f = \frac{C_i V_i}{V_f} = C_i \left(\frac{V_i}{V_f} \right) = C_i (DF) \tag{3.10}$$

Another way to think of the DF is the ratio of the final and initial concentrations for each step:

$$DF = \frac{C_f}{C_i} \tag{3.11}$$

Some conventions that apply to the term "dilution factor" are as follows:

1. DF always refers to the *forward* direction (initial mass or volume divided by final mass or volume for each step).
2. The term DF is usually used even if the step is a concentration; less commonly, the term "concentration factor" may be used.
3. Since DF is a multiplier of the initial concentration:
   (a) If the step is a dilution, DF < 1 ($C_f < C_i$).
   (b) If the step is a concentration, DF > 1 ($C_f > C_i$).

The "fold" or "X" of a dilution is defined as follows:

$$\text{dilution "fold" or "X"} = \frac{1}{DF} \tag{3.12}$$

Example A3

For the thiamine solution described earlier, if 1 mL of the solution is added to a 10 mL volumetric flask and diluted to volume, the ratio of the final to initial volumes is 10:1, or a "tenfold" or "10X" dilution, and therefore, the final concentration is tenfold (10X) lower than, or one/tenth of, the initial concentration:

$$C_f = \frac{C_i V_i}{V_f} = C_i \left(\frac{V_i}{V_f} \right) = \frac{1.12 \text{ mg thiamine}}{\text{mL solution}} \left(\frac{1 \text{ mL}}{10 \text{ mL}} \right)$$

$$= \frac{0.112 \text{ mg thiamine}}{\text{mL solution}}$$

For a tenfold dilution:

$$\text{dilution "fold" or "X"} = \frac{1}{DF}$$

Therefore, $$DF = \frac{1}{\text{dilution "fold" or "X"}}$$

$$DF = \frac{1}{10} = 0.1$$

$$C_f = C_i(DF) = \frac{1.12 \text{ mg thiamine}}{\text{mL solution}}(0.1)$$

$$= \frac{0.112 \text{ mg thiamine}}{\text{mL solution}}$$

As discussed above, note that this is different than adding 10 mL water to 1 mL of the thiamine solution, which makes the final volume 11 mL for an 11-fold or 11X dilution:

$$C_f = C_i\left(\frac{V_i}{V_f}\right) = \frac{1.12 \text{ mg thiamine}}{\text{mL solution}}\left(\frac{1 \text{ mL}}{1 \text{ mL} + 10 \text{ mL}}\right)$$

$$= \frac{1.12 \text{ mg thiamine}}{\text{mL solution}}\left(\frac{1 \text{ mL}}{11 \text{ mL}}\right)$$

$$= \frac{1.102 \text{ mg thiamine}}{\text{mL solution}}$$

For concentration processes, the calculation is the same except the analyte concentration increases from initial to final. For example, suppose that 12.8 mL of the thiamine solution is reduced (by boiling, freeze drying, rotary evaporation, etc.) to 3.9 mL. The final concentration is:

$$C_f = C_i\left(\frac{V_i}{V_f}\right) = \frac{1.12 \text{ mg thiamine}}{\text{mL solution}}\left(\frac{12.8 \text{ mL}}{3.9 \text{ mL}}\right)$$

$$= \frac{3.68 \text{ mg thiamine}}{\text{mL solution}}$$

Note again that it is critical to understand that the amount of thiamine did not change during concentration; only the volume of sample containing that amount changed.

Note that masses can be used instead of volumes:

$$C_i m_i = C_f m_f \qquad (3.13)$$

Furthermore, both masses and volumes can be used:

$$C_i m_i = C_f V_f \quad \text{or} \quad C_i V_i = C_f m_f \qquad (3.14 \text{ and } 3.15)$$

Example A4

Suppose that soybean oil contains 175 mg oleic acid per g, and 2.5 g of the oil is diluted with hexane to a final volume of 75 mL. The concentration of oleic acid in the final solution can be determined as follows:

$$C_i m_i = C_f V_f$$

$$C_f = C_i\left(\frac{m_i}{V_f}\right) = \frac{175 \text{ mg oleic acid}}{\text{g oil}}\left(\frac{2.5 \text{ g oil}}{75 \text{ mL}}\right)$$

$$= \frac{5.83 \text{ mg oleic acid}}{\text{mL}}$$

3.4.4 Back Calculations

The examples worked up to this point involve "forward" calculations (calculating the final concentration from a known initial concentration, see *Practice Problems 3 and 4*). However, many food analysis calculations will involve "back" calculations (calculating the initial sample concentration from a final concentration obtained by analysis of the diluted or concentrated sample, see *Practice Problems 1 and 6*). This calculation is simply the reverse of the "forward" calculation, for which the initial concentration is solved for as opposed to the final concentration:

$$C_i V_i = C_f V_f \quad \text{or} \quad C_i m_i = C_f m_f, \qquad \text{therefore:}$$

$$C_i = \frac{C_f V_f}{V_i} = C_f\left(\frac{V_f}{V_i}\right) \quad \text{or} \quad C_i = \frac{C_f m_f}{m_i} = C_f\left(\frac{m_f}{m_i}\right)$$

$$(3.16)$$

Note that for back calculations, we flip the DF:

$$C_i = \frac{C_f V_f}{V_i} = C_f\left(\frac{V_f}{V_i}\right) = C_f\left(\frac{1}{DF}\right)$$

$$\text{or} \quad C_i = \frac{C_f m_f}{m_i} = C_f\left(\frac{1}{DF}\right) \qquad (3.17)$$

Example A5

Suppose that 50 mL of water is added to a 35 mL sample of apple juice (therefore, initial volume = 35 mL and final volume = 85 mL), and titration indicates that the concentration of malic acid in the diluted sample is 0.24% w/v. What is the concentration of malic acid in the undiluted juice? This can be determined using Eq. 3.17 above:

$$C_i = C_f\left(\frac{V_f}{V_i}\right) = 0.24\% \text{ w/v}\left(\frac{85 \text{ mL}}{35 \text{ mL}}\right) = 0.583\% \text{ w/v}$$

A special case of concentration is when a solution is completely evaporated or dried (so that there is no liquid left, only nonvolatile solutes) and then reconstituted with a volume of liquid smaller than the initial volume. This is technically composed of two distinct steps: a concentration step (drying the sample or solution until only solutes, usually a very small mass, are left) followed by a dilution step (diluting the residual dry mass to a define volume).

As long as the entire residue is reconstituted, the calculation can be treated as a single step.

Example A6

Suppose that 10 mL apple juice, containing 0.025 g malic acid/mL, is freeze-dried, leaving a residue of 0.09 g. This residue is then reconstituted to a final volume of 3 mL with water. What is the concentration of malic acid in the final solution? For the first step, the problem is set up as:

$$C_f = C_i \left(\frac{V_i}{V_f} \right) = \frac{0.025 \text{ g malic acid}}{\text{mL juice}} \left(\frac{10 \text{ mL juice}}{0.09 \text{ g residue}} \right)$$

$$= \frac{2.78 \text{ g malic acid}}{\text{g residue}}$$

For the second step, the problem is set up as:

$$C_f = C_i \left(\frac{V_i}{V_f} \right) = \frac{2.78 \text{ g malic acid}}{\text{g residue}} \left(\frac{0.09 \text{ g residue}}{3 \text{ mL}} \right)$$

$$= \frac{0.0833 \text{ g malic acid}}{\text{mL}}$$

However, in actual laboratory practice, this is typically treated as a single step (starting with the initial volume and ending with the final diluted volume) for two reasons. First, the whole dried sample is typically reconstituted, so the final mass for the first step and the initial mass of the second step are the same and thus cancel each other out in the calculation. Second, for most concentrations involving solutions, the mass of the solutes remaining after complete drying is much too small to be measured accurately with balances commonly found in most food analysis laboratories, and it is difficult to get all of the residue out accurately for weighing. Therefore, these problems are typically solved in a single step as follows:

$$C_f = C_i \left(\frac{V_i}{V_f} \right) = \frac{0.025 \text{ g malic acid}}{\text{mL juice}} \left(\frac{10 \text{ mL juice}}{3 \text{ mL}} \right)$$

$$= \frac{0.0833 \text{ g malic acid}}{\text{mL}}$$

3.4.5 Multiple-Step Dilutions and Concentrations

Up to this point, we have dealt exclusively with single-step dilutions and concentrations. However, real-world analytical practice often requires that several dilutions and/or concentrations be performed (see *Practice Problems 1, 2, and 6*). There are two ways to approach this problem. Suppose that three dilutions are used as follows:

$$\text{Step1} : C_1 \rightarrow C_2 \text{ by performing } V_1 \rightarrow V_2$$

$$\text{Step2} : C_2 \rightarrow C_3 \text{ by performing } V_3 \rightarrow V_4$$

$$\text{Step3} : C_3 \rightarrow C_4 \text{ by performing } V_5 \rightarrow V_6$$

The obvious approach is to solve for the final (or initial) concentration of the first step, use the final (or initial) concentration of the first step as the initial (or final) concentration of the second step, and so forth:

$$C_2 = C_1 \left(\frac{V_1}{V_2} \right) \quad \text{then} \quad C_3 = C_2 \left(\frac{V_3}{V_4} \right) \quad \text{then} \quad C_4 = C_3 \left(\frac{V_5}{V_6} \right)$$

This can be time-consuming. Furthermore, performing multiple calculations can introduce rounding errors and increases the probability of making other errors. Notice that we can substitute the formula used to calculate a concentration in one step for that value in the next step:

$$C_2 = C_1 \left(\frac{V_1}{V_2} \right)$$

$$C_3 = C_2 \left(\frac{V_3}{V_4} \right) = \left[C_1 \left(\frac{V_1}{V_2} \right) \right] \left(\frac{V_3}{V_4} \right)$$

$$C_4 = C_3 \left(\frac{V_5}{V_6} \right) = \left[C_1 \left(\frac{V_1}{V_2} \right) \left(\frac{V_3}{V_4} \right) \right] \left(\frac{V_5}{V_6} \right)$$

Therefore, as long as we have the volumes (or masses) from each step, we can reduce this multistep calculation to a single-step calculation that goes directly from the initial to the final concentration (or vice versa), keeping in mind that when going forward (from initial sample to final solution), the initial concentration is multiplied by a series of dilution factors for each step, with the starting mass/volume divided by the final mass/volume for each step. If we have a dilution or concentration scheme with *n* steps, we can set up the calculations so that:

$$\text{Step1} : C_i \rightarrow C_2 \text{ by performing } V_1 \rightarrow V_2$$

$$\text{Step2} : C_2 \rightarrow C_3 \text{ by performing } V_3 \rightarrow V_4$$

and so forth until the last step, where:

$$\text{Step } n : C_3 \rightarrow C_f \text{ by performing } V_{k-1} \rightarrow V_k$$

Then, a general formula for multistep forward calculations is:

$$C_f = C_i \left(\frac{m \text{ or } V_1}{m \text{ or } V_2} \right) \left(\frac{m \text{ or } V_3}{m \text{ or } V_4} \right) \cdots \left(\frac{m \text{ or } V_{k-1}}{m \text{ or } V_k} \right)$$

$$= C_i \left(DF_1 \right) \left(DF_2 \right) \cdots \left(DF_n \right) \qquad (3.18)$$

We can simplify this by calculating the "overall DF" (DF_Σ), the product of DFs from each step:

$$DF_\Sigma = \left(DF_1 \right) \left(DF_2 \right) \cdots \left(DF_n \right) \qquad (3.19)$$

$$C_f = C_i \left(DF_\Sigma \right) \qquad (3.20)$$

When going from final solution to initial sample, the calculation is reversed: the initial concentration is multiplied by dilution factors for each step, with the final mass/volume divided by the initial mass/volume for each step. A general formula for multistep back calculations is:

$$C_i = C_f \left(\frac{m \text{ or } V_2}{m \text{ or } V_1} \right) \left(\frac{m \text{ or } V_4}{m \text{ or } V_3} \right) \cdots \left(\frac{m \text{ or } V_k}{m \text{ or } V_{k-1}} \right)$$

$$= C_f \left(\frac{1}{DF_1} \right) \left(\frac{1}{DF_2} \right) \cdots \left(\frac{1}{DF_n} \right) \qquad (3.21)$$

$$C_f = C_i \left(\frac{1}{DF_\Sigma} \right) \qquad (3.22)$$

$$DF_\Sigma = \frac{C_f}{C_i} \qquad (3.23)$$

This approach of solving dilution or concentration problems in a single step can be used for dilutions, concentrations, and mixed procedures involving both dilutions and concentrations.

We can set the problem up like this to calculate C_i (in ug/g):

Step1: $Ci = 0.56 \text{ g} \ (m_1)$ diluted to 500 mL (V_2)

Step2: 25 mL (V_3) diluted to 125 mL $(V_4, 25 \text{ mL} + 100 \text{ mL})$

Step3: 75 mL (V_5) concentrated to 50 mL (V_6)

Step4: 50 mL (V_7) diluted to 100 mL $(V_8, 50 \text{ mL} + 50 \text{ mL})$

$$C_f = \frac{12.9 \ \mu g}{mL}$$

The calculation is a back calculation of the initial concentration from the final calculation:

$$C_i = C_f \left(\frac{m \text{ or } V_2}{m \text{ or } V_1} \right) \left(\frac{m \text{ or } V_4}{m \text{ or } V_3} \right) \cdots \left(\frac{m \text{ or } V_k}{m \text{ or } V_{k-1}} \right)$$

$$= C_f \left(\frac{1}{DF_1} \right) \left(\frac{1}{DF_2} \right) \cdots \left(\frac{1}{DF_n} \right)$$

There are four steps, so there must be four individual DFs (this is a good way to make sure that you are doing the problem correctly; does the number of DFs you used equal the number of steps?).

$$C_i = C_f \left(\frac{m \text{ or } V_2}{m \text{ or } V_1} \right) \left(\frac{m \text{ or } V_4}{m \text{ or } V_3} \right) \left(\frac{m \text{ or } V_6}{m \text{ or } V_5} \right) \left(\frac{m \text{ or } V_8}{m \text{ or } V_7} \right)$$

$$C_i = \frac{12.9 \ \mu g \text{ caffeine}}{mL} \left(\frac{500 \text{ mL}}{0.56 \text{ g green tea extract}} \right)$$

$$\left(\frac{125 \text{ mL}}{25 \text{ mL}} \right) \left(\frac{50 \text{ mL}}{75 \text{ mL}} \right) \left(\frac{100 \text{ mL}}{50 \text{ mL}} \right)$$

$$= \frac{76,800 \ \mu g \text{ caffeine}}{g \text{ green tea extract}}$$

Another way to make sure that this calculation has been done correctly is to assign each solution a letter or number, and those should "cancel each other out" along with the units. If the units and/or solution identities do not cancel out properly, one or more concentrations, volumes, or masses have been used incorrectly. For this problem, the solutions could be identified as follows:

Step1: $0.56 \text{ g} \ (m_1)$ diluted to 500 mL

$$(V_2, \text{Solution A})$$

Step 2 : 25 mL $(V_3,$ Solution A$)$ diluted to 125 mL

$(V_4,$ Solution B, 25 mL + 100 mL$)$

Step 3 : 75 mL $(V_5,$ Solution B$)$ concentrated to 50 mL

$(V_6,$ Solution C$)$

Step 4 : 50 mL $(V_7,$ Solution C$)$ diluted to 100 mL

$(V_8,$ Solution D, 50 mL + 50 mL$)$

$$C_i = \frac{12.9\,\mu g}{mL}$$

The problem would then be solved the same way with the solutions identified:

$$C_i = \frac{12.9\,\mu g \text{ caffeine}}{mL \text{ Solution D}} \left(\frac{500 \text{ mL Solution A}}{0.56 \text{ g green tea extract}} \right)$$

$$\left(\frac{125 \text{ mL Solution B}}{25 \text{ mL Solution A}} \right) \left(\frac{50 \text{ mL Solution C}}{75 \text{ mL Solution B}} \right)$$

$$\left(\frac{100 \text{ mL Solution D}}{50 \text{ mL Solution C}} \right) = \frac{76,785.7\,\mu g \text{ caffeine}}{\text{g green tea extract}}$$

Note that there are various ways to solve this problem, but the principles are the same.

3.5   SPECIAL CASES

3.5.1   Extraction

There are several special cases regarding dilution that occur frequently enough in food analysis to warrant specific mention: extraction and homogenization/mixing. Regarding **extraction**, often, we use a solvent for which an analyte has a high affinity (e.g., lipids are more soluble in nonpolar solvents such as hexane or chloroform than the foods they are in, so these solvents "pull" the lipid out of the food) to extract that analyte from a sample. The solvent is typically immiscible with the sample as a whole. This means that the sample and the solvent do not mix and can be easily separated by centrifugation or by using a separatory funnel. Most of the analyte is transferred to the solvent, while most of the sample remains behind. For dilution and concentration purposes, we assume that 100% of the desired analyte is quantitatively transferred from the sam-

ple into the solvent (see *Practice Problem 1*). While this assumption is never actually true, it is necessary for our purposes. Further dilutions and/or concentrations may then be performed on the extraction solvent.

In some cases, we use a single extraction step. In this situation, the sample is mixed with the solvent to achieve transfer (extraction) of the analyte from the solvent. The sample and solvent are then separated. This is different than the simple dilution problems we have done previously, in that the sample remains behind while only the analyte is transferred to the solvent. However, the principle is still the same: 100% of the analyte that was in the sample is assumed to now reside in the solvent, and we can use the equations previously discussed (Eqs. 3.16 and 3.17).

Example B1

Suppose that peanut butter (PB) contains 0.37 g fat/g, and 1.4 g peanut butter is extracted with 100 mL hexane. What is the concentration of fat in the hexane (the assumption here is that the volume of the hexane does not change)?

In this case, the fat is transferred from the peanut butter to the solvent, as opposed to being diluted into a total mass/volume of peanut butter + solvent. This problem can be solved using the standard procedure:

$$C_f = C_i \left(\frac{m_i}{V_f} \right) = \frac{0.37 \text{ g fat}}{\text{g PB}} \left(\frac{1.4 \text{ g PB}}{100 \text{ mL}} \right) = \frac{0.00518 \text{ g fat}}{mL}$$

Note that this is the same as if the entire sample was diluted into the solvent, but for an extraction only a negligible part of the sample (but presumably all of the analyte) is actually transferred.

In reality, only a fraction (generally unknown) of the analyte is transferred from the sample to the solvent. In order to overcome this, repeated extractions are often performed. These are the same as single extractions, except the extraction is repeated with fresh solvent each time and the solvent volumes from each extraction are combined into a single extract. In this case, we can treat this as a single extraction where the combined solvent volume is the final volume, and again we assume that 100% of the analyte is transferred to the solvent.

Example B2

Suppose that the peanut butter example above is adjusted so that the peanut butter is extracted three times with 75 mL of fresh hexane, and the three extracts are combined. What is the concentration of fat in the hexane?

In this case, the final volume would be 3 × 75 mL = 225 mL:

$$C_f = C_i\left(\frac{m_i}{V_f}\right) = \frac{0.37\text{ g fat}}{\text{g PB}}\left(\frac{1.4\text{ g PB}}{225\text{ mL}}\right) = \frac{0.00230\text{ g fat}}{\text{mL}}$$

Extraction procedures often specify that the extraction solvent be brought up to a known final volume after extraction. This accounts for potential loss of solvent, or significant changes in volume due to extraction of sample components, during extraction and also increases the precision and accuracy of the extraction. This is handled by simply making this final volume the final volume for the dilution calculation.

Example B3

Suppose that the peanut butter example above is adjusted so that the peanut butter is extracted 2X with 100 mL hexane, and the pooled hexane extract is diluted to a final volume of 250 mL. What is the concentration of fat in the hexane?

In this case, the final volume would simply be 250 mL, as the pooled extracts (~200 mL) are further diluted with ~50 mL to an exact total volume of 250 mL. The problem would then be solved as shown above, using 250 mL as the final volume.

In summary, extraction can be thought of as a dilution (or concentration) in which the final volume is the total volume of solvent. Regardless of the number of extractions, it can be viewed as a single step as long as all the solvent extracted from the sample is kept and the final volume into which the sample is extracted is known. After extraction, dilutions or concentrations of the extract may be performed; these are handled as any other dilution or concentration.

3.5.2 Homogenization/Blending/Mixing

The other scenario that warrants consideration is **homogenization, blending,** and **mixing**. Often, samples need to be dispersed or blended into a solvent for sample preparation. This is typically done using a lab blender, food processor, Polytron homogenizer, stomacher, sonicator, etc. These processes cannot typically be done in volumetric glassware. Furthermore, the problem with these processes, from a calculation standpoint, is that homogenization, blending, and mixing of solids with liquids typically change the volume of a sample.

Example B4

Suppose you homogenize 2.5 g cheese (of an unknown volume) in 100 mL phosphate buffer. What is the final volume?

The final volume of the liquid will be >100 mL and is not easily measured by volumetric glassware. Therefore, in this example, the final volume is unknown. This poses a challenge for accurate dilution calculations. This process is usually overcome similarly to the extractions described above: the sample is homogenized, quantitatively transferred to a volumetric flask, and then diluted to a known final volume with the same solvent used for homogenization.

Example B5

Suppose that 22.7 g swordfish fillet is homogenized in a lab blender with 150 mL denaturing buffer. Following homogenization, the liquid is decanted into a 500 mL flask. Fresh denaturing buffer (~25 mL) is used to rinse the blender, and the rinse liquid is also decanted into the flask. The flask is brought to volume with denaturing buffer. The concentration of methylamines in the diluted homogenate is 0.92 µmol/mL. What is the concentration of methylamines in the fish?

In this case, the sum of the homogenization process is that 22.7 g fish is diluted into 500 mL final volume. Therefore, the calculation is performed as follows:

$$C_i = C_f\left(\frac{V_f}{V_i}\right) = \frac{0.92\,\mu\text{mol methylamines}}{\text{mL}}\left(\frac{500\text{ mL}}{22.7\text{ g fish}}\right)$$

$$= \frac{20.3\,\mu\text{mol methylamines}}{\text{g fish}}$$

In summary, mixing processes can be thought of as a dilution (or concentration) step for which the final volume is the total volume of solvent. Regardless of the number of steps involved, it can be viewed as a single step as long as all the homogenate from whole sample is kept and the final volume into which the sample is diluted is known. Keep in mind that, after mixing and quantitative dilution, subsequent dilutions or concentrations of the homogenate may be performed, and these are handled as any other dilution or concentration.

3.6   STANDARD CURVES

It is very common for analytical procedures to require a **standard curve**, which is a set of solutions containing a range of different concentrations of the analyte. Typically, a **stock solution** containing a high concentration of the analyte is prepared, and various dilutions are performed to get the desired standard curve solutions (see *Practice Problems 3, 4, 5, and 7*). These dilutions can be done with either sequential (i.e., serial) dilutions (Fig. 3.1a) or parallel dilutions (Fig. 3.1b).

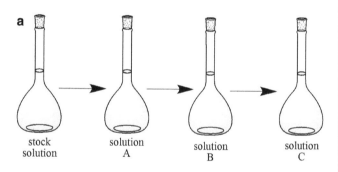

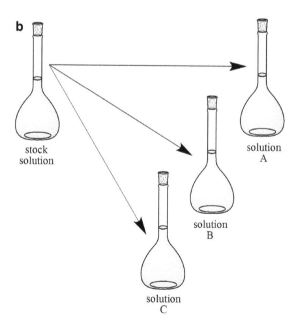

Fig. 3.1 Example standard curve dilution schemes for sequential dilutions (**a**) and parallel dilutions (**b**)

3.6.1   Sequential vs. Parallel Dilutions

In the case of **sequential dilutions**, each dilution is used to prepare the next dilution in the series:

stock solution → solution 1 → solution 2 → solution 3, etc.

Example C1
Suppose a 1 M solution of citric acid is serially diluted as follows: 1 mL is diluted with 2 mL water, and the resulting solution is diluted the same way (1 mL + 2 mL water) until four diluted solutions are obtained. What are the concentrations in each solution?

We can solve this problem by assigning each solution a name (Solutions A–D) and then calculating the concentration in each. For each step, the initial volume is 1 mL and the final volume is 3 mL, since 2 mL water is combined with each solution, and the initial concentration is the concentration of the previous solution (stock when making Solution A, Solution A when making Solution B, etc.):

$$\text{Solution A}: \quad C_f = C_i\left(\frac{V_i}{V_f}\right) = 1M\left(\frac{1\,\text{mL}}{3\,\text{mL}}\right) = 0.333\,M$$

$$\text{Solution B}: \quad C_f = C_i\left(\frac{V_i}{V_f}\right) = 0.333\,M\left(\frac{1\,\text{mL}}{3\,\text{mL}}\right)$$
$$= 0.111\,M$$

$$\text{Solution C}: \quad C_f = C_i\left(\frac{V_i}{V_f}\right)$$
$$= 0.111\,M\left(\frac{1\,\text{mL}}{3\,\text{mL}}\right)$$
$$= 0.0370\,M$$

$$\text{Solution D}: \quad C_f = C_i\left(\frac{V_i}{V_f}\right)$$
$$= 0.0370\,M\left(\frac{1\,\text{mL}}{3\,\text{mL}}\right)$$
$$= 0.0123\,M$$

For **parallel dilutions**, different DFs are applied to the stock to obtain the desired solutions:

$$\text{stock solution} \xrightarrow{\text{DF1}} \text{solution 1}$$

$$\text{stock solution} \xrightarrow{\text{DF2}} \text{solution 2}$$

$$\text{stock solution} \xrightarrow{\text{DF3}} \text{solution 3}$$

Example C2

Suppose the 1 M stock solution of citric acid is diluted as follows: 1 mL of stock solution is diluted to volume in a 5 mL, 10 mL, 25 mL, and 50 mL volumetric flask. What are the citric acid concentrations in each solution? We can solve this problem by assigning each solution a name (Solutions A–D) and then calculating the concentration in each. For each dilution, the starting concentration is the stock concentration (since they were made in parallel), and the final volume is the volume of each volumetric flask:

$$\text{Solution A}: \quad C_f = C_i \left(\frac{V_i}{V_f}\right) = 1\,M\left(\frac{1\text{ mL}}{5\text{ mL}}\right) = 0.200\,M$$

$$\text{Solution B}: \quad C_f = C_i \left(\frac{V_i}{V_f}\right) = 1\,M\left(\frac{1\text{ mL}}{10\text{ mL}}\right) = 0.100\,M$$

$$\text{Solution C}: \quad C_f = C_i \left(\frac{V_i}{V_f}\right) = 1\,M\left(\frac{1\text{ mL}}{25\text{ mL}}\right) = 0.040\,M$$

$$\text{Solution D}: \quad C_f = C_i \left(\frac{V_i}{V_f}\right) = 1\,M\left(\frac{1\text{ mL}}{50\text{ mL}}\right) = 0.020\,M$$

Preparing standards in parallel is typically more accurate than in series, as each dilution is performed in a single step from the stock. This reduces error associated with each solution by reducing the perpetuation of errors.

Example C3

If preparing a dilution involves an error of ~1% (a factor of 0.01), what is the error in the third solution if the solutions are prepared in parallel?

Performing three dilutions in parallel will result in each solution being off by ~1%.

$$\text{stock solution} \xrightarrow{\text{DF1}} \text{Solution 1}\left(X \pm 1\% \text{ or } 1.01\,X\right)$$

$$\text{stock solution} \xrightarrow{\text{DF2}} \text{Solution 2}\left(X \pm 1\% \text{ or } 1.01\,X\right)$$

$$\text{stock solution} \xrightarrow{\text{DF3}} \text{Solution 3}\left(X \pm 1\% \text{ or } 1.01\,X\right)$$

If dilutions are performed in series rather than in parallel, the error compounds because each solution is used to prepare the next solution.

Example C4

For the same three dilutions, what is the error in the third solution if the dilutions are prepared in series?

When prepared in series, the error increases with each subsequent dilution:

$$\text{stock solution} \rightarrow \text{Solution 1}\left(\pm 1\% \text{ or } 1.01X\right)$$

$$\rightarrow \text{Solution 2}\left(\pm 1\% \text{ or } 1.01X\right)$$

$$\rightarrow \text{Solution 3}\left(\pm 1\% \text{ or } 1.01X\right)$$

$$\text{error in Solution 1} = \pm 1.01X = \pm 1\%$$

$$\text{error in Solution 2} = \pm(1.01)(1.01)X$$

$$= \pm 1.0201X = \pm 2.01\%$$

$$\text{error in Solution 3} = \pm(1.01)(1.01)(1.01)X$$

$$= \pm 1.030301X = \pm 3.0301\%$$

Parallel dilutions are often preferred over series dilutions for accuracy and improved quantification for standard curves. Regardless of how they are performed, the principles of dilution calculations apply. As with any dilution, these dilutions can be performed as "dilute to" or "dilute with" procedures, and the calculations should be performed accordingly.

3.6.2 Designing Dilution Schemes

Occasionally, you may be given the stock solution concentration and the needed concentrations of standard curve solutions and asked to design a dilution scheme. The way to do this is to calculate the needed DFs for each solution and then design a scheme to achieve the DFs.

Example C5

Suppose you have a 0.2% w/v NaCl solution and you need standard curve solutions of 0.1, 0.04, 0.02, 0.01, and 0.002% w/v. How would you make these?

First, start by calculating each DF from the stock solution:

$$DF = \frac{C_f}{C_i}$$

$$DF_A = \frac{C_f}{C_i} = \frac{0.1\%}{0.2\%} = 0.5 \left(i.e., \frac{1}{2} \right)$$

$$DF_B = \frac{C_f}{C_i} = \frac{0.04\%}{0.2\%} = 0.2 \left(i.e., \frac{1}{5} \right)$$

$$DF_C = \frac{C_f}{C_i} = \frac{0.02\%}{0.2\%} = 0.1 \left(i.e., \frac{1}{10} \right)$$

$$DF_D = \frac{C_f}{C_i} = \frac{0.01\%}{0.2\%} = 0.05 \left(i.e., \frac{1}{20} \right)$$

$$DF_E = \frac{C_f}{C_i} = \frac{0.002\%}{0.2\%} = 0.01 \left(i.e., \frac{1}{100} \right)$$

From this point, there are several options. You can dilute in series or in parallel. And you can "dilute to" or "dilute with" to achieve the desired concentrations. For simplicity, the parallel "dilute to" example is shown in Table 3.1.

Table 3.1 Dilution example for a standard curve

Solution	C_i (% w/v)	Dilute (mL)	To final volume (mL)	DF	C_f (% w/v)
A	0.2	1	2	0.5	1
B	0.2	1	5	0.2	0.04
C	0.2	1	10	0.1	0.02
D	0.2	1	20	0.05	0.01
E	0.2	1	100	0.002	0.002

You should be able to design a scheme to dilute a stock solution to a wide array of diluted solutions using both parallel and series dilutions and perform either with "dilute to" or "dilute with" procedures. One consideration to keep in mind

is that Class A volumetric glassware comes in discreet volumes, and only those volumes can be used for dilutions where maximum accuracy and precision are desired. The most common volumes available are:

1. Class A volumetric flasks: 5, 10, 25, 50, 100, 200, 250, 500, 1000, and 2000 mL.
2. Class A volumetric transfer pipettes: 1, 2, 3, 4, 5, 10, 20, 25, 50, and 100 mL.

Therefore, your designed dilution scheme should use only those volumes for transferring volumes (pipettes) and bringing to volume (flasks), if possible, to optimize accuracy and precision. Some schemes may require use of an adjustable pipettor (to transfer or add volumes such as 0.1 mL, 950 µL, etc.), particularly when small volumes are used. This is often the case when preparing standard curves (see *Practice Problems 3 and 4*).

Example C6

Suppose that you have a stock solution of 2000 ppm Fe(II) in 0.1 N HCl. You need to create 1 mL of 1000, 750, 500, 250, and 100 ppm Fe(II) by combining different volumes of the stock solution and 0.1 N HCl. What volumes should be used?

We know the starting (2000 ppm) and final concentration (1000–100 ppm) and final volume (1 mL). Therefore, we must calculate the initial volume of stock solution for each solution:

$$C_i V_i = C_f V_f \quad \text{and} \quad V_i = \frac{C_f V_f}{C_i}$$

$$\text{for 1000 ppm,} \quad V_i = \frac{C_f V_f}{C_i} = \frac{(1000 \text{ ppm})(1 \text{ mL})}{2000 \text{ ppm}}$$

$$= 0.05 \text{ mL}$$

$$\text{for 750 ppm,} \quad V_i = \frac{C_f V_f}{C_i} = \frac{(750 \text{ ppm})(1 \text{ mL})}{2000 \text{ ppm}}$$

$$= 0.375 \text{ mL}$$

and so forth. Once the volume of stock solution is known for each dilution, we then calculate the amount of 0.1 N HCl needed to dilute to 1 mL:

$$\text{for 1000 ppm, } 0.5 \text{ mL} \rightarrow 1 \text{ mL} - 0.5 \text{ mL}$$

$$= 0.5 \text{ mL } 0.1 \text{ N HCl}$$

$$\text{for 750 ppm, } 0.375 \text{ mL} \rightarrow 1 \text{ mL} - 0.375 \text{ mL}$$

$$= 0.625 \text{ mL } 0.1 \text{ N HCl}$$

and so forth. The dilution scheme would thus be as shown in Table 3.2:

Table 3.2 Dilution example for a standard curve

Fe(II) (ppm)	Stock (mL)	0.1 N HCL (mL)	Total volume (mL)
1000	0.5	0.5	1.0
750	0.375	0.625	1.0
500	0.25	0.75	1.0
250	0.125	0.875	1.0
100	0.05	0.95	1.0

In cases when an adjustable pipettor is used, accuracy and precisions should be optimized by frequently calibrating and maintaining the pipettors. Furthermore, using proper (and consistent) manual pipetting technique is critical in these scenarios.

3.7 UNIT CONVERSIONS

Up to this point, we have examined calculations for which the units of the dilutions and the units of the concentrations that you have been given "match up" so that no conversions are required to get the correct answer. However, this will not always be the case in actual lab practice. This will require you to perform unit conversions to get an answer that is correct and makes sense.

Example D1

Suppose that you dilute 14.4 g of yogurt to a final volume of 250 mL with water. HPLC analysis of the diluted sample indicates a riboflavin concentration of 0.0725 ng/µL. What is the riboflavin content in the yogurt?

This is a straightforward dilution problem that can be solved as follows:

$$C_i = C_f \left(\frac{V_f}{V_i} \right)$$

However, if the calculation were performed without unit conversions, the answer obtained would be as follows:

$$C_i = C_f \left(\frac{V_f}{V_i} \right) = \frac{0.0725 \text{ ng riboflavin}}{\mu L \text{ diluted sample}}$$

$$\left(\frac{250 \text{ mL diluted sample}}{14.4 \text{ mL yogurt}} \right) = \frac{1.26 \text{ ng riboflavin} \times \text{mL}}{\text{g yogurt} \times \mu L}$$

Obviously, (ng × mL)/(g × µL) is not an acceptable unit for expressing the riboflavin concentration in a product. Therefore, we must change either µL to mL or

mL to µL for the units to be expressed correctly in the final answer. The calculation is performed correctly as follows:

$$C_i = C_f \left(\frac{V_f}{V_i} \right) = \frac{0.0725 \text{ ng riboflavin}}{\mu L \text{ diluted sample}} \left(\frac{1000 \ \mu L}{\text{mL}} \right)$$

$$\left(\frac{250 \text{ mL diluted sample}}{14.4 \text{ mL yogurt}} \right) = \frac{1,260 \text{ ng riboflavin}}{\text{g yogurt}}$$

If the units are not converted, the magnitude (number only) of the final answer is off by a factor of 1000 in this case, in addition to having the wrong units.

By keeping track of the units (often called "dimensional analysis" in chemistry textbooks), you easily catch an incorrect value and adjust the calculation.

3.8 AVOIDING COMMON ERRORS

The most common errors associated with dilutions and concentrations are:

1. Setting up the calculation incorrectly (incorrectly using the information provided).
2. Lack of needed unit conversion or incorrect unit conversion.

Three strategies can be employed to avoid these mistakes or catch them if they have been made:

1. Draw a picture of the dilution or concentration scheme.
2. Perform **unit analysis** (also referred to as **dimensional analysis** or the **factor-label method**) and assign "names" to each solution in a scheme.
3. Perform the "sniff test."

While these may seem unnecessary or laborious, these simple steps will improve the ease of calculations and the quality of your data.

3.8.1 Draw a Picture

Drawing a picture can be very useful for making sure the problem has been set up correctly. This allows you to visualize the dilution or concentration scheme, which is often helpful. Drawing a picture or diagram can help clarify the meaning of the information provided and assist in setting up the calculation.

Drawing a picture can be particularly useful for visualizing complex multistep processes and correctly assigning values to each step from a large set of provided data.

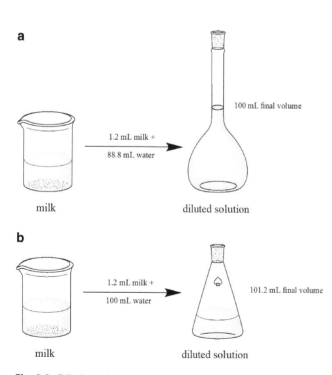

a

1.2 mL milk +
88.8 mL water

100 mL final volume

milk

diluted solution

b

1.2 mL milk +
100 mL water

101.2 mL final volume

milk

diluted solution

Fig. 3.2 Dilution schemes for a "dilute to" (**a**) and "dilute with" (**b**) scenario

Figure 3.3 shows a fairly complex scheme, and organizing the information in your head can be challenging. Drawing this scheme allows you to assign names to various solutions and organize the information so that it can easily be plugged into the calculation. Very experienced analytical chemists draw pictures to visualize processes and calculations, so this is a valuable skill to develop early on.

3.8.2 Unit Analysis

We have already discussed unit (or "dimensional") analysis, a key step in verifying that the calculation has been set up correctly. Along with this, assigning names to each solution in a scheme and incorporating those into the calculation can assure that each step has been set up correctly in the calculation. If the problem has been set up correctly, each solution name for the intermediate steps should be "cancelled out," leaving only the final solution or sample when the calculation is complete.

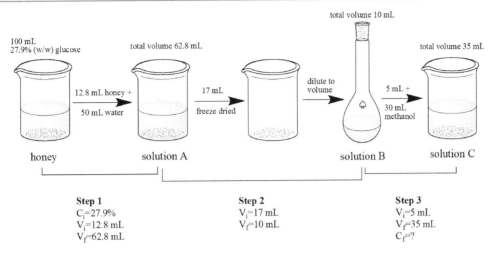

Fig. 3.3. Diagram of a complex multistep dilution and concentration scheme

3.8.3 "Sniff Test"

Finally, use the "sniff test" on your calculations to detect any obvious errors. For example:

1. If the procedure is an overall dilution, then the final concentration should be lower than the initial concentration.
2. If the procedure is an overall concentration, then the final concentration should be higher than the initial concentration.
3. With very few exceptions (such as % moisture expressed on a dry weight basis for some samples), the % of an analyte in a sample will never be $\geq 100\%$.
4. Does the calculated value make sense? For example, one would not expect the following calculated values to be true:
    (a) % moisture > a few % in very dry products
    (b) % solids > a few % in dilute solutions
    (c) % fat, protein, ash, or carbohydrates that don't make sense given what is known about the product (such as vegetable oil containing 35% moisture or 12% fat)
    (d) Analyte levels that do not make sense given what is known about typical product composition (particularly ≥ 1 order of magnitude greater than expected).

For the sniff test, it is important that you know a little about the product you are analyzing, if possible. If the "sniff test" suggests an error, carefully reexamine how you set up the problem and how the calculation was performed. The value of an experienced analyst is being able to catch errors early without simply reporting them blindly.

3.9 PRACTICE PROBLEMS

(Note: Answers to problems are in the last section of the laboratory manual.)

The following example problems demonstrate real-world calculations involving dilutions and concentrations:

1. You randomly select a cup of applesauce (containing 113 g applesauce) from the processing line. You extract 10.3 g of the applesauce 3X with 50 mL ethyl acetate, pool the extracts, and dilute to 250 mL with ethyl acetate. You evaporate 25 mL of the extract to dryness under a stream of N_2 and redissolve the residue to 5 mL with methanol. GC analysis of the methanol indicates a methoxyfenozide concentration of 0.00334 µg/mL. What is the concentration of methoxyfenozide in the applesauce (µg/g)? What is the total amount of methoxyfenozide in the applesauce cup (µg)?

2. You are given a stock solution of (−)-epicatechin (EC, MW = 290.26 g/mol) in water (0.94 mg/mL). You prepare a series of solutions by serial dilution as follows: (1) diluting 0.5 mL of the stock solution to 10 mL with water (Solution A), (2) diluting 1.5 mL of Solution A with 4 mL water (Solution B), and (3) diluting 3 mL Solution B with 9 mL water (Solution C). What is the concentration of Solution C (mg/mL and µM)? What is the overall DF? How many "fold" has the stock solution been diluted to Solution C?

3. You are performing a spectrophotometric assay for riboflavin using a commercial kit. The kit comes with 2 mL of a solution of riboflavin (1.45 mg/mL). The instructions

tell you to make a stock solution by diluting the riboflavin with 25 mL water. Then, the instructions tell you to make a set of standard solutions by diluting 100 μL of the stock solution with 0.5 mL (A), 0.75 mL (B), 1 mL (C), 2 mL (D), and 5 mL (E) water. What is the riboflavin concentration (mg/mL) in the stock solution and the five standards?

4. You are using a colorimetric method to measure anthocyanin pigments in raspberry juice. The method requires a total sample volume of 2 mL. You have a stock solution of 160 g/L anthocyanins, and you need solutions of 0, 0.1, 0.2, 0.3, and 0.5 mg/mL. Design a dilution scheme employing commonly available volumetric flasks, volumetric pipettes, and/or an adjustable pipettor (but do *not* use any volumes less than 0.2 mL, and the final volume of each solution must be at least 2 mL) to obtain the desired concentrations.

5. As discussed previously, samples often need to be diluted or concentrated in order to obtain analyte concentrations in the range of the standard curve. Ideally, the sample concentration would be near the middle of the standard curve values. This can be challenging, because you often do not know the approximate sample concentration prior to the analysis. For the standard curve described in *Practice Problem 4*, the standards cover the range of 0–0.5 mg/mL. Suppose you get a sample of a new type of juice into the quality assurance lab and you have no idea how to dilute the juice for analysis. Scientific literature suggests that this juice can have anthocyanin concentrations of anywhere from 750 to 3000 μg/mL. Based on this information, design a dilution scheme that will likely yield diluted juice samples within the range of the standard curve.

6. You are analyzing Ca content of milk using atomic absorption spectroscopy (AAS). This analysis requires dry ashing to isolate the minerals from a sample (dry ashing is essentially a concentration and extraction step: The sample is incinerated to remove all organic mineral and leave only the minerals). You dry ash a 2.8 mL sample of milk, dissolve the ash in 12 mL 1 N HCl, and dilute the solution to 50 mL with 1 N HCl. You further dilute 7 mL of this solution by adding 13 mL 1 N HCl. You then analyze 0.5 mL sample and 0.5 mL of standards (10–100 ppm Ca) by AAS. You find that the diluted sample contains 28.2 ppm Ca. What is the Ca content of the undiluted milk in ppm?

7. You are measuring caffeine (194.2 g/mol) in drinks by high-performance liquid chromatography (HPLC). You prepare standard solutions of 0–100 μM caffeine. Your method calls for dilution of the sample to 250 mL with water prior to analysis. You know that a particular drink contains 170 mg caffeine per 400 mL. How many mL should be diluted to 250 mL total to obtain a concentration in the middle of your standard curve?

Acknowledgments The author thanks Sean F. O'Keefe (Department of Food Science and Technology, Virginia Polytechnic Institute and State University, Blacksburg, VA, USA) for his input on this chapter in the previous edition of the book.

Statistics for Food Analysis

4

Andrew P. Neilson

Contents

4.1 INTRODUCTION

This chapter is a review of basic statistics and will demonstrate how statistics is used in the context of food analysis. It is meant to be a "survival guide," primarily for reference. Students taking a food analysis course should have already taken an undergraduate statistics course, or be taking one concurrently. A foundation for this chapter is Chap. 4, Evaluation of Analytical Data, in *Nielsen's Food Analysis* textbook. The learning objectives for this chapter are to be able to:

1. Calculate a mean, standard deviation, Z-score and t-score for a sample dataset.

2. Determine whether a population parameter is significantly different from a given value using a one-sample t-test.
3. Calculate a confidence interval for a sample mean using t-scores.
4. Determine if two populations are significantly different using a two-sample t-test.

A significant component of a typical food analysis course and the use of food analysis in a professional setting (industry, research, regulatory/government) is evaluation of obtained analytical data in order to answer questions based on the purpose of the analysis. In food analysis laboratory experiments, you first collect data (values), or you are given data to work with (i.e., for laboratory exercises, homework, exams, etc.). Then, you evaluate and manipulate the data, answering questions such as these:

1. Are the values significantly different from a desired value or not?

A. P. Neilson (✉)
Department of Food, Bioprocessing and Nutrition Sciences, North Carolina State University, Raleigh, NC, USA
e-mail: aneilso@ncsu.edu

2. Are two values significantly different from each other or not?

In your food analysis course, you will use statistics to solve problems typically encountered in the food industry, research, and regulatory or government scenarios. Data analysis concepts will be used in this chapter.

4.2   POPULATION DISTRIBUTIONS

The **distribution** of a given value in a **population** refers to how many members of a population have each value of the parameter measured. You can examine distributions by plotting a **histogram,** which is a graph with parameter values on the x-axis and number of members of the population having that value of the parameter (i.e., the "frequency" of each value) on the y-axis. Suppose there were 100 bottles of water in a population and you knew the sodium content of all of them. The histogram of this population might look like Fig. 4.1. Many populations are "normally distributed" (also known as **normal** populations). Normal populations have histograms that look roughly like Fig. 4.2. Note that the y-axis of a histogram could be the absolute number of population members with a given value, the proportion (fraction or decimal) of population members with a given value, or the percentage of population members with a given value. Some populations are not normally distributed, but that is a topic beyond the scope of this chapter. For the purposes of this chapter, we will assume a normal distribution. Normal population distributions are defined by two parameters (Fig. 4.3):

1. Population **mean** (μ): center of the plot, the average value.
2. Population **standard deviation** (σ): a measure of the spread of the plot, the variability of the values.

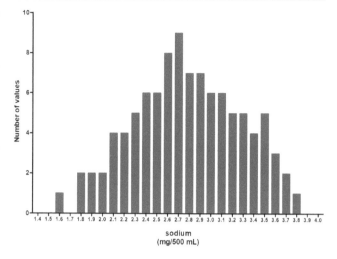

Fig. 4.2 Example population histogram

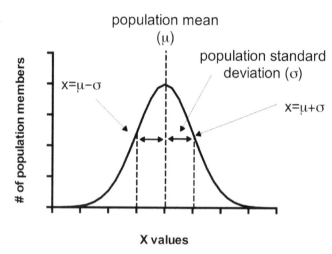

Fig. 4.3 Shape of a normal population

For the example sodium data, you can plot the histogram of the data with the population mean (μ) and population standard deviation (σ). If you define sodium content as (x), you can calculate the mean and standard deviation of x:

$$\text{Population mean of } x = \mu_x = \frac{\sum x_i}{n} = \frac{x_1 + x_2 + \ldots + x_{n-1} + x_n}{n}$$

(4.1)

$$\text{Standard deviation of } x = \sigma_x = \sqrt{\frac{\sum (x_i - \bar{x})^2}{n}}$$ (4.2)

$$\text{Population mean} = \mu_x = 2.783$$

$$\text{Standard deviation} = \sigma_x = \sqrt{\frac{\sum (x_i - 2.783)^2}{100}} = 0.492$$

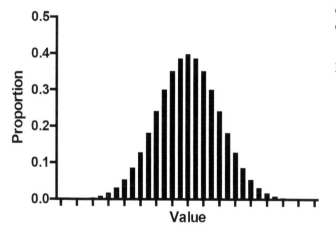

Fig. 4.1 Example normal distribution

You can use these values to determine the center (average) and spread (average variation from center) of the data, as shown in Fig. 4.4 for the example sodium data. Normal distributions can have different shapes but are still always defined by N(μ, σ). The notation "N(μ, σ)" indicates a normally distributed (N) population with center μ and standard deviation σ. The shape of the curve is dictated by the standard deviation. Figure 4.5 shows three populations, each with the same mean but with different standard deviations. For a normally distributed population, if you randomly pick (or **sample**) a member from the population, the randomly selected member can have any value in the population. Since most of the population is clustered around the mean, the randomly selected value is most likely to be close to the mean. The further a value is from the mean, the less often it occurs in the population and the less likely it is to be randomly selected from the population. If we know μ and σ, we can predict the probability that any given value will be selected at random from the population, as shown in Fig. 4.6. Normal distributions have:

1. 50% of values > μ and 50% < μ
2. 68%, 95%, and 99.7% of values within ±1, 2, and 3 standard deviations (σ) of μ.

Assume you randomly sample four values from the population of 100 water bottles, and bottles with sodium contents of 1.9, 2.7, 2.9, and 2.9 mg/500 mL were selected. As seen in Fig. 4.7, three of the randomly chosen values (2.7, 2.9, and 2.9 mg/500 mL) were close to the mean, and one value (1.9 mg/500 mL) was relatively far from the mean. Remember, any of the 100 values could have been selected at random. However, the values closest to the mean have the highest probability of being chosen based purely on chance.

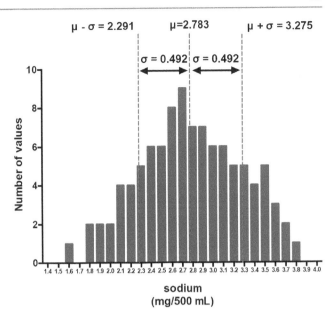

Fig. 4.4 Example population histogram with mean and standard deviation

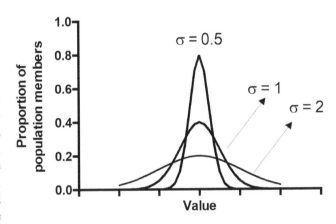

Fig. 4.5 Normal populations with the same mean but distinct standard deviations

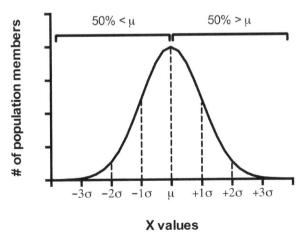

Fig. 4.6 Densities of normal distributions

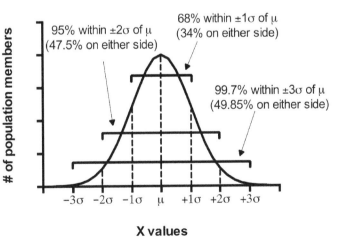

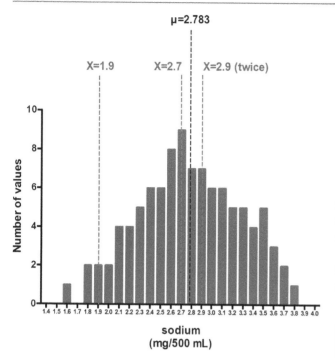

μ=2.783

X=1.9   X=2.7 X=2.9 (twice)

Fig. 4.7 Random values from the example distribution

4.3   Z-SCORES

Each normal population has a different μ, σ combination. However, this is inconvenient for statistical calculations. To make normal distributions easier to work with, you can transform, or "standardize," all normally distributed populations by converting each possible population value (x) to a standardized variable, called Z:

$$Z = \frac{x - \mu}{\sigma} \qquad (4.3)$$

The variable x is normally distributed with mean μ and standard deviation σ [x ~ N(μ, σ)]. An infinite number of possible x population distributions exist. The distribution of the **Z-scores** is Z ~ N(0, 1), that is, " the standard normal distribution" (Fig. 4.8):

1. The center of the Z-distribution is μ = 0.
2. The spread of the Z-distribution is defined by σ = 1.

The standard normal distribution applies regardless of the specific μ, σ values. Each x value in a population has a corresponding Z-score. For any x value, the corresponding Z-score is the number of standard deviations that X is away from μ. The more standard deviations (i.e., multiples of σ or 1) that x (or Z) is from μ (or 0), the less likely that value is to be randomly selected from the population. You can calculate a Z-score for any x value.

Example A1

For the sodium data, calculate the Z-scores for a hypothetical *x* value close to the mean (2.45) and one that is far from the mean (3.8)

$$x = 2.45, \quad Z = \frac{x - \mu}{\sigma} = \frac{2.45 - 2.783}{0.492} = \frac{-0.333}{0.492} = -0.677$$

$$x = 3.8, \quad Z = \frac{x - \mu}{\sigma} = \frac{3.8 - 2.783}{0.492} = \frac{1.017}{0.492} = 2.067$$

Therefore, for X ~ N(2.783, 0.492), *x* values of 2.45 and 3.8 are 0.68 standard deviations below and 2.07 standard deviations above the mean, respectively.

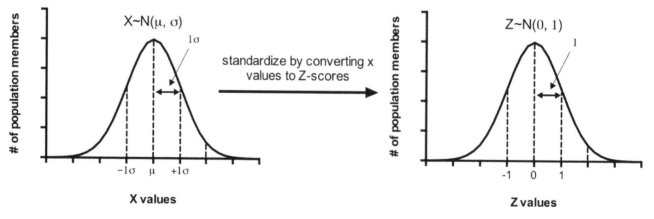

Fig. 4.8 Transformation of *x* values to Z-scores

The same distribution probabilities that apply to $X \sim N(\mu, \sigma)$ also apply to $Z \sim N(1, 0)$:

1. 0% of Z values are >0 and 50% are <0
2. 68%, 95%, and 99.7% of Z values are within ±1, 2, and 3 of 0.

Some other important properties of normal populations:

1. The population curve = 100% of population values.
2. The area under the population curve = 100% (percent) or 1 (fraction).
3. The % area under the distribution curve between any two points = probability of randomly selecting a value in that range from the population (see below).

Based on these properties of normal populations, statisticians have developed a "Z-table" (www.normaltable.com), which contains the **probability** (P) of getting a Z-score smaller than the observed Z-score if $Z \sim N(1, 0)$ (Fig. 4.9). To use the Z-table:

1. Find your observed x value.
2. Calculate a Z-score (Z_{obs}) from the x value.
3. Find the corresponding table value = $P(Z < Z_{obs})$ if $Z \sim N(1,0) = P(x < x_{obs})$ if $x \sim N(\mu, \sigma)$.

For ease of use, the Z-table is organized with the first two digits of Z on the left and the third digit (second decimal place: 0.00, 0.01, 0.02, etc.) across the top. Therefore, to find $P(Z < 1.08)$, you would find the row with 1.0 on the left, then go across that row to the 0.08 column. The steps listed above can be applied to analysis of the sodium data.

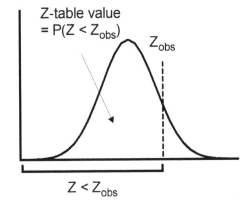

Fig. 4.9 Meaning of Z-table values

Example A2
Calculate the probability of observing sodium values of *less* than 2.783 mg/500 mL (the mean) and *less* than 3.5 mg/500 mL

Calculate Z-scores for both:

$$\text{When } x = \mu = 2.783, \ z = \frac{x - \mu}{\sigma}$$
$$= \frac{2.783 - 2.783}{0.492} = \frac{0}{0.492} = 0$$

$$\text{When } x = 3.5, \ z = \frac{x - \mu}{\sigma} = \frac{3.5 - 2.783}{0.492} = \frac{0.717}{0.492} = 1.4$$

Find table values for those Z-scores:

$$P(Z < 0) = \text{table value for } Z = 0 \rightarrow 0.50 (50\%)$$

$$P(Z < 1.46) = \text{table value for } Z$$
$$= 1.46 \rightarrow 0.9279 (92.79\%)$$

For the sodium population (normally distributed, $\mu = 2.783$, $\sigma = 0.492$), the probability of observing sodium values <2.783 and 3.5 is 50% and 92.79%, respectively.

Standardization to Z-scores is convenient because it eliminates the need for a different Z-score table for each population (which would be impossible, given the infinite number of possible populations, most of which would have unknown μ, σ). Although the Z-table contains P values of getting a Z-score smaller than the observed Z-score, you can calculate the P value of getting a Z-score larger than the observed Z-score. The sum of the area = 1 (100%). Thus, the P value of getting a Z-score smaller than the observed Z-score plus the P value of getting a Z-score larger than the observed Z-score equals 100%:

$$P(Z > Z_{obs}) + P(Z < Z_{obs}) = 1(100\%) \quad (4.4)$$

Therefore:

$$P(Z > Z_{obs}) = 1 - P(Z < Z_{obs}) \quad (4.5)$$

The full expression then becomes:

$$P(Z > Z_{obs}) + P(Z < Z_{obs}) + P(Z = Z_{obs}) = 1(100\%) \quad (4.6)$$

This equation covers all possible Z-scores. Thus, the P value of getting a Z-score larger than the observed Z-score is simply 1 − P value of getting a Z-score larger than the observed Z-score. Therefore, to find the P value of getting a Z-score larger than the observed Z-score (Fig. 4.10):

1. From the x value, calculate a Z-score (Z_{obs}).
2. From Z_{obs}, find table value = P(Z < Z_{obs}).
3. From P(Z < Z_{obs}), calculate P(Z > Z_{obs}) = 1 − table value for Z_{obs} = P(X > X_{obs})if X~N(μ,σ).

Example A3
Calculate the probability of observing sodium values of greater than 2.9 mg/500 mL, again following the steps listed previously
  Calculate the Z-score:

When $x = 2.9$, $z = \dfrac{x-\mu}{\sigma} = \dfrac{2.9-2.783}{0.492} = \dfrac{0.117}{0.492} = 0.24$

  Find the table value for the Z-score:
  P(Z < 0.24) = table value for Z = 0.24 → 0.5948(59.48%).
  Convert to P(Z > Z_{obs}) by subtracting from 1:

$$P(Z > Z_{obs}) = 1 - P(Z < Z_{obs})$$
$$= 1 - 0.5948 = 0.4052 = 40.52\%$$

So, for the sodium population, normally distributed with $\mu = 2.783$ and $\sigma = 0.492$, the probability of observing a sodium value >2.9 is 40.52%.

How do you calculate the probability of getting an x value (and corresponding Z-score) between two x values (with corresponding Z-scores)? The probability that a Z-score lies between two given Z-scores is simply the difference between

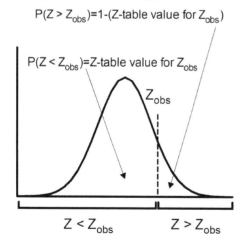

P(Z > Z_{obs})=1-(Z-table value for Z_{obs})

P(Z < Z_{obs})=Z-table value for Z_{obs}

Z_{obs}

Z < Z_{obs} Z > Z_{obs}

Fig. 4.10 Probability of observing Z smaller than Z_{obs}

the two table values for P(Z < Z_{obs}) (i.e., subtract the smaller table value from larger value) (Fig. 4.11):

$$P(Z_1 < Z < Z_2) = P(Z < Z_1) - P(Z < Z_2) \qquad (4.7)$$

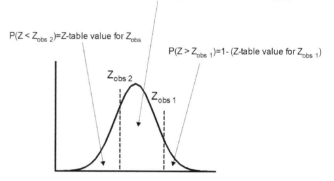

P($Z_{obs\ 1}$ < Z < $Z_{obs\ 2}$)=(Z-table value for $Z_{obs\ 1}$) - (Z-table value for $Z_{obs\ 2}$)

P(Z < $Z_{obs\ 2}$)=Z-table value for Z_{obs}

P(Z > $Z_{obs\ 1}$)=1- (Z-table value for $Z_{obs\ 1}$)

$Z_{obs\ 2}$

$Z_{obs\ 1}$

Fig. 4.11 Probability of observing Z between two Z_{obs} values

Example A4
Calculate the probability of observing sodium values between 1.9 and 3.1 mg/500 mL
  Calculate the Z-scores:

When $x = 1.9$, $z = \dfrac{x-\mu}{\sigma} = \dfrac{1.9-2.783}{0.492} = \dfrac{-0.883}{0.492} = -1.79$

When $x = 3.1$, $z = \dfrac{x-\mu}{\sigma} = \dfrac{3.1-2.783}{0.492} = \dfrac{0.317}{0.492} = 0.64$

  Find the table values for the Z-scores:
  (Z < 0.64) = table value for Z = 0.64 → 0.7389
  P(Z < 0.64) = table value for Z = − 1.79 → 0.0376
  Calculate the area between the two Z-scores:

$$P(Z_1 < Z < Z_2) = P(Z < Z_2) - P(Z < Z_1)$$
$$= 0.7389 - 0.0376$$
$$= 0.7013 = 70.13\%$$

So, for the sodium population which is normally distributed with $\mu = 2.783$ and $\sigma = 0.492$, the probability of observing a sodium values between 1.9 and 3.1 mg/500 mL is 70.13%.

4.4   SAMPLE DISTRIBUTIONS

You almost never know the actual population values. This is due to several factors. First, the population is often too large to sample all members. Second, sampling often means that the product is no longer available for use. It would be impractical for various reasons to produce a population of 10,000 cans of soup and then analyze all 10,000 cans to measure the sodium content. Instead, the analyst typically samples a few representative members (n) and estimates population parameters from sample parameters. Population parameters (the characteristics of the members of the population) are as follows:

$$\text{Population}(x)\text{mean} = \mu_x \qquad (4.8)$$

$$\text{Population}(x)\text{standard deviation} = \sigma_x \qquad (4.9)$$

Sample parameters (the characteristics of each possible sample of n members of the population) are as follows:

$$\text{Sample size} = n \qquad (4.10)$$

$$\text{Mean of } \bar{x} \text{ distribution} = \mu_{\bar{x}} \approx \mu_x \qquad (4.11)$$

$$\text{SD of } \bar{x} \text{ distribution} = \sigma_{\bar{x}} = \frac{\sigma_x}{\sqrt{n}} \qquad (4.12)$$

Sample parameters are related to population parameters:

1. Sample mean ≈ population mean.
2. Sample standard deviation < population standard deviation.

This implies that sampling can be used to estimate population parameters.

Example B1
Determine the mean and standard deviation of the sample mean distribution if you sample five or ten members of the population
  For both means:

$$\mu_{\bar{x}} \approx \mu_x = 2.783$$

The standard deviations of the sample means:

$$\sigma_{\bar{x}} = \frac{\sigma_x}{\sqrt{n}}$$

$$\text{for } n = 5, \ \sigma_{\bar{x}} = \frac{\sigma_x}{\sqrt{n}} = \frac{0.492}{\sqrt{5}} = 0.220$$

$$\text{for } n = 10, \ \sigma_{\bar{x}} = \frac{\sigma_x}{\sqrt{n}} = \frac{0.492}{\sqrt{10}} = 0.156$$

You can see that:

1. The sample mean standard deviation < population standard deviation.
2. Larger n size decreases sample mean standard deviation.

Sampling from a normal population results in a normally distributed population of all possible sample means ($\bar{x}$). The population distribution refers to the individual population values of x. The sample mean distribution refers to the values of sample means ($\bar{x}$) generated by sampling x an infinite number of times. Because the sample mean is an *average* of the sampled population members:

1. The average sample mean is the same as the average population member.
2. The sample mean is less widely distributed than the population (outliers in the population get diluted by taking a sample *mean*).

The larger the sample size (n), the tighter (SD) the sample mean distribution becomes:

$$\text{SD of } \bar{x} \text{ distribution} = \sigma_{\bar{x}} = \frac{\sigma_x}{\sqrt{n}}$$

$$\text{Therefore, as } n \uparrow, \sigma_{\bar{x}} \downarrow$$

Everything you learned about Z-scores for population values applies to sample mean values. You can transform the sample mean distribution to the Z-distribution:

$$\bar{x} \sim N\left(\mu, \frac{\sigma}{\sqrt{n}}\right) \rightarrow Z \sim N(0,1) \qquad (4.13)$$

You can transform an observed sample mean into the corresponding Z-score:

$$Z = \frac{\bar{x} - \mu}{\frac{\sigma}{\sqrt{n}}} \qquad (4.14)$$

You can use the Z-table to calculate the probability of getting a sample mean smaller or greater than an observed mean or between two observed means (same math as before).

Example B2
How likely would it be to have a sample mean of greater than 3.0 mg/500 mL if you sampled four bottles?
  Transform the $\bar{x}$ into a Z-score:

$$Z = \frac{\bar{x} - \mu}{\frac{\sigma}{\sqrt{n}}} = \frac{3 - 2.783}{\frac{0.492}{\sqrt{4}}} = \frac{0.217}{0.246} = 0.88$$

Find the table value for the Z-score (area to the left of Z_{obs}):
  P(Z < 0.88) = table value for Z = 0.88 : 0.8106
  Find P(Z > Zobs):

$$P(Z > 0.88) = 1 - P(Z < 0.88)$$
$$= 1 - 0.8106 = 0.1894 = 18.94\%$$

So, for the sodium data, the probability of sampling n = 4 bottles and getting a mean sodium content of >3.0 mg/500 mL is 18.9%.

4.5 CONFIDENCE INTERVALS

How confident are you that the observed sample mean is close to the actual population mean? And how close? You can calculate the probability of getting a sample mean as extreme as the one you observe, *if* you assume the population is normally distributed and the observed sample mean equals the actual population mean.

Procedure:

1. Calculate Z-score for $\bar{x}$.
2. Find area between $+Z$ and $-Z = P(Z < | Z_{obs} |) = C$.

The area (C) is the probability of getting a sample mean as extreme as the one you observed (Fig. 4.12) for an assumed actual population mean. You can calculate the probability of getting a sample mean more extreme than the one you observed (α) on either end, or getting a sample mean more extreme on a single end ($\alpha/2$).

$$C = P\left(-Z_{obs} < Z < +Z_{obs}\right)$$
$$= P\left(Z < +Z_{obs}\right) - P\left(Z < -Z_{obs}\right) \quad (4.15)$$

$$\alpha = 1 - C \quad (4.16)$$

$$\frac{\alpha}{2} = \frac{1-C}{2} \quad (4.17)$$

Example C1

Suppose you sample n = 3 bottles and get a mean of 2.4 mg/500 mL. Calculate the probabilities of getting a sample mean this extreme, a sample mean smaller than 1.8 mg/500 mL, and a sample mean larger than 1.8 mg/500 mL.

Calculate the Z-score for the observed population mean:

$$Z_{obs} = \frac{\bar{x} - \mu}{\frac{\sigma}{\sqrt{n}}} = \frac{2.4 - 2.783}{\frac{0.492}{\sqrt{3}}} = \frac{-0.383}{0.284} = -1.35$$

Find the $+Z$ and $-Z$ values:

$Z = -1.35$, so you are interested in the range between

$$Z = 1.35 \text{ and } -1.35$$

Find $C = P(-Z_{obs} < Z < +Z_{obs})$ from the table:

$$C = P\left(-Z_{obs} < Z < +Z_{obs}\right)$$
$$= P\left(Z < +Z_{obs}\right) - P\left(Z < -Z_{obs}\right)$$
$$C = P\left(Z < +Z_{obs}\right) - P\left(Z < -Z_{obs}\right)$$
$$= \text{table value}\left(+Z_{obs}\right) - \text{table value}\left(-Z_{obs}\right)$$
$$C = 0.9115 - 0.0885 = 0.823 = 82.3\%$$

For the sodium population which is N(2.783, 0.492), there is an 82.3% chance that you would observe a sample mean within 1.35 standard deviations ($-1.35 < Z < +1.35$) on either side of the mean if you sampled 3 bottles from the population.

Now, find α:

$$\alpha = 1 - C = 1 - 0.823 = 0.177 = 17.7\%$$

For the sodium population which is N(2.783, 0.492), there is a 17.7% chance that you would observe a sample mean > 1.35 standard deviations ($Z < -1.35$ or $Z > +1.35$, i.e., $Z > |1.35|$) on either side of the mean if you sampled three bottles from the population.

Now, find $\alpha/2$:

$$\frac{\alpha}{2} = \frac{1-C}{2} = \frac{1-0.823}{2} = 0.0885 = 8.85\%$$

For the sodium population which is N(2.783, 0.492), there is an 8.85% chance that you would observe a sample mean > 1.35 standard deviations ($Z > +1.35$) above the mean if you sampled 3 bottles from the population. This either indicates that you randomly chose a sample that was highly unlikely simply by chance OR suggests that the true population mean and standard deviation are not 2.783 and 0.492, respectively.

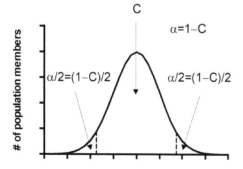

Fig. 4.12 Probability of getting a sample mean as extreme as the observed mean

Usually you do not know the actual population mean (μ) or standard deviation (σ). You sample and calculate sample mean (observed $\bar{x}$) and sample standard deviation (sample SD = $\sigma_{\bar{x}}$) to estimate these parameters (observed $\bar{x} \approx \mu_{\bar{x}} = \mu_x$ and SD $\approx \sigma_{\bar{x}} = \sigma_x / \sqrt{n}$). It is highly unlikely that the observed sample mean is *exactly* equal to the population mean. However, based on the variability of the data sample, you can calculate a range (i.e., an "interval") around your observed sample that is likely to cover the true population mean with a given level of statistical **confidence**. The **confidence interval** (CI) gives a margin of error around our sample mean:

$$CI : \bar{x} \pm \text{margin of error} \qquad (4.18)$$

Suppose that you want to generate a CI that is within a specified number of standard deviations (i.e., number of Z-scores) from the mean. Using the Z-score formula:

$$Z = \frac{\bar{x} - \mu}{\frac{\sigma}{\sqrt{n}}} \rightarrow Z\left(\frac{\sigma}{\sqrt{n}}\right) = \bar{x} - \mu$$

Since it could be on either side of the mean, change to:

$$\pm Z\left(\frac{\sigma}{\sqrt{n}}\right) = \bar{x} - \mu$$

Now, you have to choose how wide the sample margin of error should be. You do this by selecting the number of standard deviations you desire to be within on each side, which gives a maximum absolute value of Z. You could decide that you want our CI to cover the true mean with a given percent confidence (C). This means that the probability that the CI does *not* cover the true mean would be $1 - C$ (which was defined previously as α):

$$\alpha = 1 - C \rightarrow C = 1 - \alpha$$

Then, you would find the "critical" Z-scores where:

1. $P(Z_{crit} < Z < Z_{crit}) = C$ (the desired confidence level)
2. $P(Z > |Z_{crit}|) = \alpha = 1 - C$ (the probability that the CI does *not* cover the true mean).

The easiest way to do this is to find $\alpha/2$ and then find the corresponding Z-score:

First, determine $\alpha/2$ from the desired confidence (C):

$$\frac{\alpha}{2} = \frac{1-C}{2} = \frac{1-0.823}{2} = 0.0885 = 8.85\%$$

$\alpha/2$ is an area to the *right* of the Z_{crit} needed to determine the area to the left of Z_{crit} (aka $Z_{1-\alpha/2}$) i.e., $P(Z < Z_{1-\alpha/2})$ (Fig. 4.13):

$$P(Z < Z_{crit}) = P\left(Z < Z_{1-\frac{\alpha}{2}}\right) = 1 - \frac{\alpha}{2} = 1 - \frac{1-C}{2} = \frac{1+C}{2}$$

$P(Z < Z_{1-\alpha/2})$ is the table value for $Z_{1-\alpha/2}$. Use this to find $Z_{1-\alpha/2}$.

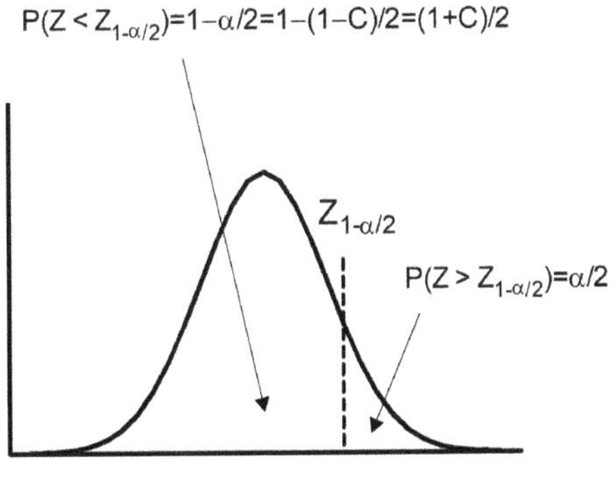

$$P(Z < Z_{1-\alpha/2}) = 1 - \alpha/2 = 1 - (1-C)/2 = (1+C)/2$$

Fig. 4.13 Relationship between C, α/s, and Z-table values

Now you can calculate the "margin of error" for the CI:

$$\text{Margin of error} : \pm Z_{1-\frac{\alpha}{2}} x \frac{\sigma}{\sqrt{n}} \qquad (4.19)$$

The confidence interval, with confidence level C, is:

$$CI : \bar{x} \pm \text{margin of error} \rightarrow CI : \bar{x} \pm Z_{1-\frac{\alpha}{2}} x\left(\frac{\sigma}{\sqrt{n}}\right) \qquad (4.20)$$

Do not get flustered by the statistics details. You just need to be able to do the following:

1. Determine the desired C.
2. Calculate $\alpha/2$.
3. Calculate $1- \alpha/2$.
4. Use this value to find $Z_{1-\alpha/2}$.
5. Calculate the CI.
6. Get from $C \rightarrow \alpha/2 \rightarrow 1 - \alpha/2 \rightarrow Z_{1-\alpha/2}$.
7. Plug these values into the formula and calculate the CI.

Example C2

Suppose you sample n = 5 cans of soda, and the mean caffeine content is 150 mg/can. You know that the population has a standard deviation (σ) of 15 mg caffeine/can. What is the 95% confidence interval for the mean caffeine content per can?

$$\text{Calculate } \alpha/2 : \frac{\alpha}{2} = \frac{1-C}{2} = \frac{1-0.95}{2} = \frac{0.05}{2} = 0.025$$

$$\text{Calculate } 1-\alpha/2 : P\left(Z < Z_{1-\frac{\alpha}{2}}\right) =$$

$$1 - \frac{\alpha}{2} = 1 - 0.025 = 0.975$$

$1 - \alpha/2 = P(Z < Z_{1-\alpha/2}) = $ table value for $Z_{1-\alpha/2}$...use this value to find $Z_{1-\alpha/2}$:

$$\text{Table value} = 0.975 \rightarrow Z_{1-\frac{\alpha}{2}} = 1.96$$

$$\text{Calculate the CI:} \bar{x} \pm Z_{1-\frac{\alpha}{2}} x \frac{\sigma}{\sqrt{n}}$$

$$150\frac{mg}{can} \pm 1.96 x \frac{15\frac{mg}{can}}{\sqrt{5}} \rightarrow 150 mg/can \pm 13.1 mg/can$$

$$\text{Lower and upper limits} = 150\frac{mg}{can} \pm 13.1\frac{mg}{can}$$

$$= 136.9 \text{ and } 163.1 mg/can$$

Therefore, you estimate that the true population mean is somewhere between 136.9 and 163.1 mg caffeine/can with 95% confidence.

Typically, you use 90, 95, or 99% confidence in food analysis. We can choose any desired level of confidence that we want. However:

1. The more confidence you want, the broader the interval gets.
2. The less confidence you are willing to accept, the narrower the range gets.

The confidence and the range of the interval are directly proportional to each other. In other words, you can be very certain that your confidence interval covers the true mean, but it will be a very broad interval. You can alternatively have a very tight interval, but a much lower confidence that it contains the true mean.

Example C3
For the soda data, calculate the 90% CI of the sample mean

$$\text{Calculate}\,\alpha/2 : \frac{\alpha}{2} = \frac{1-C}{2} = \frac{1-0.90}{2} = \frac{0.10}{2} = 0.05$$

$$\text{Calculate}\,1-\alpha/2 : 1 - \frac{\alpha}{2} = 1 - 0.05 = 0.95$$

$1 - \alpha/2 = P(Z < Z_{1-\alpha/2}) = $ table value for $Z_{1-\alpha/2}$. Use this value to find $Z_{1-\alpha/2}$:

$$\text{Table value} = 0.95 \rightarrow Z_{1-\frac{\alpha}{2}} = 1.645$$

$$\text{Calculate the CI}: \bar{x} \pm Z_{1-\frac{\alpha}{2}} \, x \frac{SD}{\sqrt{n}}$$

$$\frac{150\,\text{mg}}{\text{can}} \pm 1.645\,x\,\frac{15\frac{\text{mg}}{\text{can}}}{\sqrt{5}} \rightarrow 150\,\text{mg}/\text{can} \pm 11.0\,\text{mg}/\text{can}$$

$$\text{Lower and upper limits} = 150\frac{\text{mg}}{\text{can}} \pm 11.0\frac{\text{mg}}{\text{can}}$$
$$= 139.0\,\text{and}\,161.0\,\text{mg}/\text{can}$$

Therefore, you estimate that the true population mean is between 139.0 and 161.0 mg/can with 90% confidence. You see that our interval is tighter, but you are therefore less confident that it contains the true population mean.

4.6 T-SCORES

Usually you do not have large enough n and do not know σ, which are requirements for using the Z-distribution. Therefore, statisticians developed the **t-score**:

$$t = \frac{\bar{x} - \mu}{\frac{SD}{\sqrt{n}}} \tag{4.21}$$

Recall that the sample standard deviation (SD) is calculated as follows:

$$SD_n = \sqrt{\frac{\sum(x_i - \bar{x})^2}{n}} \tag{4.22}$$

or

$$SD_{n-1} = \sqrt{\frac{\sum(x_i - \bar{x})^2}{n-1}} \tag{4.23}$$

You should use the "n − 1" formula for SD when n < 25–30. The t-score is similar to the Z-score. However, there are some critical differences. The t-score:

1. Uses the SD of the sample (known) instead of the population σ (usually unknown).
2. Is more conservative than Z-distribution (for a given mean and SD, t is larger than Z).
3. Is typically only presented for a few selected values of $\alpha/2$ (typically 0.1, 0.05, 0.025, 0.01, and 0.005, which correspond to the commonly used confidence values of 80, 90, 95, 98, and 99%, respectively).
4. Is only listed in positive values, not negative values.

Typically, you should use the t-distribution (and sample SD) for n < 25–30 and the Z-distribution and sample SD for n > 25–30. The t-distribution is presented in a table (www.normaltable.com), similar to Z. One major difference between t and Z is that the t is based on a value called **degrees of freedom** (df). This value is based on the sample size:

$$df = n - 1 \tag{4.24}$$

Therefore, for each sample size value (n), there is a unique t-score (whereas Z is independent of n). Note that the t-table is divided into columns for each $\alpha/2$ values, and within those columns, for each df value. Also, note that the t-table gives the area to the *right* of t (Fig. 4.14), whereas the Z-table gives the area to the *left* of Z. Although this is confusing at first, it actually makes t easier to use for CIs than Z. First, determine $\alpha/2$ from the desired confidence (C):

$$\frac{\alpha}{2} = \frac{1-C}{2}$$

This is the table value for t. From the table value, find $t_{\alpha/2}$, df = n − 1. Then, use this value to calculate the CI:

$$CI : \bar{x} \pm t_{\frac{\alpha}{2}, df=n-1} \, x \frac{SD}{\sqrt{n}} \tag{4.25}$$

In a food analysis course, t-scores (as opposed to Z-scores) are used almost exclusively, and they have practical use in

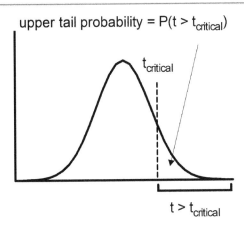

upper tail probability = P(t > $t_{critical}$)

$t_{critical}$

$t > t_{critical}$

Fig. 4.14 Interpretation of t-table values

the food industry. Because t is more conservative for Z, the same confidence level will give a wider interval if calculated using t.

Example D1
Suppose that you sample n = 7 energy bars and obtain a mean total carbohydrate content of 47.2% with a sample standard deviation of 3.22%. Calculate the 99% confidence interval
  Use the t-distribution because n is relatively small.

Determine $\alpha/2$: $\dfrac{\alpha}{2} = \dfrac{1-C}{2} = \dfrac{1-0.99}{2} = \dfrac{0.01}{2} = 0.005$

Therefore, you would look in the 0.005 column on the t-table.
  Then, find $t_{\alpha/2}$, df = n − 1 : $df = n − 1 = 6 − 1 = 5$.
  Therefore, within the 0.005 column on the t-table, you will look at the row for df = 5:

$\dfrac{\alpha}{2} = 0.005$ and $df = 5 \rightarrow t_{0.005, df=5} = 4.032$

Calculate the CI:

$$CI : \bar{x} \pm t_{\frac{\alpha}{2}, df=n-1} \; x \dfrac{SD}{\sqrt{n}} \quad \rightarrow \quad 47.2\% \pm 4.032 \, x \dfrac{3.22\%}{\sqrt{7}}$$

$$\rightarrow \qquad 47.2\% \pm 4.91$$

$$CI : \bar{x} \pm t_{\frac{\alpha}{2}, df=n-1} \; x \dfrac{SD}{\sqrt{n}} \quad \rightarrow \quad 47.2\% \pm 4.032 \, x \dfrac{3.22\%}{\sqrt{7}}$$

$$\rightarrow 47.2\% \pm 4.91$$

Upper limit = 47.2% + 4.91% = 52.1%

Lower limit = 47.2% − 4.91% = 42.3%

  Therefore, you estimate that the true population mean is between 42.3% and 52.5% with 99% confidence.

The take-home message is that a sample mean is a **single point estimate** of the true population mean that is probably not exactly correct (there is essentially zero chance that a sample mean calculated from a sample of n members of a population will be exactly the same as the true population mean). Therefore, a single point estimate is almost useless in food analysis because it is almost always incorrect. So, why even sample and test if it is almost always wrong?

Example D2
Suppose your company's specification says that the cheese powder to be used in your products must contain 17.0% protein by weight. You measure n = 10 bags of the cheese powder and calculate a mean of 16.8% by weight and a SD of 0.3%. The observed sample mean does not equal the specification value (and it is unlikely to ever fall exactly on 17.0%). Do you reject the lot because it is not exactly 17.0%? Do you keep sampling until the observed mean is 17.0%? However, suppose your company specifies that the raw ingredient must meet the specification, from n = 10 sampled units, using a 95% CI
  This CI is:

$$\bar{x} \pm t_{\frac{\alpha}{2}, df=n-1} \; x \dfrac{SD}{\sqrt{n}} \rightarrow 16.8\% \pm t_{0.025, df=4} \, x \dfrac{0.3\%}{\sqrt{10}} \rightarrow$$

$$16.8\% \pm 2.776 \, x \dfrac{0.3\%}{\sqrt{10}}$$

  CI : 16.8 % ± 0.263 % or 16.537 % − 17.063%
  Therefore, although the point estimate does not equal the specification value, you are 95% confident that the true mean lies somewhere between 16.537% and 17.063%, which contains the specification value (17.05). Therefore, based on this, you would accept the raw material.

The example D2 demonstrates the utility of a CI. It is a range of values that have a specified likelihood of containing the true population mean, whereas the point estimate has a very small likelihood of being the true mean. You may wonder what use a CI is if it is a range of values. However, you can often get very "tight" CIs (i.e., a narrow range that is useful) by random sampling of a population and choosing an appropriate level of confidence. As discussed in the textbook Chap. 4, you can use the concept of a CI to determine how large the n size should be to obtain a CI of a specific width. In addition to CIs, another technique that is often used to express the observed sample mean as a range of values is to calculate the sample mean ± SD or the mean ± SEM. The SEM, or **standard error of the mean**, is simply:

$$SEM = \frac{SD}{\sqrt{n}} \qquad (4.26)$$

You will notice that the SEM term appears in the t-test and CI formulae Eq. 4.23 and 4.27. When you express the observed sample mean ± SD or SEM, you are not giving a probability-based estimate (that would require the t-term multiplier used for t-tests and CIs) but rather simply presenting the data as an estimate of the mean based on the sample variability. The range calculated based on SEM will always be narrower, and hence less conservative, than the t-test and CI. This may not be true for the range calculated based on SD, depending on the sample size. For the cheese powder data (Example D2), you could express the estimate as:

$$Mean \pm SD = 16.8\% \pm 0.3\% = 16.5 - 17.1$$

$$Mean \pm SEM = 16.8\% \pm \frac{0.3\%}{\sqrt{10}}$$
$$= 16.8\% \pm 0.0949 = 16.705 - 16.895\%$$

Therefore, based on these ranges, we would accept the raw material based on mean ± SD but reject it based on mean ± SEM. Therefore, it is very important to know how to use your analytical data once it is obtained to inform statistics-based decisions. What are the specific rules you will use to evaluate data and make decisions? Typically, these are specified by company lab quality assurance/quality control manuals, industry standards, or regulatory agencies.

4.7 T-TESTS

Sometimes you need to determine if the observed sample mean indicates that the population is statistically the same, or different from, a chosen value. You can use a procedure called a t-test if you have the sample mean ($\bar{x}$), standard deviation (SD), sample sizes (n), and a desired population mean (μ) that you want to compare the same mean to. First, determine $\alpha/2$ from the desired confidence (C) and the df value, as before:

$$\frac{\alpha}{2} = \frac{1-C}{2} \text{ and } df = n-1$$

From these values, find the table value $t_{\alpha/2,\ df = n-1}$. This is the "critical value" for a t-test, labeled as "$t_{critical}$." Then, calculate the observed t-score ("t_{obs}") for the sample mean, SD and n:

$$t = \frac{\bar{x} - \mu}{\frac{SD}{\sqrt{n}}}$$

Then, the $t_{critical}$ and t_{obs} values are compared.

If $|t_{obs}| > t_{\frac{\alpha}{2}, df=n-1}$ → sample mean is significantly different from μ with confidence (C)

If $|t_{obs}| < t_{\frac{\alpha}{2}, df=n-1}$ → sample mean is not significantly different from μ with confidence (C)

A few notes are worth mentioning:

1. The more confidence you want that you have the correct answer, the larger $t_{critical}$ becomes, and the larger t_{obs} must be to provide evidence that the sample mean is significantly different from μ.
2. The absolute value of t_{obs} is compared vs. $t_{critical}$, as the table only lists positive t-scores.
3. You select μ to compare to the sample mean. You typically do not know the population mean, but you may have a "target value" that you are trying to reach or think the population should have, so you use that as μ. This is often an internal specification for a product, or a regulature value, etc.

Example E1

Suppose that your company makes a multivitamin with a label value of 35 mg vitamin E/capsule. Per FDA requirements, the label value needs to be within a certain amount of the actual value. You sample eight capsules and measure the vitamin E content. The observed sample mean is 31.7 mg/capsule, and the sample standard deviation is 3.1 mg/capsule. Can you say that the sample mean includes the label value with 99% confidence?

Determine $\alpha/2$ and df:

$$\frac{\alpha}{2} = \frac{1-C}{2} = \frac{1-0.99}{2} = \frac{0.01}{2}$$
$$= 0.005 \text{ and } df = n-1 = 8-1 = 7$$

From these values, find the $t_{critical}$ value ($t_{\alpha/2,\ df = n-1}$):
$t_{0.005,\ df=7} = 3.499$.

Calculate t_{obs} from the sample data:

$$t_{obs} = \frac{\bar{x} - \mu}{\frac{SD}{\sqrt{n}}} = \frac{31.7 - 35}{\frac{3.1}{\sqrt{8}}} = -3.01 \text{ and } |t_{obs}| = 3.01$$

Since $|t_{obs}|$ (3.01) $< t_{\alpha/2,\, df = n-1}$(3.499), you can say the sample mean is not significantly different from the label value (35 mg/capsule) with 99% confidence. If $|t_{obs}|$ had been $> t_{\alpha/2,\, df = n-1}$(3.499), you would say that you have evidence the sample mean was significantly different from 35 mg/capsule. Note here that you are concerned with the absolute value of t_{obs}($|t_{obs}|$)vs. $t_{critical}$. Given your chosen confidence, you have a $t_{critical}$ value of 3.499, so as long as your $|t_{obs}|$ is <3.499 you can conclude that the observed sample mean is not significantly different from the chosen μ value. Therefore, t_{obs} could be anywhere on the interval (−3.499, +3.499) and still be considered to be not significantly different from the label value. So, the observed mean could be above or below the chosen μ, as long as it is not "too far" in either direction.

Confidence intervals and t-tests provide the same information for a chosen confidence level. Either procedure may be used to compare the sample mean and chosen μ value. Sometimes you need to determine if the means of two samples are "different" (as opposed to one sample and a chosen μ). Sample means are "point estimates" that do not have any probability associated with them. Sample means themselves cannot typically be used to make decisions. Just because two sample means are not exactly the same (which would be almost impossible, even if the populations sampled from were in fact identical, or even the same population) does not mean that the populations are significantly different. You need a statistical basis for determining if the sample means are far enough apart such that you can say they are different with some level of confidence (or not). Given two sample means ($\bar{x}_1$ and $\bar{x}_2$), standard deviations (SD$_1$ and SD$_2$), and sample sizes (n_1 and n_2), first we decide on a confidence level (C), and we calculate $\alpha/2$ as before. Then, we determine df. For two sample means, df is calculated as follows:

$$df = n_1 + n_2 - 2 \tag{4.27}$$

Then you find the critical t-value as before, using the $\alpha/2$ and df values:

$$t_{\frac{\alpha}{2}, df = n_1 + n_2 - 2}$$

Once you have $t_{critical}$, you calculate t_{obs} as follows:

$$t_{obs} = \frac{|\bar{x}_1 - \bar{x}_2|}{\sqrt{s_p^2 \left(\frac{1}{n_1} + \frac{1}{n_2} \right)}} \tag{4.28}$$

This formula differs from the one-sample t-test in that both n_1 and n_2 are used, and you employ a value known as the **pooled variance** (s_p^2) instead of the sample SD. The pooled variance is a weighted average of two sample SD values. The pooled variance is calculated as follows:

$$\text{Pooled variance} = s_p^2 = \frac{(n_1 - 1)SD_1^2 + (n_2 - 1)SD_2^2}{n_1 + n_2 - 2} \tag{4.29}$$

Variance is simply standard deviation squared. Therefore, although you will not use this value, it may be useful to understand that the "pooled standard deviation" (s_p), or the weighted average of the two sample SD values, is simply the square root of the pooled variance:

$$\text{Pooled standard deviation} = \sqrt{\text{pooled variance}}$$
$$= \sqrt{s_p^2} = \sqrt{\frac{(n_1 - 1)SD_1^2 + (n_2 - 1)SD_2^2}{n_1 + n_2 - 2}} \tag{4.30}$$

Once you have obtained the $t_{critical}$ and t_{obs} values, you compare them as for the one-sample t-test:

$$t_{obs} > t_{\frac{\alpha}{2}, df = n_1 + n_2 - 2} \rightarrow \text{means } are \text{ significantly different}$$
with specified confidence (C)

$$t_{obs} < t_{\frac{\alpha}{2}, df = n_1 + n_2 - 2} \rightarrow \text{means } are\ not \text{ significantly}$$
different with specified confidence (C)

Example E2
Suppose your company is concerned that two production lines are producing spaghetti sauces with different acid contents. You measure the titratable acidity in samples from both lines (Line 1: 2.1; 2.0; 2.1; 2.2; 2.3; and 2.4%; Line 2: 2.7; 2.3; 2.2; 2.2; 2.4; 2.6; and 2.5%). Do the two production lines appear to be producing significantly different acidity levels with 90% confidence?
First, calculate the n, $\bar{x}$, and SD value for each sample:

$$\text{Line 1}: n = 6, \bar{x} = 2.183 \text{ and SD} = 0.147$$

$$\text{Line 2}: n = 7, \bar{x} = 2.4, \text{ and SD} = 0.195$$

What is the confidence level (C): 90% or 0.90?

$$\text{Next, calculate } \alpha/2: \frac{\alpha}{2} = \frac{1 - C}{2} = \frac{1 - 0.90}{2} = \frac{0.1}{2} = 0.05$$

For the t-table, df: $df = n_1 + n_2 - 2 = 6 + 7 - 2 = 11$
Find the critical t-value on the table:

$$t_{\frac{\alpha}{2}, df = n_1 + n_2 - 2} = t_{0.05, 11} = 1.796$$

Calculate a pooled variance (s_p^2):

$$s_p^2 = \frac{(n_1 - 1)SD_1^2 + (n_2 - 1)SD_2^2}{n_1 + n_2 - 2}$$

$$= \frac{(6-1)(0.147)^2 + (7-1)(0.195)^2}{6 + 7 - 2}$$

$$= \frac{0.1083 + 0.2286}{11} = 0.03063$$

Calculate t_{obs}:

$$t_{obs} = \frac{|\bar{x}_1 - \bar{x}_2|}{\sqrt{s_p^2\left(\frac{1}{n_1} + \frac{1}{n_2}\right)}} = \frac{|2.183 - 2.4|}{\sqrt{0.03063\left(\frac{1}{6} + \frac{1}{7}\right)}}$$

$$= \frac{|-0.217|}{0.09737} = 2.23$$

Decision:

$$t_{obs} = 2.23 \text{ and } t_{\frac{\alpha}{2}, df = n_1 + n_2 - 2}$$

$$= 1.796 \rightarrow t_{obs} > t_{\frac{\alpha}{2}, df = n_1 + n_2 - 2}$$

Thus, Lines 1 and 2 produce sauces with significantly different acidities (with 90% confidence).

4.8 PRACTICAL CONSIDERATIONS

Here are some practical considerations to help you use statistics in a food analysis course and in a career in which decisions are based on analytical data:

4.8.1 Sample Size

1. A larger sample size enables more accurate approximation of the true population values.
2. Larger sample sizes decrease the sample SD.
3. Use t for $n < 25$–30; use Z for n > 25–30.

4. Sample size determines df, which determines $t_{critical}$ for a specified $\alpha/2$ value.
5. Sampling more units is often useful but not always practical or cost-effective.
6. A "happy medium" needs to be reached that provides "adequate" approximation of the mean with a fixed amount of confidence. This is done by calculating the minimum needed sample size (this is covered in the textbook chapter on sampling).

4.8.2 Confidence

1. Confidence values are chosen based on the acceptable risk to consumers (safety, poor quality, etc.), the company (e.g., quality, profit margin, formulation accuracy), or government regulatory agencies (e.g., labeling accuracy).
2. Confidence determines $\alpha/2$, which determines $t_{critical}$ within a specified df value.
3. The more confidence is required (or desired):
   (a) The wider a CI will be.
   (b) The larger $t_{critical}$ will be.
   (c) The larger t_{obs} must be to be > $t_{critical}$.
   (d) The larger the n size must be to show significance.
4. Increasing desired confidence increases the statistical burden required to shown significant differences; 90–99% confidence is typically used (95% is fairly standard).
5. Increasing the desired confidence is not always practical or cost-effective.

4.8.3 What Test to Use?

Table 4.1 gives "rules of thumb" for what types of tests to use for various scenarios:

4.9 PRACTICE PROBLEMS

(Note: Answers are at end of laboratory manual.)

Table 4.1 Examples of when specific statistical tools should be used for food analysis

Question	Test/calculation	Examples
Is an observed sample mean statistically similar to or different from a chosen value?	One-sample t-test	• Is the actual value in agreement with the label value? • Does the permitted level differ from the allowed level? • Does the raw material meet company specifications? • Is the composition of the finished product acceptable based on company specifications? • Is a pipette delivering the desired volume? • Is a packaging machine filling packages to the desired level?
What is the actual population value, based on the sample?	One-sample CI	• What is the actual concentration of a compound of interest?
Based on samples from two populations, are these populations different?	Two-sample t-test ($\mu = 0$)	• Are two lots different? • Are two lines, plants, etc., producing products with different composition? • Is your product the same or different from a competitor's product?

1. As quality assurance manager for a canned soup manufacturer, you need to make sure that chicken noodle soup has the sodium content indicated on the label (343 mg/cup). You sample six cans of soup and measure the sodium content: 322.8, 320.7, 339.1, 340.9, 319.2, and 324.4 mg/cup. What is the mean of the observations (mg/cup)? What is the standard deviation of the observations (mg/cup)? Calculate the 96% confidence interval for the true population mean, and determine both the upper and lower limits of the confidence interval. Determine if the sample mean provides strong enough evidence that the population is "out of spec" with 99% confidence.

2. You perform moisture analysis on sweetened condensed milk using a forced-draft oven. The following data are obtained. Lot A: 86.7, 86.2, 87.9, 86.3, and 87.8% solids; Lot B: 89.1, 88.9, 89.3, 88.8, and 89.0%. Determine if the lots are statistically different with 95% confidence.

4.10 TERMS AND SYMBOLS

Confidence (C). Statistical probability of being correct based on chance alone.

Confidence interval (CI). A range of values (based on a point estimate plus a statistically determined margin of error) used to predict information about a population.

Degrees of freedom (df). The number of values that is free to vary independently.

Distribution. All values in a population and the relative or absolute occurrence of each value.

Histogram. A plot of all values in a population and the relative occurrence of each value.

Mean (μ). Average value.

Normal (N). A population distribution whose shape is defined by a mean and standard deviation.

Pooled variance (s_p^2). A measure of variability for two individual samples, incorporating both the sample sizes and sample standard deviations of both.

Population. All individuals of interest.

Population mean (μ_x). The average value of all the members of a population.

Population standard deviation (σ). The average difference between the value of each population member and the population mean.

Probability (P). Statistical likelihood of an event or state occurring by chance alone.

Sample. Selected members of a population, from which inferences are made about the entire population.

Sample mean ($\mu_{\bar{x}}$). The average value of all the members of a sample.

Sample standard deviation (SD, $\sigma_{\bar{x}}$). The average difference between the value of each sample member and the sample mean.

Single point estimate. A single value (mean or single data point) used to predict information about a population.

Standard deviation (σ). The average difference between the value of each population member and the population mean.

Standard error of the mean (SEM). A measure of variability of a sample, incorporating the sample standard deviation and the sample size.

t-score (t). A normalized value describing how many standard deviations a value is from the mean, but more conservative than the Z-score.

Z-score (Z). A normalized value describing how many standard deviations a value is from the mean.

Acknowledgments The author thanks Sean F. O'Keefe (Department of Food Science and Technology, Virginia Polytechnic Institute and State University, Blacksburg, VA, USA) for his input on this chapter in the previous edition of the book.

Part II

Laboratory Exercises

Nutrition Labeling Using a Computer Software Program

5

Ann M. Roland

Contents

5.1 INTRODUCTION

5.1.1 Background

The 1990 Nutrition Labeling and Education Act mandated nutritional labeling of most foods. As a result, a large portion of food analysis is performed for nutrition labeling purposes. Interpretation of these regulations (21 CFR 101.9) and the appropriate usage of rounding rules, available nutrient content claims, reference amounts, and serving size can be challenging. Additionally, during the product development process, the effect of formulation changes on the nutrition label may be important. As an example, a small change in the amount of an ingredient may determine if a product can be labeled low fat. As a result, the ability to immediately approximate how a formulation change will impact the nutritional label can be valuable. In some cases, the opposite situation may occur and a concept called reverse engineering is used. In reverse engineering, the information from the nutrition label is used to determine a formula for the product. Caution must be used during reverse engineering. In most cases, only an approximate formula can be obtained and additional information not provided by the nutrition label may be necessary.

The use of nutrient databases and computer software programs designed for preparing and analyzing nutrition labels can be valuable in the situations described earlier. In this laboratory, you will use a computer software program to prepare a nutrition label from a product formula, determine how changes in the formula affect the nutrition label, and observe an example of reverse engineering.

5.1.2 Reading Assignment

Nielsen SS (2024) Nutrition labeling. Ch. 3. In: Ismail BP, Nielsen SS (eds). *Nielsen's food analysis*, 6th edn. Springer, New York.

A. M. Roland (✉)
Owl Software, Columbia, MO, USA
e-mail: aroland@owlsoft.com

5.1.3   Objective

Prepare a nutrition label for a yogurt formula, determine how formulation changes will affect the nutrition label, and observe an example of reverse engineering a formula based on its nutrition facts values.

5.1.4   Materials

TechWizard™ – Formulation and Nutrition Labeling Software for Microsoft® Excel® for Windows®.

Owl Software. TechWizard™ Software Manual, Columbia, MO. www.owlsoft.com

5.1.5   Notes

Instructions on how to receive and install the software used for this laboratory are located online at www.owlsoft.com, under the *Academic* link located in the heading. *It is possible that the TechWizard™ program has been updated since the publication of this laboratory manual and any changes in the procedures described below will also be found on this web page.*

*Install the software prior to the laboratory session to ensure that it works properly with your PC. TechWizard™ will not run on a Mac® with Microsoft® Excel®.

5.2   PREPARING NUTRITION LABELS FOR SAMPLE YOGURT FORMULAS

5.2.1   Procedure

1. Start the TechWizard™ Application.
   (a) Double-click the TechWizard™ icon on the computer desktop, if a yellow strip appears at the top of the screen with a button *Enable Content*, click it.
   (b) Click on the *Enter Program* button.
   (c) Select the username "*Administrator*" and click *Log In*.
2. Go to the Labeling section. This section is for building a recipe and entering the information required to create a nutrition facts label.
   (a) Select the *Labeling* menu at the top of the screen.
   (b) Click on *Labeling Section*.
3. Enter the ingredients for Formula #1 listed in Table 5.1.
   (a) Click the *Add Ingredients* button located on the left side of the screen.
   (b) In the Add Ingredients window, find each ingredient name in the list or use *Search Ingredient List* at the

Table 5.1 Sample yogurt formulas

Ingredient name	Ref #	Formula #1 (lb)	Formula #2 (lb)
Milk (3.7% fat)	Twz 0077	38.201	48.201
Milk (skim), no vit. A add	NDB No. 1151	35.706	25.706
Milk (skim), condensed (35% TS)	Twz 0083	12.888	12.888
Swtnr, sugar, (liquid)	Twz 0132	11.905	11.905
Modified starch (<1 DE), Ntl. Stch., N-LiteR CL	Twz 0096	0.800	0.800
Stblzr, gelatin, dry powder, unsweetened	NDB No 19177	0.500	0.500

top of the window. Select the ingredient and click the *Add* button or double-click on the ingredient. As ingredients are added, they will appear on the right. If the wrong ingredient is added, select it in the list on the right and click the *Delete Selected* button.
   (c) When all the ingredients are added, *Close* the window.
4. Enter the amounts of the ingredients for Formula #1 listed in Table 5.1.
   (a) Click the *Labeling* menu and click *Enter Recipe* from the *Formula group*. The *Recipe Entry* window appears.
   (b) Select an ingredient from the list. Under *Enter Amount*, type in the number, then from the *Select Units* dropdown select *lb*. Click the *Update* button or press the *Enter* key on your keyboard.
   (c) Repeat to enter the amount and units for each ingredient.
   (d) To update the formula, click *OK*. Note the amounts that now populate the *%(wt/wt)* column.)
5. Review and complete ingredient information. Notice the warning at the top of the window that states, "Warning – Missing Values!" This message informs us that some of the ingredients are missing values for nutrients that appear on the label.
   (a) Select the *Labeling* menu at the top of the screen and click on *Composition*. Each row represents the composition of each ingredient. Notice that the column for the nutrient, *Potassium* has a red background. This is to warn us that some of the ingredients are missing amounts for this nutrient. The ingredient "*Swtnr, sugar, (Liquid)*" is missing a value for Potassium. The *Potassium* content of sugar is zero.
   (b) In the blue cell for the missing Potassium value, enter the number zero and press the *Enter* key on your keyboard. Notice that the red background disappears.

(c) Scroll to the right to see the column for *Sugars*. Enter the number zero for the missing value for the *Sugars* contents of the gelatin ingredient. This is a temporary solution to the issue. A better solution is to update the ingredient information in the TechWizard™ ingredient file.

6. Return to the Labeling screen.
    (a) Select the *Labeling* menu at the top of the screen.
    (b) Click on *Labeling Section*. Notice that the missing values warning is gone.

7. Determine the serving size of the yogurt.
    (a) To find the serving size, click the *Show Ref. Table* button which is located above and to the right of the *% (wt/wt)* column. This table is a copy of Reference Amounts Customarily Consumed Per Eating Occasion (21 CFR 101.12(b)). Yogurt is listed under the heading, *Dairy Products and Substitutes*. The *reference amount*, or serving size, is *170 g*. The common household amount is the amount in cups that corresponds to 170 g. In this example 170 g of yogurt, that measures 2/3 cup, is in each container.
    (b) Exit the document by clicking the X button, to return to the TechWizard™ *Labeling* section.

8. Enter the serving size information.
    (a) To the right of the *Show Ref. Table* button, click the *Serving Size* button below the *Common Household Unit* heading. When the Serving Size window appears, enter 2/3 and click *OK*. Select *cup* from the *units* dropdown to the right of the 2/3 value.
    (b) Click the *Serving Size* button below the *Equivalent Metric Quantity* heading. When the Serving Size window appears, enter *170* and click *OK*. To the right of the value, set units to "g." The *Equivalent Metric Quantity* is the amount of the formula used to calculate the Nutrition Facts information. You may have noticed the values in the *Nutrient* table changed when this value was entered.
    (c) Click the *Number of Servings* button, when the prompt appears, enter 1 and click *OK*.

9. Set label options and view the nutrition label. Yogurt contains a substantial amount of protein. On a typical nutrition facts label, the amount per serving of protein in grams is viewed but not the % daily value (DV). To include the % daily value of protein, a value is required for the Protein Digestibility Corrected Amino Acid Score (PDCAAS). This value for our yogurt is 1.
    (a) Click on the *Label Options* button which is located to the right of the *Serving Size* buttons. The *Label Options* window appears. Make sure that the *Label Type* is set to *Standard*, and the *Dual Column* option is not checked.
    (b) Check the box next to *"Protein – Show ADV"* and enter 1 for *PDCAAS*. Click the *Apply* button and

close the window. The Nutrition Facts Panel for our yogurt appears.

10. Save Formula #1.
    (a) Click the *Return* button at the top left of the screen.
    (b) Select the *File* menu at the top of the screen and click on *Save Formula As* on the right. The *Save Formula As* window appears.
    (c) In the space for *Formula Name*, enter your first and last name followed by the name of the formula, which is Yogurt Formula #1 (First Name Last Name Yogurt Formula #1). Click the *Save* button.

11. Copy and paste the nutrition label and ingredient statement for Formula #1 to a Microsoft® Word® Document.
    (a) Select the *Labeling* menu, click the *Copy* dropdown, and select *Standard Label*. The *Format Options* window appears. Click *OK* to copy the label. Click *OK* when the label copied prompt appears.
    (b) Start Microsoft ® Word ®. Open a new document by double clicking on *Blank Document*. Type the same information (First Name Last Name Yogurt Formula #1) in the Word® document that you entered when you saved the formula and press the *enter* key on the keyboard to move to a new line. Select the *Home* menu and click on *Paste*. The Nutrition Facts panel is now in the Word® document. Press the e*nter* key to move to a new line under the label.
    (c) Go back to TechWizard™ by clicking on the Excel® icon on the bottom of the screen.
    (d) Select the Labeling menu and click on *Simple Declaration Editor*. The *Declaration Window* appears. Select all the text under *Formula Declaration* and press the CTRL key and the C key on your keyboard at the same time. This will copy the ingredient statement. Close the window.
    (e) Go back to the Word® document and paste the ingredient statement under the Nutrition Facts panel. Save the Word® document.
    (f) Go back to TechWizard™.

12. Repeat the process for Formula #2 (Table 5.1).
    (a) Change the amount of each ingredient in TechWizard™ to match the amounts in the Formula #2 (%) column (Table 5.1) by using the *Recipe Entry* function explained in Step 4.
    (b) Confirm that the information remains the same as in Steps 5–9.
    (c) Save the formula with the new name (First Name Last Name Yogurt Formula #2) using *Save Formula as* explained in Step 10.
    (d) Add the Nutrition Facts label, and ingredient statement to the Word® document and save it, as described in Step 11.

5.3 ADDING NEW INGREDIENTS TO A FORMULA AND DETERMINING HOW THEY INFLUENCE THE NUTRITION LABEL

Sometimes it may be necessary to add additional ingredients to a formula. As an example, let us say you decided to add an additional source of calcium to yogurt formula #1. After contacting several suppliers, you decided to add Whey Calcium 1000, a calcium phosphate product, to the yogurt formula. This product is a dairy-based whey mineral concentrate and contains 25% calcium. You want to determine how much Whey Calcium 1000 you need to add to have 50% and 100% of the Daily Value (DV) of calcium in one serving of your yogurt. The composition of the Whey Calcium 1000 you will add is shown in Table 5.2. If we review Table 5.2, we can see that some of the values necessary to use this ingredient to create a Nutrition Facts panel are missing (Dietary Fiber, Added Sugars, Vitamin D). We will have to use our judgement as a food scientist to fill the missing values.

TechWizard™ includes an artificial intelligence (AI) feature that provides for creating a formula based on ingredient, nutrient, and composition requirements. We will review the use of the AI feature to create the yogurt formula used in the previous exercise. To watch TechWizard™ in action, we will utilize an automated example, which shows the major steps involved in a procedure and provides details on what is happening. Afterward, we will save the formula and use it to develop our own formulations with an additional ingredient.

Table 5.2 Composition of Whey Calcium 1000

Component	Amount
Total solids (%)	92
Calories (cal/100 g)	40
Total fat (g/100 g)	0.3
Saturated fat (g/100 g)	0.2
Trans fatty acid (g/100 g)	0.01
Cholesterol (mg/100 g)	2.0
Sodium (mg/100 g)	745
Total carbohydrates (g/100 g)	10
Dietary fiber	0
Total sugars (g/100 g)	10
Added sugars	0
Protein (g/100 g)	4.0
Vitamin D (D2 + D3)	0
Calcium (mg/100 g)	25,000
Iron (mg/100 g)	1.3
Potassium (mg/100 g)	988
Ash (%)	75
Lactose (%)	10
Phosphorus (mg/100 g)	13,000
Water (%)	8.0

5.3.1 Procedure

1. Run an Automated Example to create a formula.
   (a) Select the *Help* menu and click on *Cultured Products Automated Examples*. The *Cultured Products Examples* window appears. Click button 1.
   (b) Click the *Next* button as needed to continue the example.
   (c) When *EXAMPLE 1 COMPLETED* appears in the window, click the X to close the window. The data from the example remains.
2. Save the formula.
   (a) Select the *File* menu at the top of the screen and click on *Save Formula As* on the right. The *Save Formula As* window appears.
   (b) In the space provided for *Formula Name*, enter your first and last name before the current formula name, *Yogurt, Lowfat (cultured products example 1)*, (First Name Last Name *Yogurt, Lowfat (cultured products example 1)*.
   (c) Click the *Save* button.
3. Add the name of the new ingredient to the ingredient file.
   (a) Select the *Edit Ingredient File* menu. The Edit Ingredient File menu has 4 groups which appear at the bottom of the menu these are *File*, *Form*, *Ingredients*, and *Properties*.
   (b) Select *Edit Current File* from the *File* group. This opens the *Ingredient File Editor*.
   (c) Select the *Edit Ingredient File* menu and click on *Add* in the *Ingredients* group. The *Add Ingredient to Ingredient File* window appears.
   (d) In the space for *Enter Ingredient Name*, type Whey Calcium 1000 followed by your First and Last name (Whey Calcium 1000 First Name Last Name).
   (e) Click the *Add* button, answer *Yes*, then *Close* the window.
4. Enter the new ingredient composition (Table 5.2).
   (a) Select the *Edit Ingredient File* menu and click on *Use Data Entry Form*, from the *Form* group. The *Ingredient Data Entry Form* appears with *Typical nutritional information* showing under *Select Property List Appearance*.
   (b) In the *Search Ingredient List* at the top of the window, type *Whey Calcium 1000* and press the *Search* button.
   (c) Select the *Whey Calcium 1000* with your name. *Price* is selected in the middle portion of the window. The *Modified Value* box has *1.00* in it. Click on the *Enter* button next to *Modified Value* to accept the *price/lb* and move to the next property.
   (d) We will leave *Reference Number* blank. Click the *Enter* button to move to the next property.

(e) Enter the amounts found in Table 5.2. For *Total Solids*, type *92* in the *Modified Value* box and click *Enter*. Repeat the process to enter the values for *Calories, Total Fat, Saturated Fat, Trans Fat, Cholesterol, Sodium,* and *Total Carbohydrates*.

(f) This ingredient is sourced from milk which does not contain dietary fiber. Enter *0* for the *Dietary Fiber* amount.

(g) This ingredient contains 10% lactose which is a part of Total Sugars. Enter 10 for *Total Sugars*.

(h) Lactose does not count towards added sugars because of the purpose of the ingredient (it is not added as a sweetener) and the type of sugar (lactose has limited sweetness). Enter *0* for *Added Sugars*.

(i) Enter *4* for *Protein*.

(j) We do not have a value for *Vitamin D (D2 + D3)*, enter zero for the amount.

(k) Enter the values for *Calcium, Iron,* and *Potassium*.

(l) Select the *Property* option in *Search Ingredient List* at the top of the screen. Enter *Ash* in the space provided and press *Search*. Enter the amount for *Ash* in the *Modified Value* space. Use the same procedure to enter the amount for *Phosphorus*.

(m) Search for the property *Ingredient Declaration*. Type "Whey mineral concentrate" in the *Modified Value* space and press the *Enter* button next to *Modified Value*.

(n) Press the *Update* button to complete the data entry. *Close* the window. Answer *Yes* when prompted to save your changes.

5. Calculate the amount of calcium (mg/100 g) required to meet 50 and 100% of the DV (see, example below):

$$\text{Calcium required} = \left(\frac{\text{DV for calcium}}{\text{serving size}} \right)$$

$$x\,100\,g \ x \ \% \text{ of DV required}$$

$$\text{Calcium required for 50\% of the DV} = \left(\frac{1300\,mg}{170\,g} \right)$$

$$x\,100\,g \ x \ 0.50$$

$$\text{Calcium required for 50\% of the DV} = 382 \left(\frac{mg}{100\,g} \right)$$

6. Open the yogurt formula created from the previous automated example.

(a) Select the *Formula Dev and Batching* menu, then *Formula Dev*.

(b) Select the *File* menu and click on *Open Formula*. The *Retrieve Formula* window appears.

(c) Select your Y*ogurt, Lowfat (cultured products example 1)* formula and press the *Open* button. Answer *Yes* if prompted to open the formula in the Formula Dev section. Answer *Yes* if prompted to continue.

7. Add your new Whey Calcium 1000 ingredient to the formula.

(a) Click the *Add Ingredients* button. Enter *Whey Calcium 1000* in *Search Ingredient List* at the top of the window. Press the *Search* button.

(b) Select the ingredient *Whey Calcium 1000 your first name your last name*. Press the *Add* button. *Close* the window.

8. Add the property Calcium and constraints to the formula.

(a) Press the *Add Properties* button. Enter *Calcium* in *Search Property List* at the top of the window. Press the *Search* button. The property is selected. Press the *Add* button. *Close* the window.

(b) In the blue cells to the right of *Calcium*, enter the amount of calcium (mg/100 g) needed for 50% DV/ serving, *382*, in the *Max* column then enter *382* in the *Min* column. Note the settings in the blue *Min* and *Max* cells for all ingredients and properties. These specify the desired composition of the formula.

9. Create the formula.

(a) Press the *Formulate* button above the *Add Ingredients* button. This will have the AI feature determine the ingredient amounts necessary to meet all the constraints set on ingredients and properties in the blue *Min* and *Max* cells.

(b) Record the %wt./wt. to answer Question 1 in Sect. 5.5.

10. Save the formula.

(a) Select the *File* menu at the top of the screen.

(b) Click on *Save Formula As* on the right.

(c) Change the formula name to include, "added calcium 50% DV."

11. Open the formula into the Labeling section.

(a) Select the *Formula Dev and Batching* menu.

(b) Select *Create Nutrition Label* from the *Formula* group.

(c) Answer *Yes,* then *OK* to the prompts.

12. Update the composition of ingredients missing values as in Sect. 5.2, Step 5.

13. Create a Nutrition Facts panel, ingredient statement, and update the Word® document with this information.

(a) Follow the instructions described in Sect. 5.2, Steps 6–11.

14. Produce a formula and label that has 100% of the calcium DV.
    (a) Repeat Steps 8–13 except using the calculated amount of calcium required to meet 100% of the calcium DV. You will have to perform this calculation yourself following the example in Step 5.

5.4   AN EXAMPLE OF REVERSE ENGINEERING IN PRODUCT DEVELOPMENT

This is an example in which we are using a TechWizard™ automated example to watch the application in action. This example will show how values from a nutrition facts panel and the ingredient statement can be used to create a formula using artificial intelligence. Pay close attention to the description provided for each step and the procedures TechWizard™ utilizes to create a comparable recipe.

5.4.1   Procedure

1. Select the *Help* menu and click on *Frozen Dessert Automated Examples*. The *Frozen Dessert Examples* window appears.
2. Click button *5*. Read the description in the window that appears.
3. Click the *Next* button as needed to continue the example. Observe the actions and data on the screen.
4. When *EXAMPLE 5 COMPLETED* appears in the window, click X to close the window. The data from the example remains.

5.5   QUESTIONS

1. How much Whey Calcium 1000 did you have to add to the yogurt formula to have 50 and 100% of the DV of calcium in the formula?
2. If Whey Calcium 1000 costs $2.50/lb. and you are going to have 100% of the DV for calcium in your yogurt, how much extra will you have to charge for a serving of yogurt to cover the cost of this ingredient?
3. Assume you added enough Whey Calcium 1000 to claim 100% of the DV of calcium, would you expect the added calcium to cause any texture changes in the yogurt?

Acknowledgments The author of this chapter and editors of the laboratory manual thank Katherine Klammer, Department of Food Science & Nutrition at the University of Minnesota, for reviewing this chapter multiple times.

RESOURCE MATERIALS

Nielsen SS (2024) Nutrition labeling. Ch. 3. In: Ismail BP, Nielsen SS (eds). Nielsen's food analysis, 6th edn. Springer, New York
Owl Software TechWizard™ Manual, Columbia, MO. www.owlsoft.com

Accuracy and Precision Assessment

6

S. Suzanne Nielsen

Contents

6.1   INTRODUCTION

6.1.1   Background

Volumetric glassware, mechanical pipettes, and balances are used in many analytical laboratories. If the basic skills in the use of this glassware and equipment are mastered, laboratory exercises are easier and more enjoyable, and the results obtained are more accurate and precise. Measures of accuracy and precision can be calculated based on data generated, given the glassware and equipment used, to evaluate both the skill of the user and the reliability of the instrument and glassware.

Determining mass using an analytical balance is the most basic measurement made in an analytical laboratory. Determining and comparing mass is fundamental to assays such as moisture and fat determination. Accurately weighing reagents is the first step in preparing solutions for use in various assays.

Accuracy and precision of the analytical balance are better than for any other instruments commonly used to make analytical measurements, provided the balance is properly calibrated and the laboratory personnel use proper technique. With proper calibration and technique, accuracy and precision are limited only by the readability of the balance. Repeatedly weighing a standard weight can yield valuable information about the calibration of the balance and the technician's technique.

Once the performance of the analytical balance and the technician using it has been proven acceptable, determination of mass can be used to assess the accuracy and precision of other analytical instruments. All analytical laboratories use volumetric glassware and mechanical pipettes. Mastering their use is necessary to obtain reliable analytical results. To report analytical results from the laboratory in a scientifically justifiable manner, it is necessary to understand accuracy and precision.

A procedure or measurement technique is validated by generating numbers that estimate their accuracy and precision. This laboratory includes assessment of the accuracy and precision of automatic pipettors. An example application is determining the accuracy of automatic pipettors in a

S. S. Nielsen (✉)
Department of Food Science, Purdue University,
West Lafayette, IN, USA
e-mail: nielsens@purduee.edu

research or quality assurance laboratory, to help assess their reliability and determine if repair of the pipettors is necessary. Laboratory personnel should periodically check the pipettors to determine if they accurately dispense the intended volume of water. To do this, water dispensed by the pipettor is weighed, and the weight is converted to a volume measurement using the appropriate density of water based on the temperature of the water. If replicated volume data indicate a problem with the accuracy and/or precision of the pipettor, repair is necessary before the pipettor can be reliably used again.

It is generally required that reported values minimally include the mean, a measure of precision, and the number of replicates (Smith 2024). The number of significant figures used to report the mean reflects the inherent uncertainty of the value, and it needs to be justified based on the largest uncertainty in making the measurements of relative precision of the assay. The mean value is often expressed as part of a confidence interval (CI) to indicate the range within which the true mean is expected to be found. Comparison of the mean value or the CI to a standard or true value is the first approximation of accuracy. A procedure or instrument is generally not deemed inaccurate if the CI overlaps the standard value. Additionally, a CI that is considerably greater than the readability indicates that the technician's technique needs improvement. In the case of testing the accuracy of an analytical balance with a standard weight, if the CI does not include the standard weight value, it would suggest that either the balance needs calibration or that the standard weight is not as originally issued. Accuracy is sometimes estimated by the relative error ($\%E_{rel}$) between the mean analysis value and the true value. However, $\%E_{rel}$ only reflects tendencies and in practice is often calculated even when there is no statistical justification that the mean and true value differ. Also, note that there is no consideration of the number of replicates in the calculation of $\%E_{rel}$, suggesting that the number of replicates will not affect this estimation of accuracy to any large extent. Absolute precision is reflected by the standard deviation (SD), while relative precision is calculated as the coefficient of variation (CV). Calculations of precision are largely independent of the number of replicates, except that more replicates may give a better estimate of the population variance.

Validation of a procedure or measurement technique can be performed, at the most basic level, as a single-trial validation, as is described in this laboratory that includes estimating the accuracy and precision of commonly used laboratory equipment. However, for more general acceptance of procedures, they are validated by collaborative studies involving several laboratories. Collaborative evaluations are sanctioned by groups such as AOAC International, AACC International, and the American Oil Chemists' Society (AOCS) (Nielsen 2024). Such collaborative studies are prerequisite to procedures appearing as approved methods in manuals published by these organizations.

6.1.2 Reading Assignment

Neilson AP (2024) Laboratory standard operating procedures, Ch. 1. In: Nielsen's food analysis laboratory manual, fourth edn., Ismail BP, Nielsen SS (eds), Springer, New York.

Nielsen SS (2024) Introduction to food analysis, Ch. 1. In: Ismail BP, Nielsen SS (eds) Nielsen's food analysis, sixth edn. Springer, New York.

Smith JS (2024) Evaluation of analytical data, Ch. 4. In: Ismail BP, Nielsen SS (eds) Nielsen's food analysis, sixth edn. Springer, New York.

6.1.3 Objective

Familiarize, or refamiliarize, oneself with the use of balances, mechanical pipettes, and volumetric glassware, and assess accuracy and precision of data generated.

6.1.4 Principle of Method

Proper use of equipment and glassware in analytical tests helps ensure more accurate and precise results.

6.1.5 Supplies

- Beaker, 100 mL
- Beaker, 20 or 30 mL
- Beaker, 250 mL
- Buret, 25 or 50 mL
- Erlenmeyer flask, 500 mL
- Funnel, approximately 2 cm diameter (to fill buret)
- Mechanical pipettor, 1000 µL, with plastic tips
- Plastic gloves
- Ring stand and clamps (to hold buret)
- Rubber bulb or pipette pull-up
- Standard weight, 50 or 100 g
- Thermometer, to read near room temperature
- Volumetric flask, 100 mL
- Volumetric pipettes, one each of 1 and 10 mL

6.1.6 Equipment

- Analytical balance
- Top loading balance

6.1.7   Notes

Before or during the laboratory exercise, the instructor is encouraged to discuss the following: (1) difference between dispensing from a volumetric pipette and a graduated pipette and (2) difference between markings on a 10-mL versus a 25- or 50-mL buret.

6.2   PROCEDURE

(Record data in tables that follow.)

1. Obtain ~400 mL deionized distilled (dd) H_2O in a 500-mL Erlenmeyer flask for use during this laboratory session. Check the temperature of the water with a thermometer.
2. Analytical balance and volumetric pipettes.
   (a) Tare a 100-mL beaker, deliver 10 mL of water from a volumetric pipette into the beaker, and record the weight. Repeat this procedure of taring the beaker, adding 10 mL, and recording the weight, to get six determinations on the same pipette. (Note that the total volume will be 60 mL.) (It is not necessary to empty the beaker after each pipetting.)
   (b) Repeat the procedure as outlined in Part 2a but use a 20- or 30-mL beaker and a 1.0-mL volumetric pipette. Do six determinations.
3. Analytical balance and buret
   (a) Repeat the procedure as outlined in Part 2a, but use a 100-mL beaker and a 50-mL (or 25-mL) buret filled with water, and dispense 10 mL of water (i.e., tare a 100-mL beaker, deliver 10 mL of water from the buret into the beaker, and record the weight). (Handle the beaker wearing gloves, to keep oils from your hands off the beaker.) Repeat this procedure of taring the beaker, adding 10 mL, and recording the weight, to get six determinations on the buret. (Note that the total volume will be 60 mL.) (It is not necessary to empty the beaker after each addition.)
   (b) Repeat the procedure as outlined in Part 3a but use a 20- or 30-mL beaker and a 1.0-mL volume from the buret. Do six determinations.
4. Analytical balance and mechanical pipette. Repeat the procedure as outlined in Step 2a but use a 20- or 30-mL beaker and a 1.0-mL mechanical pipette (i.e., tare a 20- or 30-mL beaker, deliver 1 mL of water from a mechanical pipettor into the beaker, and record the weight). Repeat this procedure of taring the beaker, adding 1 mL, and recording the weight to get six determinations on the same pipettor. (Note that the total volume will be 6 mL.)

(It is not necessary to empty the beaker after each pipetting.)

5. Total content (TC) versus total delivery (TD). Tare a 100-mL volumetric flask on a top loading balance. Fill the flask to the mark with water. Weigh the water in the flask. Now tare a 250-mL beaker and pour the water from the volumetric flask into the beaker. Weigh the water delivered from the volumetric flask.
6. Readability versus accuracy. Zero a top loading balance and weigh a 100-g (or 50-g) standard weight. Record the observed weight. Use gloves or finger cots as you handle the standard weight to keep oils from your hands off the weight. Repeat with the same standard weight on at least two other top loading balances, recording the observed weight and the type and model (e.g., Mettler, Sartorius) of balance used.

6.3   DATA AND CALCULATIONS

Calculate the exact volume delivered in Sect. 6.2, Parts 2–5, using each weight measurement and the known density of water (see Table 6.1). Using *volume* data, calculate the following indicators of accuracy and precision: mean, standard deviation, coefficient of variation, percent relative error, and 95% confidence interval. Use your first three measurements for $n = 3$ values requested and all six measurements for $n = 6$ values.

Data for Sect. 6.2, Parts 2, 3, and 4

| | Volumetric pipette | | | | Buret | | | | Mechanical pipettor | |
| | 1 mL | | 10 mL | | 1 mL | | 10 mL | | 1 mL | |
Rep	Wt.	Vol.	Wt.	Vol.	Wt.	Vol.	Wt.	Vol.	Wt.	Vol.
1										
2										
3										
4										
5										
6										
$n = 3$										
Mean	–		–		–		–		–	
SD	–		–		–		–		–	
CV	–		–		–		–		–	
% E_{rel}	–		–		–		–		–	
$CI_{95\%}$	–		–		–		–		–	
$n = 6$										
Mean	–		–		–		–		–	
SD	–		–		–		–		–	
CV	–		–		–		–		–	
% E_{rel}	–		–		–		–		–	
$CI_{95\%}$	–		–		–		–		–	

Data for Sect. 6.2, Part 5

	Wt.	Vol.
Water in flask =		
Water in beaker =		

Data for Sect. 6.2, Part 6

Balance	Type/model of balance	Standard weight (g)
1		
2		
3		

Table 6.1 Viscosity and density of water at various temperatures

Temp. (°C)	Density (g/mL)	Viscosity (cps)	Temp. (°C)	Density (g/mL)	Viscosity (cps)
20	0.99823	1.002	24	0.99733	0.9111
21	0.99802	0.9779	25	0.99707	0.8904
22	0.99780	0.9548	26	0.99681	0.8705
23	0.99757	0.9325	27	0.99654	0.8513

6.4   QUESTIONS

(Questions refer to parts of Sect. 6.2)

1. Theoretically, how are standard deviation, coefficient of variation, mean, percent relative error, and 95% confidence interval affected by (1) more replicates and (2) a larger size of the measurement? Was this evident in looking at the actual results obtained using the volumetric pipettes and the buret, with $N = 3$ versus $N = 6$ and with 1 mL versus 10 mL? (see table below)

	Theoretical		Actual, with results obtained	
	More replicates	Larger measurement	More replicates	Larger measurement
Standard deviation				
Coefficient of variation				
Mean				
Percent relative error				
95% confidence interval				

2. Why are percent relative error and coefficient of variation used to compare the accuracy and precision, respectively, of the volumes from pipetting/dispensing 1 and 10 mL with the volumetric pipettes and buret in Parts 2 and 3, rather than simply the mean and standard deviation, respectively?

3. Compare and discuss the accuracy and the precision of the volumes from the 1 mL pipetted/dispensed using a volumetric pipette, buret, and mechanical pipettor (Parts 2, 3, and 4). Are these results consistent with what would be expected?

4. If accuracy and precision using the mechanical pipettor are less than should be expected, what could you do to improve its accuracy and precision?

5. In a titration experiment using a buret, if you expect to use much less than a 10-mL volume in each titration, would you expect your accuracy and precision to be better using a 10-mL buret or a 50-mL buret? Why?

6. How do your results from Part 5 of this lab differentiate "to contain" from "to deliver"? Is a volumetric flask "to content" or "to deliver"? Which is a volumetric pipette?

7. From your results from Part 6 of this lab, would you now assume that since a balance reads to 0.01 gram that it is accurate to 0.01 gram?

8. What sources of error (human and instrumental) were evident or possible in Parts 2–4, and how could these be reduced or eliminated? Explain.

9. You are considering adopting a new analytical method in your lab to measure the moisture content of cereal products. How would you determine the precision of the new method and compare it to the old method? How would you determine (or estimate) the accuracy of the new method?

Acknowledgments The author of this chapter thanks Charles E. Carpenter (Department of Nutrition, Dietetics and Food Sciences, Utah State University, Logan UT, USA) for contributing material for this chapter when it was originally developed.

RESOURCE MATERIALS

Neilson AP (2024) Laboratory standard operating procedures, Ch. 1. In: Nielsen's food analysis laboratory manual, 4th ed., Ismail BP, Nielsen SS (eds), Springer, New York

Nielsen SS (2024) Introduction to food analysis, Ch. 1. In: Ismail BP, Nielsen SS (eds) Nielsen's food analysis, 6th edn. Springer, New York

Smith JS (2024) Evaluation of analytical data, Ch. 4. In: Ismail BP, Nielsen SS (eds) Nielsen's food analysis, 6th edn. Springer, New York

Preparation of Solutions and Buffers

7

B. Pam Ismail

Contents

B. P. Ismail (✉)
Department of Food Science and Nutrition, University of
Minnesota, St. Paul, MN, USA
e-mail: bismailm@umn.edu

© The Author(s), under exclusive license to Springer Nature Switzerland AG 2024
B. P. Ismail, S. S. Nielsen (eds.), *Nielsen's Food Analysis Laboratory Manual*, Food Science Text Series,
https://doi.org/10.1007/978-3-031-44970-3_7

7.1   INTRODUCTION

7.1.1   Background

Many chapters in this laboratory manual (in "Part II, Laboratory Exercises") require the use of numerous solutions that must be prepared using appropriate laboratory techniques, while ensuring accuracy and precision. Practicing the preparation of these solutions helps develop the necessary skills for work in a laboratory. This laboratory exercise consists of preparing a variety of solutions and dilutions, utilizing the calculation principles presented in Chap. 2 and 3 of this manual. Also introduced in this chapter is use of a pH meter to adjust the pH of a buffer.

Before describing the procedures to make various solutions and a buffer in this laboratory exercise, it is very important to review prior chapters in the Reading Assignment list. All procedures of this laboratory exercise involve standard practices for use of balances, mechanical pipettes, and glassware, along with proper storage of solutions (see especially Chap. 1). In each of the four major sections of this laboratory exercise, the subsection "Hazards, Precautions, and Waste Disposal" refers to "normal laboratory safety procedures." These procedures include wearing safety glasses and gloves, using concentrated acids and any other dangerous reagents in a fume hood, and clearly labeling all solutions prepared. While these specific cautions are critically important, they will not be repeated in each part of this laboratory exercise.

Note that each of the four major parts of this laboratory exercise involves calculations that should be done by the student prior to the laboratory session (see "Calculations" sections for each part). All calculations should be checked by a teaching assistant or the instructor before proceeding with the described procedures.

7.1.2   Reading Assignment

Neilson AP (2024) Laboratory standard operating procedure, Ch. 1: In: Nielsen's food analysis laboratory manual, fourth edn., Ismail BP, Nielsen SS (eds), Springer, New York.

Neilson AP (2024) Dilutions and concentration calculations, Ch. 3. In: Nielsen's food analysis laboratory manual, fourth edn., Ismail BP, Nielsen SS (eds), Springer, New York.

Tyl C, Ismail BP (2024) Reagents and buffers, Ch. 2. In: Nielsen's food analysis laboratory manual, fourth edn., Ismail BP, Nielsen SS (eds), Springer, New York.

7.2   PREPARATION OF HYDROCHLORIC ACID SOLUTION

7.2.1   Introduction

Mineral-free 0.1 N hydrochloric (HCl) acid, as prepared in this exercise, is used in the operation of the atomic absorption spectrophotometer for analyzing the mineral content in specified foods (Chap. 11, Determination of Nutrition Label Minerals by Atomic Absorption Spectroscopy).

7.2.2   Objective

Prepare 100 mL of mineral-free 0.1 N HCl.

7.2.3   Chemicals

	CAS no.	Hazards
Hydrochloric acid (HCl), conc. (37%)	7637-07-2	Toxic, highly flammable

7.2.4   Hazards, Precautions, and Waste Disposal

Adhere to normal laboratory safety procedures. Because the solution to be prepared is intended for use in the analysis of specific minerals, use only acid-washed glassware for both solution preparation and storage, since this treatment removes any minerals bound to the glass.

7.2.5   Supplies

- Deionized distilled (dd) water (stored in a glass container that has been acid washed)
- Glass storage bottle, acid washed, to hold 100 mL
- Graduated pipette, glass, acid washed, 5 mL
- Volumetric flask, acid washed, 100 mL

7.2.6 Calculations

Calculate the volume of concentrated HCl needed to prepare 100 mL of 0.1 N HCl. Note that labels of commercially available HCl often do not state the molarity/normality; instead, the percent of HCl is provided. Assume that the percentage is 37% (v/v) HCl. Given that the density of HCl is 1.18 g/mL (or kg/L), calculate the volume of HCl needed to prepare 100 mL of 0. N HCl (Hint: You need first to calculate the molarity, M, of the 37% HCl reagent; molarity of HCl equal its normality, N). Before proceeding, have a teaching assistant or the instructor check your calculation.

7.2.7 Procedure

1. Partially fill a 100 mL volumetric flask with dd water.
2. Working under the hood, use the graduated pipette to accurately measure the appropriate acid volume and dispense the acid into the 100 mL volumetric flask containing 10–20 mL of dd water.
3. Bring the flask to volume with dd water.
4. Stopper the bottle, wrap a parafilm strip around the top to help seal the flask, and invert the flask gently several times to ensure adequate mixing.
5. Transfer the acid solution you prepared to the acid-washed storage bottle labeled 0.1 N hydrochloric acid and record the date on the label.

7.3 PREPARATION OF SATURATED SODIUM CHLORIDE SOLUTION

7.3.1 Introduction

There are many instances in research when you need to prepare a saturated solution. In the extraction procedure of the Gas Chromatography laboratory exercise (Chap. 9), you will use a saturated sodium chloride (NaCl) solution to increase the ionic strength of the methylated fatty acids to drive them into the hexane phase.

7.3.2 Objective

Prepare 50 mL of a saturated NaCl solution.

7.3.3 Chemicals

	CAS no.	Hazards
Sodium chloride (NaCl)	7647-14-5	Irritant

7.3.4 Hazards, Precautions, and Waste Disposal

Adhere to normal laboratory safety procedures.

7.3.5 Supplies

- Water (dd water)
- Glass beaker, 100 mL
- Magnetic stir bar
- Storage bottle, 100 mL
- Weigh boat

7.3.6 Equipment

- Analytical balance
- Magnetic stir plate

7.3.7 Calculations

Using the Merck Index or another online reference, calculate the amount (g) of NaCl needed for saturation at 25 °C. Before proceeding, have a teaching assistant or the instructor check your calculation.

7.3.8 Procedure

1. Weigh the needed amount (g) of NaCl into a weigh boat and transfer it into a glass beaker.
2. Add 50 mL dd water and a stir bar, then place on a magnetic stir plate.
3. To assure saturation, there should be some salt at the bottom of the beaker that does not go into solution. If your solution is clear and no salt crystals are visible, add a spatula more of salt.
4. Transfer your saturated NaCl solution to a storage bottle and properly label (including the date) for later use in the gas chromatography laboratory exercise (Chap. 9).

7.4 PREPARATION OF CAFFEINE STANDARDS

7.4.1 Introduction

A standard curve is constructed by plotting the known concentrations of a series of standards against a detection measurement, e.g., absorbance of light at a particular wavelength (λ, nm). A linear regression equation is then produced to determine the unknown concentration of the compound of interest in a sample. You will use a caffeine standard curve in the High-Performance Liquid Chromatography (HPLC) laboratory exercise (Chap. 8) to determine the caffeine concentration of several beverages.

7.4.2 Objective

Prepare a stock solution of caffeine standard and a series of diluted solutions.

7.4.3 Chemicals

	CAS no.	Hazards
Caffeine	58-08-2	Harmful

7.4.4 Hazards, Precautions, and Waste Disposal

Adhere to normal laboratory safety procedures.

7.4.5 Supplies

- Glass beaker, 10 mL, 2
- Magnetic stir bar
- Mechanical pipettes, P200 and P1000
- Micro-centrifuge tubes (for dilutions), 4
- Volumetric flask, 25 mL
- Weighing paper

7.4.6 Equipment

- Analytical balance
- Magnetic stir plate

7.4.7 Calculations

7.4.7.1 Caffeine Stock Solution

Calculate how much caffeine is needed (mg) to prepare 25 mL of a 1 mg/mL caffeine stock solution. Before proceeding, ask a teaching assistant or the instructor to check your calculation.

7.4.7.2 Diluted Standard Solutions

Calculate the volume of stock solution needed to prepare 1 mL of the following caffeine standard solutions:

Caffeine concentration (mg/mL)	Volume stock solution (µL)	Volume dd water (µL)
0.050	50.	950
0.10		
0.15		
0.20		

Example calculation for 0.05 mg/mL caffeine solution:
Use the following equation and rearrange it, to calculate dilutions.

To prepare 1 mL of the 0.050 mg/mL standard from the 1 mg/mL stock solution:

$$c_{\text{stock solution}} \times v_{\text{stock solution}} = c_{\text{diluted solution}} \times v_{\text{solution}}$$

$$1\frac{mg}{mL} \times v_{\text{caffeine stock}} = 0.050\frac{mg}{mL} \times 1\,mL$$

$$v_{\text{caffeine stock}} = 0.050\,mL = 50.\mu L$$

Use 50. µL Caffeine stock +950 µL dd water
Follow the example calculation above to determine the volumes needed for the other caffeine solutions listed in the table. Complete the table above. Before proceeding, have a teaching assistant or the instructor check your calculations and the table.

7.4.8 Procedure

7.4.8.1 Caffeine Stock Solution

1. Carefully weigh the caffeine (mg) onto weighing paper then transfer the caffeine to a 10 mL beaker. Place a magnetic stir bar and approximately 10 mL of dd water in the beaker and place the beaker on a magnetic stir plate. Quantitatively transfer the caffeine solution to a 25 mL volumetric flask (i.e., transfer the solution, then use a few mL of dd water to wash the beaker, transfer to the volumetric flask, then repeat the process another time). Dilute and make up to volume with dd water. (*Note:* Remove the stir bar when bringing the solution to volume in the 25 mL volumetric flask and then return the stir bar to the flask for complete mixing.)

7.4.8.2 Diluted Standard Solutions

1. Transfer some of the stock solution to a 10 mL beaker. (This will be used to create the standard solutions.)
2. Using a mechanical pipettor, place the calculated volume of stock solution for each of the above standard concentrations in a separate micro-centrifuge tube with the corresponding amount of dd water.
3. Cap the tubes and vortex them.

4. Label the tubes with the solution identification and concentration, your last name, and date.
5. These standards should be kept frozen until use in the HPLC laboratory exercise (Chap. 8).

7.5 PREPARATION OF A BUFFERING SOLUTION

7.5.1 Introduction

Buffers consist of mixtures of weak acids and their corresponding bases, e.g., acetic acid and sodium acetate. It is crucial that both components are present in the system for buffers to exert their function, which is to stabilize a certain pH. Much research is dependent upon the correct preparation of a solution with a buffering capacity. The buffer of choice is dependent on the desired pH of the assay, but also other factors such as compatibility with other reagents. For instance, when dietary fiber is quantified, starch needs to be digested with an α-amylase solution, which requires calcium (Ca^{2+}) to work properly. Therefore, a calcium salt such as $CaCl_2$ is added to the solution. Some buffers such as phosphate buffers would not be a good choice for such an application because they may form precipitates with the Ca^{2+} ions, and therefore some other buffer needs to be selected.

7.5.2 Objective

To gain a deeper understanding of the calculations and methods of buffer preparation covered in Chap. 2, prepare 250 mL of a 200 mM acetate buffer, pH 4.15.

7.5.3 Chemicals

	CAS no.	Hazards
Glacial acetic acid	64-19-7	Corrosive
Sodium acetate	127-09-3	Irritant

7.5.4 Hazards, Precautions, and Waste Disposal

Adhere to normal laboratory safety procedures.

7.5.5 Supplies

- Water (dd water)
- Graduated cylinder, 100 mL
- Graduated cylinder, 250 mL
- Glass beaker, 500 mL
- Glass pipette, 5 mL
- Magnetic stir bar, 2
- pH calibration standards, pH 4 and pH 7
- Pipette aid
- Volumetric flask, 250 mL, 2
- Weighing paper

7.5.6 Equipment

- Analytical balance
- Magnetic stir plate
- pH meter

7.5.7 Notes

Since this laboratory exercise involves preparation of an acetate buffer using acetic acid and sodium acetate, some notes follow regarding this type of buffer.

- The molarity of a buffer indicates the **sum of the molarity of the acid and the base**. In the case of this exercise, it means that in 1 L of buffer, 200 mM of acetic acid + sodium acetate is present. It does not mean that there are 200 mM of acetic acid and 200 mM of sodium acetate. Instead, **the two components together** make up 200 mM of buffer molecules.
- The Henderson-Hasselbalch equation is used to determine the concentration of acid and base needed to result in the desired pH.
- Identification of the acid type and the base is needed when preparing a buffer. In this laboratory exercise, these questions are relatively simple because the name "acetic acid" already indicates the acid used. There are other buffers for which it is harder to know [e.g., TRIS, tris(hydroxymethyl) aminomethane], and in this case just determine which form of the molecule is protonated; the form that is protonated is the acid.
- Finally, the pK_a of acetic acid is 4.75.

7.5.8 Calculations

To make the acetate buffer, some calculations are given to you, and you will need to do other calculations. The calculations follow to determine the needed concentrations of the acid and base (using the Henderson-Hasselbalch equation, Chap. 2) to prepare a 200 mM acetate buffer. You will need to do further calculations, as described at the end of this section, to make 250 mL of this buffer.

You know that [acetic acid] + [sodium acetate] = 200 mM. You need to determine the amount (M) or number of moles you need of each to prepare the buffer. Identify one of the compounds as x, and the other one as 200 − x, where 200 mM is the desired concentration. In this calculation, designate [acetic acid] as x, and [sodium acetate] = 200 − x. (*Note:* The designation of which is x is arbitrary.) Using the Henderson-Hasselbach equation: $pH = pK_a + \log \dfrac{[\text{base}]}{[\text{acid}]}$

1. $4.15 = 4.75 + \log \dfrac{[\text{sodium acetate}]}{[\text{acetic acid}]}$

2. $-0.6 = \log \dfrac{200 - x}{x}$

   Now get rid of the logarithm:

3. $0.25119 = \dfrac{200 - x}{x}$

4. $0.25119\,x = 200 - x$

5. $0.25119\,x + x = 200$

   $1.25119\,X = 200$

   $x = \dfrac{200}{1.25119} = 159.85$

   x = the concentration of acetic acid: ≈ 160 mM

Therefore, the concentration of sodium acetate (200 − x) = 200–160 = 40 mM

There are three ways to prepare a buffer, such as an acetate buffer, as described in Chap. 2 in this laboratory manual (Sect. 2.3). This laboratory procedure to make the 200 mM acetate buffer follows the first method described. In this method, you prepare 200 mM of each of the acid and the base, then using the Henderson-Hasselbach equation you determine the volume you need to take from each.

To prepare 250 mL of a 200 mM acetate buffer, you will need x mL of 200 mM acetic acid and (250-x) mL of 200 mM sodium acetate. Solve for x using the Henderson-Hasselbach equation and determine the specific volume (mL) of 200 mM acetic acid and the volume of 200 mM sodium acetate buffer needed 20 prepare 250 mL of your buffer. Before proceeding, have a teaching assistant or the instructor check your calculations.

7.5.9 Procedure

1. Prepare 250 mL of a 200 mM acetic acid solution:
   Calculation:
   0.25 L × 0.2 Moles Acetic Acid/L = *0.05 Moles Acetic Acid*
   molecular weight (MW) of acetic acid = 60 g/mole.

   Therefore: 60 g/mole × *0.05 mole acetic acid* = 3 g acetic acid needed to prepare 250 mL of 200 mM acetic acid solution. Three g of acetic acid is equivalent to 2.86 mL of acetic acid, which has a density of 1.05 g/mL.

Preparation:
   In a 250 mL volumetric flask containing some dd water, use a glass pipette to add 2.86 mL acetic acid then bring to volume with dd water.

2. Prepare 250 mL of 200 mM sodium acetate solution:
   Calculation:
   0.25 L × 0.2 Mole Sodium Acetate/L = *0.05 Moles Sodium Acetate*
   MW of sodium acetate = 82 g/mole
   Therefore: 82 g/mole × *0.05 moles sodium acetate* = 4.1 grams sodium acetate
   Preparation:
   (a) Weigh 4.1 grams sodium acetate in a weighing boat or paper and transfer to a 250 mL volumetric flask.
   (b) Use a dd water wash bottle to wash all the powder into the flask.
   (c) Bring to volume with dd water while washing down the sides of the flask with dd water.
   (d) Place a stir bar in the flask and stir until dissolved.

3. From the Henderson-Hasselbalch equation and calculations requested in Sect. 7.5.8, you obtained the volume needed of 200 mM acetic acid and of 200 mM sodium acetate. Use graduated cylinders to measure the two volumes, and combine the separate solutions of 200 mM acetic acid and 200 mM sodium acetate in a beaker.

4. Place the beaker on a stir plate. Add a stir bar and mix.

5. Calibrate the pH meter of the solution using both a pH 4 and pH 7 calibrator. (Your teaching assistant or instructor can assist you as needed in the pH calibration.).

6. Measure the pH of your buffer and record the value. Adjust the pH as needed with 2 *N* HCl or 2 *N* NaOH. How close to the desired pH is the solution you prepared?

7. This solution is not used in any of the laboratory exercises within the laboratory manual, so it can be disposed of down the drain.

7.6 QUESTIONS

Refer to questions at the end Chap. 2 and 3 in the laboratory manual:

Chap. 2 (Reagents and Buffers): Questions 1, 6, 14

Chap. 3 (Dilutions and Concentrations): Questions 2, 3

Acknowledgments This laboratory was developed with the help of Cindy Gallaher, Department of Food Science and Nutrition, University of Minnesota, St. Paul, Minnesota.

RESOURCE MATERIAL

O'Neil, MJ, ed. (2023) The Merck Index: An encyclopedia of chemicals, drugs, and biologicals, *Online*. Royal Society of Chemistry. Whitehouse Station, New Jersey https://www.rsc.org/merck-index

High-Performance Liquid Chromatography

8

B. Pam Ismail and S. Suzanne Nielsen

Contents

8.1 INTRODUCTION

8.1.1 Background

High-performance liquid chromatography (HPLC) has many applications in food analysis. Food components that can be analyzed with HPLC include organic acids, vitamins, amino acids, sugars, nitrosamines, phytochemicals, certain pesticides, metabolites, fatty acids, aflatoxins, pigments, and cer-

tain food additives. Unlike gas chromatography, it is not necessary for the compound being analyzed to be volatile. It is necessary, however, for the compounds to have some solubility in the mobile phase. It is important that the solubilized samples for injection be free from all particulate matter, so centrifugation and filtration are common procedures. Also, solid-phase extraction is used commonly in sample preparation to remove interfering compounds from the sample matrix prior to HPLC analysis.

Many food-related HPLC analyses utilize reversed-phase chromatography in which the mobile phase is relatively polar, such as water, dilute buffer, or aqueous methanol, or aqueous acetonitrile. The stationary phase (column packing) is relatively nonpolar, usually silica particles coated with a C_8, C_{18}, or C_{40} hydrocarbon. As compounds travel through the column, they partition between the hydrocarbon stationary phase and the mobile phase. The mobile phase may be constant dur-

B. P. Ismail (✉)
Department of Food Science and Nutrition, University of Minnesota, St. Paul, MN, USA
e-mail: bismailm@umn.edu

S. S. Nielsen
Department of Food Science, Purdue University, West Lafayette, IN, USA
e-mail: nielens@purdue.edu

© The Author(s), under exclusive license to Springer Nature Switzerland AG 2024
B. P. Ismail, S. S. Nielsen (eds.), *Nielsen's Food Analysis Laboratory Manual*, Food Science Text Series,
https://doi.org/10.1007/978-3-031-44970-3_8

ing the chromatographic separation (i.e., isocratic) or changed stepwise or continuously (i.e., gradient). When the compounds are separated from each other at the end of the column, they eventually elute and move with the mobile phase through a detector for identification and quantitation. Identification often is accomplished by comparing the retention time, i.e., the time needed to elute a compound from a column to that of standards chromatographed under the same conditions. Other than retention times, spectral analysis using detectors such as a photodiode array detector or a mass spectrometer can enhance identification of compounds with similar retention times. Identification also can be accomplished by powerful detectors. Quantitation generally involves the construction of standard curves using standards at various concentrations, typically in part per million.

8.1.2 Reading Assignment

Ismail BP (2024) Basic principles of chromatography. Ch. 12. In: Ismail BP, Nielsen SS (eds) Nielsen's food analysis, sixth edn. Springer, New York

Ismail BP, Reuhs BL (2024) High-performance liquid chromatography. Ch. 13. In: Ismail BP, Nielsen SS (eds) Nielsen's food analysis, sixth edn. Springer, New York

8.2 DETERMINATION OF CAFFEINE IN BEVERAGES BY HPLC

8.2.1 Introduction

Caffeine is a bitter alkaloid $C_8H_{10}N_4O_2$ found especially in coffee, tea, and kola nuts and is used medicinally as a stimulant and diuretic. The caffeine concentration in a beverage can be determined by simple filtration of the beverage prior to separation from other beverage components using HPLC.

In addition to caffeine, other water-soluble components can be found in caffeinated beverages extracted from natural products. Some of these components have health benefits. For example, some flavonoids can have antioxidant effects that are more potent than vitamins C and E. Examples of these types of compounds are epigallocatechin gallate (EGCG) found in tea and caffinic acid in coffee. These compounds can be separated from caffeine and quantified with little additional effort.

Analytical problems can be encountered when other food ingredients are combined with caffeinated beverages to create new products. In some cases, the food additive can contain a compound that will affect the extraction efficiency and/or interfere with the chromatographic process of the compound of interest. In these cases, sample preparation step(s) must be employed to allow quantification.

8.2.2 Objective

Determine the caffeine concentrations in beverages by reversed-phase HPLC with ultraviolet spectrometric detection. Demonstrate the use of sample preparation steps to improve analytical performance.

8.2.3 Chemicals

	CAS no.	Hazards
Acetic acid (CH₃COOH)	64-19-7	Corrosive
Caffeine	58-08-2	Harmful
Acetonitrile, HPLC grade (CH₃OH)	1334547-72-6	Highly flammable, acute toxicity, irritant

8.2.4 Hazards, Precautions, and Waste Disposal

Adhere to normal laboratory safety procedures. Wear safety glasses at all times. Acetonitrile waste must be handled as hazardous waste. Other waste likely may be put down the drain using a water rinse, but follow good laboratory practices outlined by environmental health and safety protocols at your institution.

8.2.5 Reagents

- 0.1% acetic acid in ddw (v/v)
- 0.1% acetic acid in acetonitrile (v/v)
- A solution of 0.1% acetic acid, 1% acetonitrile, and 98.9% dd water (v/v/v)
- Acetic acid concentrated, and another prepared at 2%
- Caffeine standard solutions (0.05 mg/mL, 0.1 mg/mL, 0.15 mg/mL and 0.2 mg/mL in dd water), prepared as described in Chap. 7
- Mobile phase – HPLC grade acetonitrile and deionized distilled (dd) water
- Pepsin (10 mg/mL in 0.1 N HCl)

8.2.6 Supplies

Beverages:

- Black tea (Pure Leaf)
- Dairy coffee drink (e.g., Starbucks Doubleshot Espresso)
- Decaffeinated coffee
- Soft drinks (Pepsi (or Coke) and Red Bull)

Other supplies:

- C_{18} (4.6 mm i.d. × 250 mm) reversed-phase column with a C_{18} stationary phase guard column
- Disposable plastic syringes, 3 mL
- Erlenmeyer flask, 10 mL
- Filters: 0.45 μm pore size
- Marker
- Mechanical pipettes and pipette tips (1000 μL and 200 μL)
- Parafilm
- Sample vials for autosampler
- Stir bars, small
- Test tubes (5–10 mL)
- Vortexer

8.2.7   Equipment

- Analytical balance
- HPLC system, with UV-Vis detector
  - HPLC 1: Shimadzu prominence HPLC system equipped with an automated injection unit and photodiode array detector (PDA).
  - HPLC 2: Shimadzu prominence HPLC system equipped with a manual injection unit and UV-Vis detector.
- Membrane filtering and degassing system
- Water bath maintained at 37 °C

8.2.8   HPLC Conditions

Column:		
	Chemistry	C_{18}
	Endcapped	Yes
	Inner diameter (id)	4.6 mm
	Length (mm)	250 mm
	Mode	Reversed-phase
	Particle size	5 μm
	Particle substrate	Silica
	Pore size	120 Å
Mobile phase:		
	Solvent A	0.1% acetic acid in water
	Solvent B	0.1% acetic acid in acetonitrile
HPLC conditions		
	Injection volume	20 uL
	Temperature	37 °C
	Flow rate	1 mL/min
	Wavelength	272 nm

Initial conditions are 99:1 A/B with a linear gradient to 70:30 A/B over 20 minutes followed by a 5 min linear gradient back to 99:1 A/B and 5 min equilibration at initial conditions for a total chromatographic run time of 30 min. Flow rate = 1.0 mL/min.

8.2.9   Procedure for Analysis of Soft Drinks (Red Bull and Pepsi/Coke)

(Prepare samples in triplicate)

1. Sample dilution:
   (a) To a test tube provided add 3 mL of Pepsi. Then add 2 mL dd water, to dilute the Pepsi sample 1.6 times. Vortex thoroughly.
   (b) To a test tube provided add 1 mL of Red Bull. Then add 3 mL dd water, to dilute the Red Bull sample 4 times. Vortex thoroughly.
2. Filter soft drink sample:
   (a) Attach a 0.45 μm pore size filter onto the end of a 3 mL syringe.
   (b) Pull out the syringe plunger and transfer the diluted sample into the syringe.
   (c) Replace the plunger and push sample through the filter into an auto sampler vial or a test tube (for manual injection).
   (d) Label the sample vial (include student/group name, date, sample name).
3. HPLC analysis:
   (a) HPLC 1:
      (i) Place the vial in the autosampler.
   (b) HPLC 2:
      (i) Rinse the injection port using an HPLC syringe with a solution containing 0.1% acetic acid, 1% acetonitrile, and 98.9% water (teaching assistant will provide you with the solution).
      (ii) Flush the HPLC syringe with filtered sample, then take up 15–20 μL of filtered sample (try to avoid taking up air bubbles).
      (iii) With HPLC injector valve in LOAD position, insert syringe needle into the needle port all the way.
      (iv) Gently depress syringe plunger to completely fill the injector loop with sample.
      (v) Leaving the syringe in position, simultaneously turn valve to INJECT position (mobile phase now pushes sample onto the column).
      (vi) Leave valve in the INJECT position so that the loop will be continuously flushed with mobile phase, thereby preventing cross contamination.

(vii) After caffeine peak has eluted, return valve to LOAD position in preparation for next injection. Repeat the same rinsing steps between samples.

8.2.10 Procedure for Analysis of Black Tea and Decaffeinated Coffee

(Prepare samples in triplicate)

1. Sample dilution:
   (a) To a test tube provided add 2 mL of decaffeinated coffee. Then add 2 mL dd water, to dilute the decaffeinated coffee sample 2 times. Vortex thoroughly.
   (b) To a test tube provided add 3 mL of black tea. Then add 3 mL dd water, to dilute the black tea sample 4 times. Vortex thoroughly.
2. Filter black tea and decaffeinated coffee samples:
   (a) Same as Sect. 8.2.9
3. HPLC analysis:
   (a) HPLC 1:
       (i) Place the vial in the autosampler.
   (b) HPLC 2:
       (i) Same as in Sect. 8.2.9, step 3 (b).

8.2.11 Procedure for Analysis of Dairy Coffee

(Prepare samples in triplicate)

1. Digest sample with enzyme (coffee drink, pepsin enzyme solution, and 2% acetic acid will be provided).
   (a) Add 0.6 mL of pepsin enzyme solution (10 mg/mL in 0.1 N HCl) to a 0.2 mL aliquot of coffee sample in the given glass tube.
   (b) Incubate for 20 min in a water bath set at 37 °C with gentle mixing using a small stir bar.
   (c) Combine with 2% aqueous acetic acid (1:1, volume to volume) and vortex thoroughly (i.e., 0.8 mL of 2% acetic acid).

(d) Transfer quantitatively the mixture into a microcentrifuge tube using the 1 mL mechanical pipet.
(e) Centrifuge @ 13,200 rpm for 5 min (**the precipitate will be floating at the top of the micro centrifuge tube. Please be advised to carefully avoid taking the precipitate for the filtration step that follows).
   **The dilution factor for the dairy coffee sample is 8 (0.2 mL of coffee sample in 1.6 mL solution).
2. Filter dairy coffee samples:
   (a) Same as Sect. 8.2.9
3. HPLC analysis:
   (a) HPLC 1:
       (i) Place the vial in the autosampler.
   (b) HPLC 2:
       (i) Same as in Sect. 8.2.9 , step 3 (b).

8.2.12 Data and Calculations

1. Chromatograms and peak area reports will be provided to you by the laboratory assistant. Use peak areas of the standards to construct your standard curve. Use the equation of the line for your standard curve to calculate the concentration of caffeine in your sample. Use the appropriate dilution factor when calculating the caffeine concentration in your sample. Complete the tables below with all the data.

Standard curve		
Caffeine conc. (mg/ml)	Rep	Peak area (cm^2)
0.05	1	
	2	
	3	
0.10	1	
	2	
	3	
0.15	1	
	2	
	3	
0.20	1	
	2	
	3	

Sample: (Complete this table for each type of sample)

Rep		Retention time (min)	Peak area (cm^2)	Conc. based on area from std. curve (mg/mL)	Conc. after considering dilution factor (mg/mL)	Conc. (mg/100 mL)
1						
2						
3						
	$\bar{X}=$					
	SD =					
	CV =					
	%E$_{Rel}$ =					

(*Note:* With the help of the teaching assistant, find the caffeine content in the selected beverages. FDA definition of **"decaffinated coffee":** Decaffeinated coffee must have at least 97% less caffeine than its regular caffeinated counterpart. For decaffinated coffee, caffeine content may be 5 mg or less per 6 fl oz.)

2. Using the mean values, calculate the concentration of caffeine in your sample expressed in terms of milligrams caffeine in a 12-oz. can (1 mL = 0.338 oz).

8.2.13 Questions

1. Why is it important to degas the mobile phase?
2. How is the "reversed-phase" HPLC different from "normal phase" with regard to stationary and mobile phases, and order of elution?
3. How would the use of methanol instead of acetonitrile affect the retention time of caffeine? Explain your answer.
4. How is an isocratic separation different from a gradient separation?

5. Using an isocratic separation, how would you change the mobile phase to decrease the retention time for caffeine?
6. Using the data collected, which beverage contains the most caffeine and what was its concentration?
7. What component/s of milk may interfere with the quantitation of caffeine in the sample? Why is the enzyme added to the extraction method?

Acknowledgments The authors of this chapter thank Steven T. Talcott (Department of Nutrition and Food Science, Texas A&M University, College Station, TX, USA) for contributing material for a section of the chapter when it was originally developed.

RESOURCE MATERIALS

AOAC International (2023) Official methods of analysis, 22th edn., (On-line) Method 979.08. Benzoate, caffeine, and saccharin in soda beverages. AOAC International, Rockville, MD

Ismail BP (2024) Basic principles of chromatography. Ch. 12. In: Ismail BP, Nielsen SS (eds) Nielsen's food analysis, 6th edn. Springer, New York

Ismail BP, Reuhs BL (2024) High-performance liquid chromatography. Ch. 13. In: Ismail BP, Nielsen SS (eds) Nielsen's food analysis, 6th edn. Springer, New York

Gas Chromatography

B. Pam Ismail and S. Suzanne Nielsen

Contents

9.1 INTRODUCTION

9.1.1 Background

Gas chromatography (GC) has many applications in the analysis of food products. GC has been used for the determination of fatty acids, triglycerides, cholesterol, gases, water, alcohols, non-polar pesticides, flavor compounds, and many more. While GC has been used for other food components such as sugars, oligosaccharides, amino acids, peptides, and vitamins, these substances are more suited to analysis by high-performance liquid chromatography. GC is ideally suited to the analysis of volatile substances that are thermally stable. Substances such as non-polar pesticides and flavor compounds that meet these criteria can be isolated from a food and directly injected into the GC. For compounds that are thermally unstable, too low in volatility, or yield poor chromatographic separation, a derivatization step must be done before GC analysis. The lab experiment describes the analysis of fatty acids, which requires derivatization.

9.1.2 Reading Assignment

Chen B, Ellefson WC (2024) Fat analysis. Ch. 17. In: Ismail BP, Nielsen SS (eds) *Nielsen's food analysis*, 6th edn. Springer, New York.

Pike OA (2024) Fat characterization. Ch. 23. In: Ismail BP, Nielsen SS (eds) *Nielsen's food analysis*, 6th edn. Springer, New York.

Reineccius GA, Qian MC (2024) Gas chromatography. Ch. 14. In: Ismail BP, Nielsen SS (eds) *Nielsen's food analysis*, 6th edn. Springer, New York.

B. P. Ismail (✉)
Department of Food Science and Nutrition, University of Minnesota, St. Paul, MN, USA
e-mail: bismailm@umn.edu

S. S. Nielsen
Department of Food Science, Purdue University, West Lafayette, IN, USA
e-mail: nielsens@purdue.edu

B. P. Ismail, S. S. Nielsen (eds.), *Nielsen's Food Analysis Laboratory Manual*, Food Science Text Series,
https://doi.org/10.1007/978-3-031-44970-3_9

9.2   PREPARATION OF FATTY ACID METHYL ESTERS (FAMES), AND DETERMINATION OF FATTY ACID PROFILE OF OILS BY GAS CHROMATOGRAPHY

9.2.1   Introduction

Information about fatty acid profile of food is important for nutrition labeling, which involves the measurement of not only total fat but also saturated, unsaturated, and monounsaturated fat. Gas chromatography is an ideal instrument to determine (qualitatively and quantitatively) fatty acid profile or fatty acid composition of a food product. This usually involves extracting the lipids and analyzing them using capillary gas chromatography. Before such analysis, triacylglycerols and phospholipids are saponified, and the fatty acids liberated are esterified to form fatty acid methyl esters (FAMEs) so that the volatility is increased.

The boron trifluoride (BF_3) method will be used in this experiment to prepare the sample for FAME determination. In the BF_3 method, lipids are saponified, and fatty acids are liberated and esterified in the presence of a BF_3 catalyst for further analysis. This method is applicable to common animal and vegetable oils and fats and fatty acids. Lipids that cannot be saponified are not derivatized and, if present in large amount, may interfere with subsequent analysis. This method is not suitable for preparation of methyl esters of fatty acids containing large amounts of epoxy, hydroperoxy, aldehyde, ketone, cyclopropyl, and cyclopentyl groups and conjugated polyunsaturated and acetylenic compounds because of partial or complete destruction of these groups.

It should be noted that AOAC Method 996.01 is used in this laboratory exercise, rather than AOAC Methods 996.06, which is the method for nutrition labeling, with a focus on *trans* fats. Compared to AOAC Method 996.01, Method 996.06 uses a longer and more expensive capillary column, requires a longer analysis time per sample, and involves more complicated calculations. Students will be given 5 oil samples; one of them is an olive oil sample adulterated up to 50% with another oil. Students will determine the identity of all pure oils and the adulterated oil based on the fatty acid profile.

9.2.2   Objective

Prepare methyl esters from fatty acids in food oils and then determine the fatty acid profile and their concentration in the oils by gas chromatography.

9.2.3   Chemicals

	CAS no.	Hazards
Boron trifluoride (BF_3)	7637-07-2	Toxic, highly flammable
Hexane	110–54-3	Harmful, highly flammable, dangerous for the environment
Methanol	67–56-1	Extremely flammable
Sodium chloride (NaCl)	7647-14-5	Irritant
Sodium hydroxide (NaOH)	1310-73-2	Corrosive
Sodium sulfate (Na_2SO_4)	7757-82-6	Harmful
Sodium methoxide	124–41-4	Toxic, highly flammable

9.2.4   Reagents and Samples

- Boron trifluoride (BF_3)-14% in methanol
- Hexane (GC grade. If fatty acids contain 20C atoms or more, heptane is recommended.)
- Methanolic sodium hydroxide 0.5 N (Dissolve 2 g of NaOH in 100 mL of methanol.)
- Oils: pure olive oil, adulterated olive oil, safflower oil, corn oil, and canola oil
  - (*Note: The samples will be labeled, in no particular order, A, B, C, D, E.*)
- Reference standard [15A gas-liquid chromatography (GLC) Reference standard, FAME 25 mg is dissolved in 10 mL hexane (Table below) (Nu-CHEK Prep, Inc. MN)]
- Sodium chloride solution, saturated
- Sodium sulfate, anhydrous granular.

FAME 15A GLC Reference Standard

No.	Chain	Item	Weight %
1	C16:0	Methyl palmitate	6
2	C18:0	Methyl stearate	3
3	C18:1	Methyl oleate	35
4	C18:2	Methyl linoleate	50
5	C18:3	Methyl linolenate	3
6	C20:0	Methyl arachidicate	3

9.2.5   Hazards, Precautions, and Waste Disposal

Do all work with the boron trifluoride in the hood; avoid contact with skin, eyes, and respiratory tract. Wash all glassware in contact with boron trifluoride immediately after use. Otherwise, adhere to normal laboratory safety procedures. Wear safety glasses at all times. Boron trifluoride, hexane, and sodium methoxide must be disposed of as hazardous wastes. Other wastes likely may be put down the drain using a water rinse, but follow good laboratory practices outlined by environmental health and safety protocols at your institution.

9.2.6 Supplies

- 1.5-mL GC glass vials with screw caps
- 2-mL screw cap glass vials
- 10-mL disposable glass pipettes (glass)
- 50-mL screw cap test tube (glass) with Teflon-lined caps
- Pasteur pipette (glass)
- Pipette aid

9.2.7 Equipment

- Analytical balance
- Centrifuge
- Vortex mixer
- Water bath (90–100 °C)
- Gas chromatography unit (with running conditions):

Instrument	Gas chromatograph, HP 5890A
Detector	Flame ionization detector (FID)
Capillary column	Stabilwax (cross-linked Carbowax-PEG) or equivalent
Length	30 m
Internal diameter	0.25 mm
Film thickness	0.25 μm
Sample injection	1 μL
Split ratio	30:1
Column head pressure	15 psi
Injector temperature	230 °C
Detector temperature	250 °C
Temperature program	
Initial oven	170 °C
Initial time	2 min
Rate	5 °C/min
Final temperature	230 °C
Final time	6 min

9.2.8 Procedure

(Instructions are given for single sample preparation and injection, but injections of samples and standards can be replicated.)

9.2.8.1 Preparation of Methyl Esters by Boron Trifluoride

(Adapted from AOAC Method 996.01)

(Notes: Methyl esters should be analyzed as soon as possible or sealed in an ampule and stored in a freezer. You might also add equivalent 0.005% 2, 6-di-*tert*-butyl-4--methylphenol (BHT).)

1. Add 2 drops sample oil into the bottom of a 50-mL test tube. Add 4 mL 0.5 N methanolic sodium hydroxide. Tightly close tube with Teflon lined cap. Prepare duplicate samples.
2. Place tube in 90–100 °C water bath for 10 min.
3. Remove tube from water bath carefully and cool under running warm and then cool tap water. *Make sure the tube is cool before opening.* Slowly unscrew the cap.
4. Add 4 mL boron trifluoride, recap, and place into 90–100 °C water bath for an additional 10 min.
5. Again, carefully remove tube from water bath, cool under running tap water. *Make sure it is cool before opening.* Slowly unscrew the cap.
6. Add 7 mL hexane, 7 mL saturated sodium chloride solution, re-cap tube, and shake by hand for 5 min. Place tube into test tube rack and allow solution to separate into two layers.
7. Transfer approx. 1.5-mL upper hexane solution into 2-mL glass vial; add anhydrous sodium sulfate (a spatula tip-full) to remove water. Swirl solution gently; if anhydrous sodium sulfate forms clumps in the solution, this means it has absorbed water and you need to add more anhydrous sodium sulfate. If the powder remains discrete, then your solution is dry and you can move to step 8.
8. Transfer approximately 1 mL of the "dried" solution to a GC vial using a Pasteur pipette. Be careful not to pick up any of the solid material with your pipette: you want to have a clear sample solution to inject into the GC.

9.2.8.2 Injection of Standards and Samples into GC

1. Rinse syringe three times with hexane, and three times with the reference standard mixture (25 mg of 15A GLC Reference Standard FAME dissolved in 10 mL hexane). Inject 1 μL of standard solution, remove syringe from injection port, then press start button. Rinse the syringe again three times with solvent. Use the chromatogram obtained as described below. **(This step will be conducted by the teaching assistant in the interest of time).**
2. Rinse syringe three times with hexane, and three times with the sample solution prepared by Method A. Inject 1 μl of sample solution, remove syringe from injection port,

then press start button. Rinse the syringe again three times with solvent. Use the chromatogram obtained as described below.

9.2.8.3 Demonstrations: GC Equipment

1. The instructor will show and explain the function of the different parts of a GC system.
2. The instructor will explain the function of an autosampler and discuss the benefits of using an autosampler, including time savings and injection reproducibility.
3. The instructor will demonstrate how to cut a GC column.

9.2.9 Data and Calculations

1. Report retention times, peak areas, peak area %, and weighted % for the peaks in the chromatogram from the FAME reference standard mixture in the table below. Use this information and your knowledge of how compounds of similar structure, but different carbon chain lengths, would elute from a column to identify the six peaks in the chromatogram. Remember to use the chromatogram of the standard from your instrument setup.

Empirical FAME 15A reference standard, GC results

Peak	Retention time	Peak area	Peak area %	Weighted %	Tentative ID of peak
1					
2					
3					
4					
5					
6					

2. Record peak retention times, peak areas, and peak area % for each oil sample (example table below). Calculate the weighted % for each peak by adjusting for the peak area of the solvent front. Using the retention times of the peaks in standard chromatogram, tentatively identify the peaks in the chromatograms for each type of oil analyzed.
3. Summary data tables will also need to be generated (see example below).
4. Compare and contrast the fatty acid profiles of the different oils tested and compare your results to literature values (i.e., compare the fatty acid profiles you found to published research on fatty acid analysis of the same oils). One potential source is the USDA Nutrient Database for Standard Reference.

If you use this source, please note:

- For the fatty acids 18:1, 18:2, and 18:3, the values listed in the database as "18:1 undifferentiated", "18:2 undifferentiated", and "18:3 undifferentiated" should be used.
- For safflower oil, the entry "Oil, safflower, salad or cooking, high oleic (primary safflower oil of commerce)" should be used. Do NOT use the "Oil, safflower, salad or cooking, linoleic, (over 70%)" entry.
- For canola oil, the entry "Oil, canola" should be used.
- Select "Standard Reference" under the "Select Source" drop-down menu to find the oil options with a complete nutrient report.

(*Notes*: You will notice in some sample chromatograms there are more peaks than the six peaks that correspond to the standard. Include the peak areas of these extra peaks in the total area, but you will not be able to "ID" them because they were not found in the standard. You do not have to calculate the individual % peak areas for the extra peaks.)

OIL Letter		Rep 1				Rep 2			Reps 1 & 2		
Peak	Retention time	Tentative ID	Peak area	Peak area %	Weighted %	Peak area	Peak area %	Weighted %	Ave	Std. dev.	CV
1											
2											
3											
4											
5											
6											
7											
8											
		Total area									

Example Summary Tables

FAME 15A reference standard, GC results

Peak	Retention time	Weighted %	Tentative ID of peak
1			
2			
3			
4			
5			
6			

Overall oil summary table

Sample oil identification	Olive oil Pure	Olive oil Adulterated	Safflower	Corn	Canola
Oil code identification					
Compound	Avg % (SD, CV)	Avg % (SD, CV)	Avg % (SD, CV)	Avg % (SD, CV)	Avg % (SD, CV)
Methyl Palmitate (C16:0)					
Methyl Stearate (C18:0)					
Methyl Oleate (C18:1)					
Methyl Linoleate (C18:2)					
Methyl Linolenic (C18:3)					
Methyl Arachidic (C20:0)					

9.2.10 Questions

1. In this lab, the fatty acids liberated from the triacylglycerols in the oils are derivatized; why is this is necessary and what types of compounds are best analyzed by GC? List the functions of each of the below reagents used to perform this experiment.
   (a) Methanolic sodium hydroxide
   (b) Methanolic boron trifluoride
   (c) Sodium chloride
   (d) Hexane
   (e) Anhydrous sodium sulfate
2. In this experiment, a flame ionization detector was used. Describe how a FID functions, the advantages and disadvantages of it, and then list/describe another detector that could be used to detect FAMEs.
3. The approach taken in this lab provides a fatty acid profile of each oil and an approximation of the amounts of each based on peak area. Why would peak area percent be inadequate for quantification (i.e., mg fatty acid/g fat)?
4. If you want to use the results of your GC analysis to calculate the amount of saturated, mono-, and polyunsaturated fatty acids as grams per specified serving size of the oil (such as for nutritional labeling). To make this procedure sufficiently quantitative for a purpose like that just described, an internal standard must be used.
   (a) What are the characteristics required of a suitable internal standard for FAME quantification by GC?
   (b) Would the internal standard be added to the reference standard mixture and the sample, or only to one of these?

Acknowledgments The authors of this chapter thank Michael C. Qian (Department of Food Science and Technology, Oregon State University, Corvallis, OR, USA) for contributing material for this chapter when it was originally developed.

RESOURCE MATERIALS

Chen B, Ellefson WC (2024) Fat analysis. Ch. 17. In: Ismail BP, Nielsen SS (eds) Nielsen's food analysis, 6th edn. Springer, New York

Pike OA (2024) Fat characterization. Ch. 23. In: Ismail BP, Nielsen SS (eds) Nielsen's food analysis, 6th edn. Springer, New York

Reineccius GA, Qian MC (2024) Gas chromatography. Ch. 14. In: Ismail BP, Nielsen SS (eds) Nielsen's food analysis, 6th edn. Springer, New York

Mass Spectrometry with High-Performance Liquid Chromatography

10

B. Pam Ismail

Contents

10.1 INTRODUCTION

10.1.1 Background

Mass spectrometry (MS) is an analytical technique that provides information about the molecular weight and chemical characteristics of a compound. This technique is commonly linked to high-performance liquid chromatography (LC-MS), gas chromatography (GC-MS), inductively coupled plasma (ICP-MS), and other methods.

B. P. Ismail (✉)
Department of Food Science and Nutrition, University of Minnesota, St. Paul, MN, USA
e-mail: bismailm@umn.edu

A mass spectrometer has three essential functions: sample ionization, ion separation, and ion detection. Components in a sample can be ionized by various ionization techniques; regardless of the source, the charged particles that are generated are separated based on their mass-to-charge ratio. Separation occurs in the mass analyzer, which sorts the different ion masses by applying electric and magnetic fields. These ions are then detected which provide data on the abundance of each ion fragment present.

For LC-MS analysis, the compounds are first separated on a high-performance liquid chromatography (HPLC) system. After detection in the LC system by ultraviolet-visible spectroscopy (UV-VIS), the eluent enters the ion source of a mass spectrometer where the compounds get ionized and subsequently separated based on "mass-to-charge" (m/z) ratio in the analyzer. The separated ions are then detected using an electron multiplier. Ionization using electrospray ionization (ESI) often produces precursor ions with some fragmentation. The ion, which undergoes fragmentation, is the precursor ion, and the ions that are produced upon fragmentation of the precursor ion are called product ions. To obtain further structural information about the compound, tandem mass spectrometry (MS/MS) can be employed. After ionization, the ions of interest are separated and fragmented by collision-induced dissociation (CID) by using an inert gas (helium, argon, etc.) as the collision gas. The energy given to the collision gas is varied depending on the desired extent of fragmentation, which will provide valuable structural information. The mass spectrum obtained by tandem MS contains only the product ions. A case study is presented to better understand the advantages of MS in tandem.

10.1.2 Reading Assignment

Ismail BP and Reuhs BL (2024) High-performance liquid chromatography. Ch. 13. In: Ismail BP, Nielsen SS (eds) *Nielsen's food analysis*, 6th edn. Springer, New York

Yucel U, Smith JS (2024) Mass spectrometry. Ch. 11. In: Ismail BP and Nielsen SS (eds) *Nielsen's food analysis*, 6th edn. Springer, New York

10.1.3 Objectives

Determine and identify various phytochemicals (mainly isoflavones, Table 10.1 and Fig. 10.1) in soy flour, and understand how to analyze a mass spectrum to obtain mass and structural information of a compound.

10.1.4 Chemicals

	CAS no.	Hazards
HPLC grade acetonitrile irritant	1334547-72-6	Highly flammable, acute toxicity, irritant
HPLC grade methanol	67-56-1	Highly flammable, acute toxicity, serious health hazard
Formic acid	64-18-6	Flammable, toxic

10.1.5 Hazards, Precautions, and Waste Disposal

Adhere to normal laboratory safety procedures. Wear safety glasses at all times. Methanol and acetonitrile waste must be handled as hazardous waste. Other wastes likely may be put down the drain using a water rinse, but follow good laboratory practices outlined by environmental health and safety protocols at your institution.

10.1.6 Reagents

- Deionized distilled (dd) water
- HPLC grade water
- HPLC grade water with 0.1% formic acid (v/v)
- HPLC grade acetonitrile with 0.1% formic acid (v/v)
- Methanol/water mixture, 80/20

10.1.7 Supplies

- Aluminum foil
- Automatic pipettors, 1000 μL, 200 μL, 100 μL, 10 μL
- Autosampler vials, 1.5 mL, with caps
- Beakers, 125 mL, each containing either 80/20 methanol, dd water, or acetonitrile (3)
- Beaker, 125 mL, for balancing the centrifuge tubes
- Buchner funnel
- Centrifuge tubes, 50 mL centrifuge tubes (2)
- Defatted soy flour
- Erlenmeyer flask, 25 mL (rinsed prior with 80/20 methanol)
- Erlenmeyer flask, 10 mL (rinsed prior with 80/20 methanol)
- Filter paper, Whatman 42, 90 mm diameter
- Graduated cylinders, 10 mL (3)
- Glass rods
- Markers
- Parafilm
- Rotary evaporator collection flasks (125 mL or smaller) (rinsed prior with 80/20 methanol)
- Sidearm flask, 125 mL, with a nozzle to plug vacuum pump (rinsed prior with 80/20 methanol)
- Stir bars and stirrer
- Syringes, 3 mL
- Syringe filters, non-sterile, nylon; pore size 0.45 μm; diameter 25 mm
- Tape
- Vacuum pump
- Vortex

10.1.8 Equipment

- Analytical balance
- Centrifuge, for 15 mL tubes
- High-performance liquid chromatography system

Table 10.1 Twelve known isoflavones found in soybean and their molecular weight (MW)

Isoflavone	MW	Isoflavone	MW	Isoflavone	MW
Daidzein	254	Genistein	270	Glycitein	284
Daidzin	416	Genistin	432	Glycitin	446
Acetyldaidzin	458	Acetylgenistin	474	Acetylglycitin	488
Malonyldaidzin	502	Malonylgenistin	518	Malonylglycitin	532

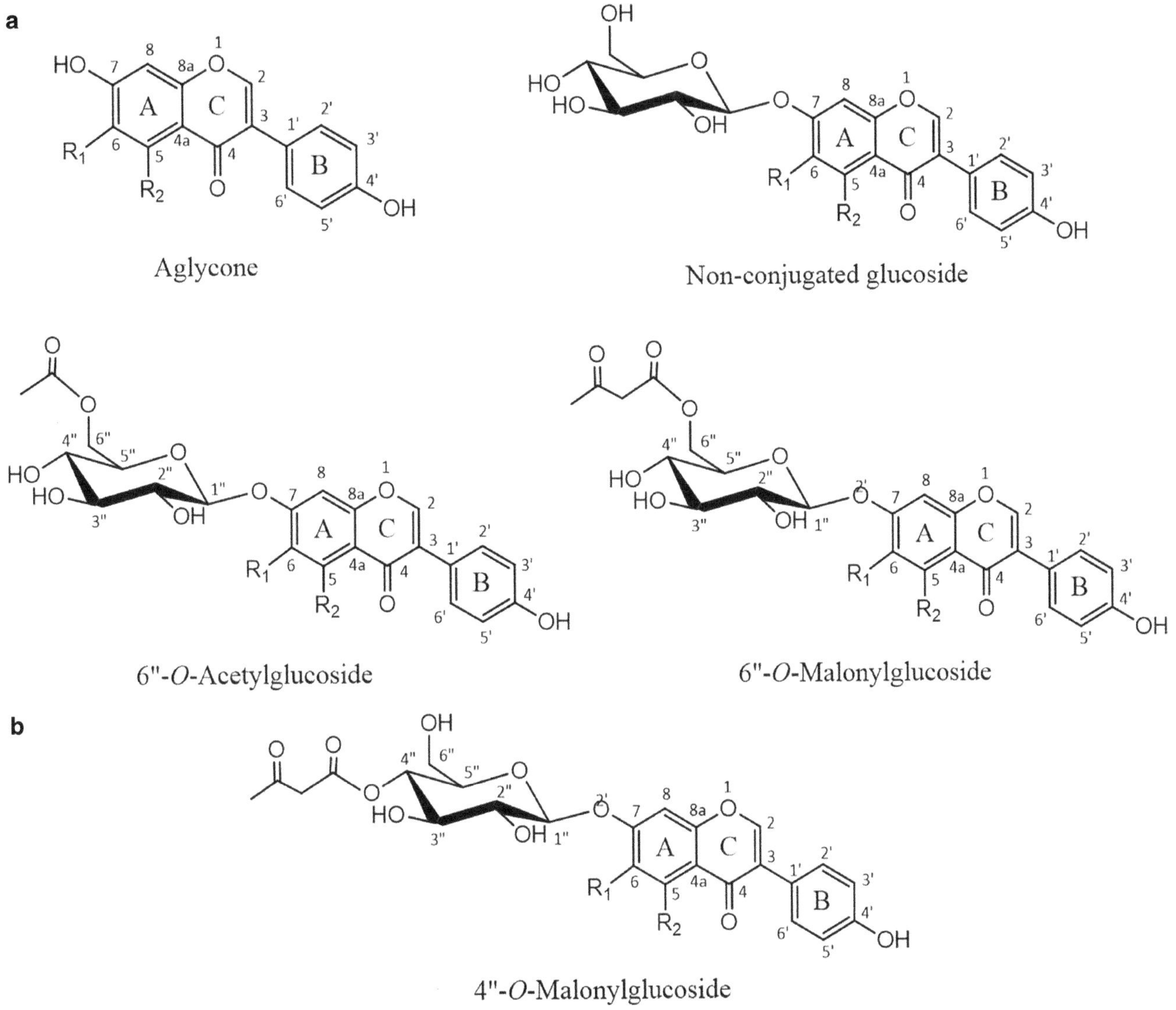

a

Aglycone

Non-conjugated glucoside

6"-O-Acetylglucoside

6"-O-Malonylglucoside

b

4"-O-Malonylglucoside

Fig. 10.1 (**a**) Structures and numbering of the 12 known isoflavones categorized as aglycone, nonconjugated glucoside, acetylglucoside, and malonylglucoside. R1 can be -H in the case of daidzin and genistin or -OCH₃ in the case of glycitin, while R2 can be -H in the case of daidzin and glycitin or -OH in the case of genistin. (**b**) Structures and numbering system of 4"-O-malonylglucosides (malonylglucoside isomers)

Table 10.2 Example combinations of ionization sources and mass analyzers coupled with chromatographic separation

Name	Ionization source	Mass analyzer	Tandem MS	Mode of separation[a]
Agilent 6150	Electrospray (ESI) and atmospheric pressure chemical ionization (APCI)	Single quadrupole (Q)	No	LC
LTQ XL, Thermo Scientific	ESI/APCI	Linear ion trap (LIT)	No	LC
Agilent 6490	ESI/APCI	Triple quadrupole (QqQ)	Yes	LC
API 6500 QTRAP, ABSciex	ESI	QqLIT	Yes	LC
MALDI-8030, Shimadzu.	Matrix-assisted laser desorption (MALDI)	Time of flight (TOF)	No	Direct injection
Xevo G2-S, waters	ESI/APCI	Q-TOF	Yes	LC
SCIEX 4800	MALDI	TOF-TOF	Yes	LC
Q-Exactive, Thermo Scientific	ESI	Q-Orbitrap	Yes	LC
Agilent 5977C	Electron impact (EI)/chemical ionization (CI)	Q	No	GC
Agilent 7000E	EI/CI	QqQ	Yes	GC
Pegasus BT GC-TOFMS, Leco	EI/CI	TOF	No	GC
Agilent 7250	EI/CI	Q-TOF	Yes	GC
Orbitrap Exploris GC 240	EI/CI	Q-Orbitrap	Yes	GC

[a]Liquid chromatography, LC; Gas chromatography, GC

- Mass-selective detector (quadrupole), equipped with electrospray ionization source
- Rotary evaporator

Various combinations of ionization sources and mass analyzers can be used for LC-MS and GC-MS analysis (see Table 10.2). This laboratory exercise as describes utilizes a LC-MS system coupled with electrospray ionization (ESI) and a quadrupole mass analyzer.

10.2 PROCEDURE

10.2.1 Sample Preparation

1. Weigh 0.05 gm of defatted soy into a labeled 25-mL Erlenmeyer flask.
2. Add 10 mL of HPLC grade acetonitrile to the sample and put in a stir bar.
3. Stir on a magnetic stir plate until the sample is thoroughly mixed (stirrer speed = 7, time = 5 min).
4. Add 9 mL dd water and stir for 2 h (stirrer speed = 7). Occasionally, scrape the surface of the plastic bottle with a glass rod to make the sample sticking to the sides go into the solution. (There will be a sample on the stirrer prepared by the TA and ready for you to take to the next step.)

5. Quantitatively transfer the sample solution from the Erlenmeyer flask to a labeled centrifuge tube. Rinse the flask three times with 1 mL acetonitrile and add the rinse to the centrifuge tube. Use the scale to balance the tube with a tube containing dd water.
6. Centrifuge the solution in a centrifuge. Example conditions with a Marathon 3200 centrifuge are as follows:
   Speed = 4000 rpm
   Time = 15 min
   Temperature = 15 °C
7. Take the centrifuge tube carefully out of the rotor so that the precipitate does not get disturbed.
8. Filter the permeate using a Buchner funnel, labeled 125-mL filter flask with a nozzle to plug vacuum pump, Whatman 42 filter paper, and vacuum pump. The setup should look like in Fig. 10.2.
9. Quantitatively transfer the filtered solution into a labeled rotary evaporator (rotovap) collection flask. Again, rinse three times with 1 mL acetonitrile and add the rinse to the rotovap collection flask.
10. Evaporate to dryness using a rotovap. Example conditions with a Buchi R-II Rotovap are as follows:
    Cooler temperature = 4 °C
    Water bath temperature = 38 °C
    Rotovap RPM = 120
    Time = 15 min (may take up to 30 min or so)

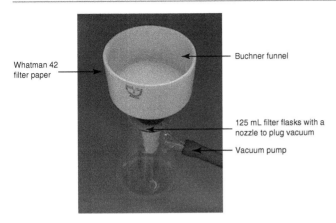

Whatman 42
filter paper

Buchner funnel

125 mL filter flasks with a
nozzle to plug vacuum

Vacuum pump

Fig. 10.2 Setup to filter sample permeate

Pressure: Since it is a two-step evaporation, evaporating of acetonitrile will be done at 115 mbar, and evaporation of water will be done at 30 mbar.

(*Note:* If the available pump is not adequate enough to use for evaporating water from the sample, shell freezing is used to freeze the sample so it can be dried using a lyophilizer (freeze dryer).)

11. There will be a sample already dried out for you and ready for the next step. Add 10 mL of 80/20 methanol/dd water and quantitatively transfer the solution into labeled 25- or 10-mL Erlenmeyer flask. This time do not rinse; just transfer using a pipette.
12. Vortex for 2 min (Speed = 5).
13. Attach a 0.45 μm filter onto the end of a 3 mL syringe.
14. Pour some of the sample into the filter assembly.
15. Push sample through the filter into an auto sampler vial.
16. Label the sample vial (include group name, date, sample name).
17. Seal the vial with parafilm, wrap with silver foil, and label adequately.

10.2.2 LC-MS Procedure

A Shimadzu LC-10 AD HPLC equipped with two solvent pumps and a CT-10A column oven will be used. The chromatographic column is a YMC AM-303 (ODS, 250 mm × 4.6 mm i.d.) comprising C18 reversed-phase packing (5 μm average pore size), equipped with a C18 guard column (4 mm × 20 mm i.d.).

The chosen flow rate is 1 mL/min and oven temp of 45 °C. A linear HPLC gradient will be used: Solvent A will be 0.1% (v/v) formic acid in HPLC grade water, and solvent B will be acetonitrile with 0.1% (v/v) formic acid. The gradient time program is described in Table 10.3.

Table 10.3 HPLC gradient time program

Solvent B concentration (%)	Time (min)
11	0 (initial)
14	30
14	35
30	40
30	50
11	52
11	60

10.2.3 MS Conditions

Mass spectrometry is performed on a Waters ZQ quadrupole instrument. Positive ionization will be employed. Masses are scanned from 200 to 600 m/z over 60 min, at a scan rate of one scans/sec.

Eight selected ions of the following m/z are monitored, $[M + H^+]$: 255, 271, 417, 433, 459, 475, 503, and 519.

The injection volume is 20 μL; the split ratio is 1:3 (so ¼ of the flow goes into the mass spectrometer, ¾ flow through the UV-detector). The source temperature is 150 °C; the desolvation temperature is 450 °C. The desolvation gas flow is 600 L/h; the cone gas flow is 75 L/h. The cone voltage is 30 eV; the capillary voltage is 3 kV.

1. Turn on the gas flow by pressing the button "API gas" on the tune page.
2. Click the button "Press to Operate" on the tune page.
3. Turn on the oven and the degasser.
4. Load the HPLC conditions on the inlet page and turn on the flow.
5. Wait until the HPLC column is conditioned and the oven has reached the desired temperature.
6. Enter the file name, MS file, inlet file, and MS tune file name, and save the data set.
7. Before you inject sample, make sure that the mass spectrometer is in the "Load" mode.
8. Press the "Start" button, inject the sample, click on "Start" again, and press the "Inject" button on the mass spectrometer.

10.3 DATA AND CALCULATIONS

Analyze the total ion chromatogram and mass spectra of the selected isoflavones provided to you at the end of the lab exercise. To help you understand how to analyze a mass spectrum to obtain mass and structural information of a compound, see the case study, Sect. 10.5. Use the resource materials provided to answer the general questions below, and analyze the ion chromatogram and mass spectra of the isoflavones.

10.4 QUESTIONS

10.4.1 General Questions

1. What is a molecular ion? Does electron impact (EI) ionization produce a single precursor ion? How about ESI? Answer by explaining the difference between soft ionization to hard ionization.
2. List the ionization interfaces that are capable of delivering soft ionization.
3. Which ion travels farthest in unit time (low m/z or high m/z) in time-of-flight mass analyzer? How does the length of time-of-flight mass analyzer affect the resolution?
4. Explain the different between a quadrupole and an ion trap.
5. What more information can MS/MS (i.e., MS in tandem) provide versus just one MS?

10.4.2 Questions Specific to Isoflavone Analysis

1. Discuss the peaks and how you identify them.
2. Indicate if there were any co-elutions and how one can determine that.
3. Discuss the presence of isomers, if any, and how useful the MS was in identifying their presence, pointing to advantages and limitations.

4. Were you able to see precursor ions? Were you able to see product ions? Use a table format (Table 10.4) to list the precursor ions identified and their product ions as in the example table below.

10.5 CASE STUDY

Malonylgenistin is a type of isoflavone abundantly found in soybeans. Epidemiological studies have shown the association of isoflavones with many health benefits such as reducing the risk of cancer, alleviation of postmenopausal symptoms, etc. Processing of soybeans into various soy products results in the conversion of malonylgenistin to various compounds. Monitoring the conversions of malonylgenistin to different compounds is important to determine if the resultant compounds have any biological significance. Recent studies have shown the formation of a new derivative upon heat treatment of malonylgenistin. The wavescans obtained using a photodiode array detector (PDA) of the new derivative and malonylgenistin were similar. Hence, LC/UV failed to provide any information about the structural differences between the two compounds. MS analysis of the two compounds revealed that the compounds had same mass (m/z = 519 Dalton) indicating that the new derivative is an isomer of malonylgenistin (Fig. 10.3). Tandem mass spectrometry was employed to probe structural differences between the two isomers. There were two noticeable differences in the product ion spectra (Fig. 10.4).

1. Isomer fragmented more at a lower collision level as compared to malonylgenistin.

Table 10.4 Identified isoflavones and precursor and product ions m/z values

Retention time	Isoflavone	Precursor ion m/z	Product ion (name and m/z)

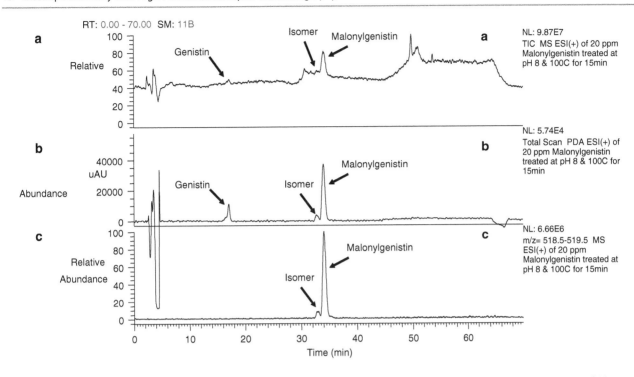

Fig. 10.3 ESI-MS analysis malonylgenistin and its isomer: (**a**) total ion chromatogram, (**b**) PDA view of malonylgensitn solution, and (**c**) reconstructed single ion chromatogram of m/z 519 ion (potonated molecule of a malonylgenistin)

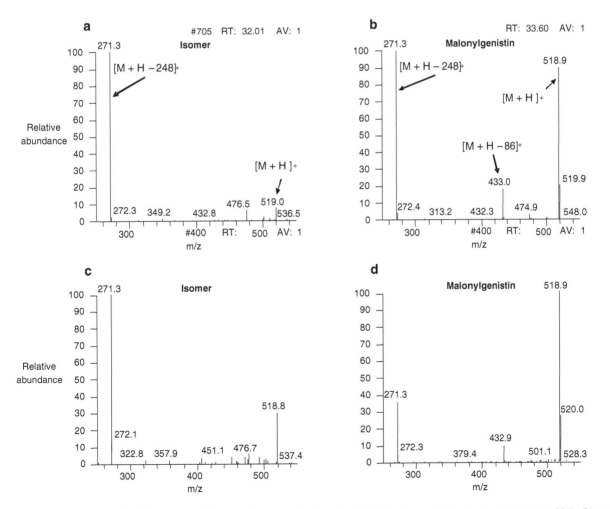

Fig. 10.4 ESI-MS/MS analysis of the protonated forms of isomer and malonylgenistin at various collision levels: (**a**) isomer at 20%, (**b**) malonylgenistin at 20%, and (**c**) isomer at 17%; and (**d**) malonylgenistin at 17% collision

2. Malonylgenistin product ion spectra had an ion with m/z = 433, which was absent in the isomer product ion spectra.

These differences highlight the structural differences between the two compounds. Nuclear magnetic resonance has to be employed to obtain further structural details.

Acknowledgments Fernanda Furlan Goncalves Dias, Department of Food Science and Nutrition, University of Minnesota, St Paul, MN, is acknowledged for her contribution to the content of Table 10.2.

RESOURCE MATERIALS

Ismail BP and Reuhs BL (2024) High-performance liquid chromatography. Ch. 13. In: Ismail BP, Nielsen SS (eds)

Yucel U, Smith JS (2024) Mass spectrometry. Ch. 11. In: Ismail BP and Nielsen SS (eds) Nielsen's food analysis, 6th edn. Springer, New York

Determination of Minerals on Nutrition Label by Atomic Absorption Spectroscopy

11

B. Pam Ismail and S. Suzanne Nielsen

Contents

11.1 INTRODUCTION

11.1.1 Background

The concentration of specific minerals in foods can be determined by a variety of methods. The aim of this lab is to acquaint you with the use of atomic absorption spectroscopy (AAS) for mineral analysis. This experiment specifies the preparation of standards and samples for determining the content of the minerals required on the nutrition label, sodium (Na), calcium Ca), iron (Fe), and potassium (K), by AAS. The samples suggested for analysis include one liquid product (a clear sports drink that does not require ashing), and two of six choices of solid food products, which will require dry or wet ashing before analysis, and which are the same products suggested for the proximate analysis laboratory exercises (Chaps. 13, 14, 15 and 16).

Sodium results from this AAS experiment can be compared with sodium results from analysis of the same products in the experiment that uses the more rapid methods of analy-

B. P. Ismail (✉)
Department of Food Science and Nutrition, University of Minnesota, St. Paul, MN, USA
e-mail: bismailm@umn.edu

S. S. Nielsen
Department of Food Science, Purdue University, West Lafayette, IN, USA
e-mail: nielsns@purdue.edu

sis of ion selective electrodes, the Mohr titration, and Quantab® test strips (Chap. 21). If your department has access to an inductively coupled plasma-optical emission spectrophotometer (ICP-OES), the Na, Fe, Ca, and K contents of the food samples can also be tested by this method, and results compared to AAS results. Note that when large analytical laboratories analyze for multiple minerals in food samples, they typically use ICP-OES rather than AAS. As described in Yeung et al., 2024, AAS relies on absorption while atomic emission spectroscopy (AES) relies on emission. When AES is combined with ICP, the technique and instrument are referred to as ICP-OES. In AAS, one mineral is analyzed at a time, while with ICP-OES multiple minerals can be analyzed at a time. However, because an ICP-OES unit is much more expensive and requires greater technical expertise to operate as compared to AAS, this laboratory exercise describes mineral analysis by AAS.

Procedures for both wet ashing and dry ashing of the solid samples are described. Experience can be gained with both types of ashing and results can be compared between the two methods of ashing. AAS results also will be compared to values per serving reported on the label. Each student or a group of students will be assigned one clear liquid product and two solid products from two different categories (e.g., a cheese and a cereal) to compare the different mineral contents.

11.1.2 Reading Assignment

Harris GK (2024) Ash analysis. Ch. 16. In: Ismail BP and Nielsen SS (eds) *Nielsen's food analysis*, 6th edn. Springer, New York

Yeung CK, Miller DD, Rutzke MA (2024) Atomic absorption spectroscopy, atomic emission spectroscopy, and inductively coupled plasma mass spectrometry. Ch. 9. In: Ismail BP and Nielsen SS (eds) *Nielsen's food analysis*, 6th edn. Springer, New York

11.1.3 Objective

The objective of this experiment is to determine the sodium, calcium, iron, and potassium contents of food products using atomic absorption spectroscopy.

11.1.4 Principle of Method

Atomic absorption is based on atoms *absorbing* energy, once heat energy from a flame has converted molecules to atoms. By absorbing the energy, atoms go from ground state to an excited state. The energy absorbed is within the allowable spectra emitted by the cathode (made up of the metallic form of the mineral) within the hollow cathode lamp. A photomul-

tiplier tube detector measures absorption at the selected wavelength, usually λ_{max} (wavelength at maximum absorption). According to Beer's law, absorption is linearly related to concentration.

11.1.5 Chemicals

	CAS no.	Hazards
Calcium (crystalline)	7440-72-21	Irritant
Hydrochloric acid (HCl)	7647-01-1	Corrosive
Hydrogen peroxide, 30% (H_2O_2)	7722-84-1	Corrosive
Iron	7439-89-6	Irritant
Lanthanum chloride ($LaCl_3$)	10025-84-0	Irritant, corrosive
Nitric acid (HNO_3)	7697-37-2	Corrosive
Potassium chloride (KCl) (for K std. solution)	7447-40-7	Irritant
Sodium chloride (NaCl) (for Na std. solution)	7647-14-5	Irritant

11.1.6 Reagents

(**It is recommended that these solutions be prepared by the laboratory assistant before class.)

- 6 *N* HCl
- 0.1 *N* HCl
- Double distilled (dd) water
- 1% Lanthanum chloride in dd water
- Sodium, calcium, iron, and potassium standard solutions, 1000 ppm **
  - Used to prepare 100-mL solutions of each of the concentrations listed in Table 11.1. Each standard solution must be prepared in 0.1 N HCl. For calcium standards, $LaCl_3$ must be added to a final concentration of 0.1%.

11.1.7 Hazards, Precautions, and Waste Disposal

Adhere to normal laboratory safety procedures. Wear safety glasses and gloves during sample preparation. Use acids in a hood.

Table 11.1 Concentrations (ppm) of Na, Ca, Fe, and K standard solutions

Sodium	Calcium	Iron	Potassium
0	0	0	0
50	4	0.5	50
100	8	2	100
200	12	5	200
300	16	10	300

11.1.8 Supplies

Food samples:

- Almonds
- Cheddar cheese
- Cheerios cereal
- Clear sports drink
- Parmesan cheese
- Process cheese
- Rice baby cereal

Other supplies:

- Beaker, 25 mL or 50 mL (1 per student/group)
- Desiccator, with dry desiccant
- Digestion tubes (for wet ashing; size to fit digestion block)
- Filter paper, ashless
- Funnels, small (to filter samples)
- Glass rods
- Mechanical pipettes, 200 μL, 1 mL, and 5 mL (3 of each per student or group)
- Parafilm
- Plastic bottles, with lids, to hold 25 mL (3 per sample and per student/group)
- Porcelain ashing crucibles, previously cleaned and heated at 550°C in a muffle furnace for 18 h (for dry ashing) (3 per sample for each mineral analysis per student or group)
- Volumetric flasks, 10 mL (3 per sample per student or group)
- Volumetric flask, 100 mL (3 for liquid sample and a blank per student or group)
- Weighing boats/paper

(*Note:* All glassware is acid washed with 0.1 *N* HCl to remove any mineral contaminants.)

11.1.9 Equipment

- Analytical balance
- Atomic absorption spectroscopy unit
- Digestion block (for wet ashing; set to 175°C)
- Muffle furnace (for dry ashing; set to 550°C)
- Hotplate
- Water bath, heated to boil water (for dry ashing)

11.2 PROCEDURE

11.2.1 Sample Preparation: Liquid Samples

1. Put an appropriate volume of liquid sample in a 100-mL volumetric flask. For the clear sports drink, use 0.2 mL for analysis by AAS or an amount predetermined by your teaching instructor to reach an appropriate dilution factor to fit within the standard curve.
2. Make up to volume using 0.1 *N* HCl.
3. Shake well. (If there is any particulate matter present, the sample will need to be filtered through ashless filter paper.)
4. Make appropriate dilution and analyze (for calcium analysis by AAS, add LaCl₃ to final conc. of 0.1%).
5. Liquid Blank: Prepare a liquid blank sample to be assayed, following the sample preparation procedure but excluding the sample.

11.2.2 Sample Preparation: Solid Samples

All samples need to be wet ashed and separate samples need to be dry ashed prior to the AAS laboratory session. Samples could be either prepared and ashed by a teaching instructor, or they could be prepared by students in a prior laboratory session, then ashed by a teaching instructor prior to this AAS lab. The teaching instructor will also need to determine the appropriate dilution of each type of sample for analysis by AAS.

11.2.2.1 Wet Ashing

(*Note:* Digestion procedure described is a wet digestion with nitric acid and hydrogen peroxide. Other types of digestion can be used instead.)

1. Label one digestion tube per sample plus one tube for the reagent blank (control).
2. Accurately weigh out 300-400 mg of each sample and place in a digestion tube. Prepare samples in duplicate or triplicate.
3. Pipette 5 mL concentrated nitric acid into each tube, washing the sides of the tube as you add the acid.
4. Set tubes with samples and reagent blank in digestion block. Turn on the digestion block and set to 175°C to start the predigestion.
5. Swirl the samples gently once or twice during the nitric acid predigestion, using tongs and protective gloves.
6. Remove tubes from digestion block when brown gas starts to elute (or when solution begins to steam, if there is no brown gas) and set in the cooling rack. Turn off the digestion block.
7. Let the samples cool for at least 30 min. (Samples can be stored at this point for up to 24 h.)
8. Add 4 mL of 30% hydrogen peroxide to each tube, doing only a few tubes at one time. Gently swirl the tubes. Put the tubes back in the digestion block. Turn on the digestion block still set to 175°C.
9. Watch tubes closely for the start of the reaction, indicated by the appearance of rapidly rolling bubbles. As

soon as the reaction starts, remove the tubes from the block and let the reaction continue in the cooling rack. (Caution: Some sample types will have a vigorous reaction, and for some the sample is lifted to the top of the tube, with the risk of boiling over.)

10. Repeat Steps 8 and 9 for all the samples and the reagent blank.

11. Put all tubes in the digestion block, and leave until ca. 1–1.5 mL remains, and then remove each tube from the digestion block. Check the tubes every 10–15 min during this digestion. (If the tubes are left on the digestion block too long and they become dry, remove, cool, and *carefully* add ca. 2 mL concentrated nitric acid and continue heating.) Turn off the digestion block when all the tubes have been digested and removed.

12. Make appropriate dilution of samples with dd water in a 10-mL volumetric flask as predetermined by your teaching instructor. (To the sample for AAS, add $LaCl_3$ to final conc. of 0.1%.)

13. If necessary, filter samples using Whatman hardened ashless filter paper into container appropriate for analysis by AAS as described in Step 10 under the dry ashing section below.

11.2.2.1.1 Dry Ashing

1. Accurately weigh out shredded, blended, or ground ca. 2-3-g sample into crucible, in duplicate.

2. Pre-dry sample over boiling water bath.

3. Complete drying of samples in vacuum oven at 100°C, 26-in. Hg, for 16 h.

4. For high fat samples (e.g., cheese) place crucibles on a hotplate until the smoking ceases. This allows the fat to burn off before the sample is placed in the muffle furnace.

5. All crucibles (including a designated blank crucible for each sample/mineral type) will be ashed in a muffle furnace at 550°C for up to 8 hr. Samples that appear dark will receive an additional treatment that consists of wetting the ash with a small amount of dd water + 0.5–3 mL concentrated nitric acid followed by heating and drying on a hotplate in the hood. These crucibles will be ashed a second time for up to 2 hours.

6. Once ashed, the samples should be placed into a desiccator to cool to room temperature. Dissolve ash in appropriate amount (mL) of 6N HCl per Table 11.2.

7. Place the crucibles on a hot plate and stir the sample with a glass rod (at least 2 min, scrape the edges with the rod) to facilitate dissolution.

8. Obtain a clean 10-mL volumetric flask, label it appropriately, and pipet 1 mL of 1% $LaCl_3$ solution for calcium analysis by AAS or dd into the bottom.

9. Make appropriate dilution of samples with 0.1 N HCl in a volumetric flask as indicated in Table 11.2.

Table 11.2 Dilution of samples for Ca and Fe analysis by AAS using dry ashing[a, b]

Sample	Calcium	Iron
All cheese samples	Ashed sample dissolved in 10 mL of 6 N HCl and 0.05 mL is diluted to 10 mL	Ashed sample dissolved in 1 mL of 6 N HCl and 0.5 mL diluted to 10 mL
Baby cereal	Ashed sample dissolved in 2 mL of 6 N HCl and 0.03 mL is diluted to 10 mL	Ashed sample dissolved in 4 mL of 6 N HCl and 0.2 mL is diluted to 10 mL
Cheerios	Ashed sample dissolved in 10 mL of 6 N HCl and 0.2 mL is diluted to 10 mL	Ashed sample dissolved in 5 mL of 6 N HCl and 1 mL is diluted to 10 mL
Almonds	Ashed sample dissolved in 10 mL of 6 N HCl and 0.5 mL is diluted to 10 mL	Ashed sample dissolved in 1 mL of 6 N HCl and 0.5 mL diluted to 10 mL

[a]Use ca. 2-3-g sample
[b]Teaching instructor will determine appropriate dilutions for Na and K analysis

10. Repeat Steps 6–9 for the remaining samples and associated blanks. Cover the flasks with parafilm, shake well, and filter the contents into the provided sample tubes as described below:

    (a) Obtain a filter paper and fold it into fourths.

    (b) Place a funnel into a 25-mL sample tube and put filter paper into funnel. Label a 25-mL beaker as "waste." Make sure to label your tubes appropriately (i.e., cheese, calcium, rep 1).

    (c) Pour a small amount of sample from the volumetric flask into the filter. Allow the first few drops to dispense into the waste beaker. Transfer funnel to the labeled tube and finish filtering the sample.

    (d) All samples and blanks will be prepared as above.

11.2.3 Analysis

Follow manufacturer's instructions for start-up, use, and shutdown of the AAS. Take appropriate caution with the acetylene and flame in using AAS. Use an appropriate hollow cathode lamp for each mineral. Analyze in this order, reagent blanks, standards, and samples. Record absorbance of each blank, standard, and sample onto your data sheet.

11.3 DATA AND CALCULATIONS

Report standard curve data and sample data in tables below in Sects. 11.3.1 and 11.3.2, respectively. Do all calculations for each duplicate sample individually, before determining a mean value on the final answer.

11.3.1 Standard Curve Data

Sodium		Calcium		Iron		Potassium	
ppm	Absorption	ppm	Absorption	ppm	Absorption	ppm	Absorption
0		0		0		0	
50		4		0.5		50	
100		8		2		100	
200		12		5		200	
300		16		10		300	

11.3.2 Sample Data

(*Note*: You will need to fill out sample data for each mineral separately. To get µg in total sample weight, µg in 100 g sample, mg in 100 g sample, and mg per serving follow the instruction under Sect. 11.3.3.)

Sample	Rep	Sample size (g or mL)	Absorption	Corrected absorbance	µg in total sample weight	µg in 100 g sample	mg in 100g sample	mg per serving
Liquid blank	1							
	2							
Liquid product	1							
	2							
Solid blank	1							
	2							
Solid product #1	1							
	2							
Solid product #2	1							
	2							

11.3.3 Calculations

1. Using the standard concentrations of the four mineral standards and their respective absorbances, construct a standard curve for each mineral, and calculate the linear regression equation of the line.
2. Using the linear regression line equation for each mineral, and the absorbance readings of the samples, calculate the mineral concentration present in each replicate sample (ppm). (Note that you need to subtract the liquid blank absorbance from the liquid sample values, and the solid blank absorbance values from the sample of solid foods.)
3. Convert ppm (µg/mL) to total µg in 100 g of original sample.

    Conc. (µg) of mineral (Na, Ca, Fe, or K) in X grams sample =

$$\left[\text{Conc in } \frac{\mu g}{\text{mL}} \times 10\,\text{mL}\right] \times \left(1\,\text{mL}\right) \text{or} \left(2\,\text{mL}\right) \text{or}$$
$$\left(10\,\text{mL}\right)/\left[0.04\,\text{mL}\right] \text{or} \left[0.1\,\text{mL}\right] \text{or}$$
$$\left[0.2\,\text{mL}\right] \text{or} \left[0.5\,\text{mL}\right]$$

(*Note:* If you dissolved your ash in 1 mL then use 1 mL in the equation above; if you dissolved your ash in 10 mL then you would use 10 mL in the equation above, and so on. If you transferred 0.05 mL of sample dissolved in 6N HCl into the 10 ml volumetric flask, then you use 0.05 mL in the equation above; if you transferred 0.2 mL of sample dissolved in 6N HCl into the 10 mL volumetric flask, then you use 0.2 mL in the equation above, and so on.)

µg of mineral (Na, Ca, Fe, or K) in 100 g samples = [Conc (µg) of mineral in sample/ Total sample weight (g)] × 100

4. Calculate the average concentration of the 3 replications and the SD as well as CV.
5. Compare your mean values to the food label.
6. Report in the table below the values for mineral analysis of the foods analyzed and compare this with values reported on the nutrition label for each food. Show examples of all calculations.
   (a) Convert the values for samples in ppm to mg/mL for the liquid and to mg/100 g for the solid samples (simply divide by 1000). Include the average amount

(mg), coefficient of variation, and % E_{Rel} for each mineral per 100 g of the liquid product and the two solid products analyzed. Covert to mg per serving based on the nutrition label.

(b) Report the values (both amount and % Daily Value) from the nutrition label of the foods you analyzed.

(c) Compare the Na, Ca, Fe, and K values for the specific foods you analyzed to those on the nutrition label and to those reported in the US Department of Agriculture

Nutrient Database for Standard Reference (http://ndb.nal.usda.gov/).

(*Note:* Remember that the baby cereal daily values (DV) are for children. Also, iron DV is calculated for the label using the requirement for women, ages 19–50 years (18 mg/day as recommended dietary allowance). If you collect data from both wet ashing and dry ashing, make sure to compare the results of Fe content.)

Results for mineral analysis of foods:

Sample	Sodium					
	Average concentration (mg/100g) (CV, %Erel)	Average concentration per serving (mg/X g)	Reported amount (mg/100g)	Reported amount per serving	Calculated percent DV per serving (% Erel)	Reported percent DV per serving

Sample	Calcium					
	Average concentration (mg/100g) (CV, %Erel)	Average concentration per serving (mg/X g)	Reported amount (mg/100g)	Reported amount per serving	Calculated percent DV per serving (% Erel)	Reported percent DV per serving

Sample	Iron					
	Average concentration (mg/100g) (CV, % Erel)	Average concentration per serving (mg/X g)	Reported amount (mg/100g)	Reported amount per serving	Calculated percent DV per serving (% Erel)	Reported percent DV per serving

Sample	Potassium					
	Average concentration (mg/100g) (CV, % Erel)	Average concentration per serving (mg/X g)	Reported amount (mg/100g)	Reported amount per serving	Calculated percent DV per serving (% Erel)	Reported percent DV per serving

11.4   QUESTIONS

1. Based on the comparison of the Na, Ca, Fe, and K values for the specific foods you analyzed to those on the nutrition label and to those reported in the US Department of Agriculture Nutrient Database for Standard Reference (http://ndb.nal.usda.gov/), briefly discuss your observations.

2. If the Fe content of your samples prepared by wet ashing differed from the Fe content of samples prepared by dry ashing, describe that difference and explain why it likely occurred.

3. For one of the minerals, describe how you would prepare the standard solutions to create the standard curve, using the commercially available 1000-ppm solution of the mineral. If possible, all solutions for points in the standard curve should be made using different volumes of the same stock solution. Do not use volumes of less than 0.2 mL. Make all standards to the same volume of 100 mL. Note that each standard solution must contain 10 mL conc. HCl/100 mL final volume, as described under Reagents.

4. Describe how you would prepare a 1000-ppm Na solution, starting with commercially available solid NaCl.

5. Why was lanthanum chloride added for calcium analysis and not for iron analysis?

6. Why were the samples and standards prepared in a final concentration of 0.1 N HCl? In other words, what is the role of the acid?

7. Were your values consistent with the reported values? Explain potential reasons for deviating from reported values, while listing in general potential reasons that may result in under estimation or over estimation of your mineral content.

8. What is the role of the flame vs. the lamp in AAS?

9. Why was it necessary to change the lamps in the AAS instrument between the various minerals analyzed? Justify your answer.

RESOURCE MATERIALS

Harris GK (2024) Ash analysis. Ch. 16. In: Ismail BP and Nielsen SS (eds) Nielsen's food analysis, 6th edn. Springer, New York

Yeung CK, Miller DD, Rutzke MA (2024) Atomic absorption spectroscopy, atomic emission spectroscopy, and inductively coupled plasma mass spectrometry. Ch. 9. In: Ismail BP and Nielsen SS (eds) Nielsen's food analysis, 6th edn. Springer, New York

Mid-Infrared Spectroscopy Analysis

12

Luis Rodriguez-Saona, Siyu Yao,
and Shreya Madhav Nuguri

Contents

12.1 INTRODUCTION

12.1.1 Background

Driven by recent advances in semiconductor and photonic techniques, mid-infrared (MIR) techniques have evolved from bulky benchtop instrumentation to field-deployable and ruggedized devices with an equivalent spectral resolution, allowing on-site and real-time chemical identification in food quality and safety monitoring. MIR spectroscopy typically measures the light absorbance of a sample in the region of 4000-650 cm^{-1}. When the frequency of the IR radiation matches the natural frequency of the vibration (or rotation), an IR photon is absorbed, and the amplitude of the vibration increases. The absorption bands in MIR spectroscopy correspond to the vibrational modes of specific functional groups, which makes it very useful in the study of organic compounds. Moreover, the positioning and intensity of the bands are related to their environment, energy, and concentration in a sample matrix. Accordingly, MIR can be used for performing quantitative and qualitative applications for rapid food analysis and in-plant control.

12.1.2 Reading Assignment

Rodriguez-Saona L, Ayvaz H (2024) Infrared and Raman spectroscopy. Ch. 8. In: Ismail BP and Nielsen SS (eds) Nielsen's Food analysis, sixth edn. Springer, New York.

L. Rodriguez-Saona (✉) · S. M. Nuguri
Department of Food Science and Technology, The Ohio State
University, Columbus, OH, USA
e-mail: rodriguez-saona.1@osu.edu;
nuguri.2@buckeyemail.osu.edu

S. Yao
Department of Nutrition and Food Hygiene, School of Public
Health, Southeast University, Nanjing, China
e-mail: siyuyao@seu.edu.cn

© The Author(s), under exclusive license to Springer Nature Switzerland AG 2024
B. P. Ismail, S. S. Nielsen (eds.), *Nielsen's Food Analysis Laboratory Manual*, Food Science Text Series,
https://doi.org/10.1007/978-3-031-44970-3_12

12.1.3 Objectives

Determine the *trans*-fat content in commercial butter and margarine products using MIR spectroscopic technique, explore the quantification analysis in IR using a calibration curve and a developed chemometrics model (an R script is provided), and become familiar with MIR techniques in food quality assurance to learn its benefits for use in the food industry.

12.2 DETERMINATION OF *TRANS*-FAT CONTENT IN BUTTER AND MARGARINE BY MID-INFRARED SPECTROSCOPY

Trans-fat can be found in food made with sources from ruminant animals or partially hydrogenated vegetable oils (PHOs). The *trans*-fat content in butter typically ranges from 2% to 7%, while in PHOs ranges from 10 to 60% [1]. Due to the health concerns of *trans* fat, in 2015 the US Food and Drug Administration released a statement that PHOs are not GRAS (generally recognized as safe), which came in effect after June 18, 2018. Gas chromatography [American Oil Chemists' Society (AOCS) Method Cd 14c-94] is the traditional technique to determine the *trans*-fat content; however, it requires lengthy procedures [2]. MIR, on the other hand, is a robust and fast method that can be applied for the analysis of numerous food products including butter and margarine. MIR calibration of the target compound in a known matrix can be developed based on a defined MIR band height or area. However, in most food matrix samples, defined bands are difficult to be identified directly due to the predominant interference from the chemical and physical information of the matrix. Therefore, chemometrics methods are becoming popular in efficiently extracting the spectral differences and training a calibration model for predicting the properties/ classes of samples in the applications.

12.2.1 Principle of Method

The rapid determination of *trans*-fat content by MIR spectroscopy is a widely used methodology [3]. This approach is based on the measurement of the height/area of the band at around 966 cm^{-1}, which is the C–H out-of-plane deformation band specific for the isolated double bonds with *trans* configuration (Fig. 12.1) [3]. Two approaches are designed for calculating *trans*-fat levels. The first approach (Sect. 12.2.6.2) requires the generation of a standard curve based on the height of the *trans* band at around 966 cm^{-1} in the second derivative transformed spectrum. The second approach (Sect. 12.2.6.3) requires

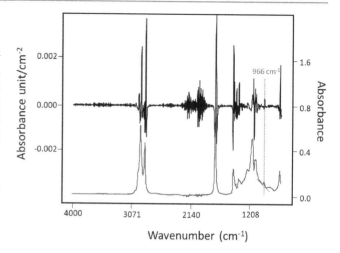

Fig. 12.1 Representative second derivative (black) and absorption spectra (blue) for a sample with a *trans*-fat level of 3.3%. (A unique band associated with isolated *trans* double bands at around 966 cm^{-1} is marked)

the use of a chemometrics model, provided as an R script within the text.

12.2.2 Chemicals

	CAS No.	Hazards
Acetone	67-64-1	Highly flammable, irritant
Ethanol (95 or 100%)	64-17-5	Highly flammable

12.2.3 Reagents

- Acetone, Fisher Scientific, Inc. (Waltham, MA, USA)
- Ethanol, 70%, 100 mL
  Use 95% or 100% ethanol (Fisher Scientific, Inc., Waltham, MA, USA) and deionized distilled water to prepare 100 mL of a 70% ethanol solution
- Trielaidin (TE, trans 18:1, absorption at 966 cm^{-1}), Fisher Scientific, Inc. (Waltham, MA, USA)
- Tripalmitin (TP, 16:0, absorption at 956 cm^{-1}), Fisher Scientific, Inc. (Waltham, MA, USA)

12.2.4 Supplies

- A compressed folder with spectra examples for Sect. 12.2.6
- Amber vials, 2 mL
- Butter and margarine products
- Centrifuge tubes, 15 mL
- Cotton pads
- Mechanical pipettors (100 uL)
- Micropipettes

12.2.5 Equipment

- Analytical balance
- Centrifuge (for 15 mL tubes)
- Fourier Transform Infrared (FTIR) spectrophotometer (experiment was developed on an Agilent 4500 series portable FTIR (4500a) with a diamond ATR crystal interface
- Heating block (use 70 °C and 80 °C)
- Oven, 65 °C, or hotplate
- Vortexer

12.2.6 Procedure and Data Handling

(*Note:* A compressed folder is available to your course instructor via the Food Analysis textbook website with teaching materials, comprising raw and second derivative transformed spectra of standards and five commercial samples (three butter and two margarine samples) in GRAMS (.spc) and EXCEL (.xls) formats. It also includes an excel sheet of spectra for multivariate analysis. The data can give students a hands-on experience of this experiment (Sect. 12.2.6.1, 12.2.6.2, and 12.2.6.3) if MIR is not accessible in the institution. Also, answers to the questions below are available through the same website, accessible only to course instructors. Access to such website can be requested by the course instructor by contacting the lab manual co-editor B. P. Ismail)

12.2.6.1 Sample Spectral Data Acquisitions Steps

1. Place ~3 g butter or margarine samples in 15 mL centrifuge tubes.
2. Heat the tubes with samples in a 65 °C oven for 1 h to separate oil and aqueous phases (Fig. 12.2). (*Note:* Alternatively, you can place the samples in boiling water on a hotplate for 10 min to achieve phase separation.

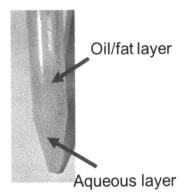

Fig. 12.2 Clear phase separations of the oil/fat layer and the aqueous layer of a butter sample

Margarine samples sometimes need centrifugation to better separate the phases. If so, centrifuge the tubes in a laboratory centrifuge at a speed of 4000 rpm for 10 min.)

3. Use cotton pads to clean the attenuated total reflection (ATR) crystal gently, first with acetone and then with the 70% water: ethanol solution.

   (*Note:* Using the specified FTIR unit, it is best to collect spectral data from 4000 to 700 cm^{-1} with co-added 64 scans at a resolution of 4 cm^{-1} to achieve suitable signal-to-noise ratio. A background spectrum must be collected before the spectral collection of each sample to eliminate the variances from environmental factors)

4. Remove the warmed sample from the oven and pipette ~ 20 μL (depending on your ATR crystal area) of the oil layer and dispense onto the MIR crystal (Fig. 12.3).

   (*Note:* Do not touch the crystal directly with the pipette tip. The amount of oil pipetted onto the MIR crystal needs to cover the whole attenuated total reflection (ATR) crystal, which will depend on the specifications of the FTIR spectrometer utilized for this lab)

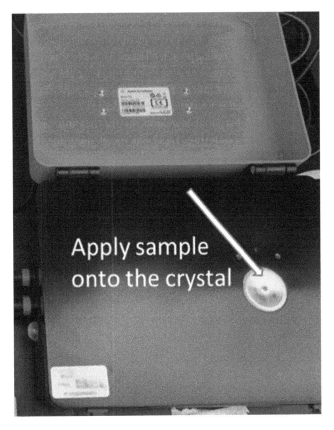

Fig. 12.3 Demonstration of spectral data acquisition on the Agilent 4500 series compact FTIR spectrometer (4500a)

5. "Read' and "save" the spectrum on the instrument software. A characteristic *trans*-fat band can be identified at around 966 cm⁻¹.

6. Use a cotton pad to absorb the oil sample and clean the crystal as in Step 3 above.

12.2.6.2 Development of the Trans-Fat Calibration Curve and Calculation of Trans Fat in Commercial Samples

The *trans*-fat calibration curve is developed following the method reported by Mossoba et al. [3]. TE (trans-18:1, absorption at 966 cm⁻¹) and TP (16:0, absorption at 956 cm⁻¹) are used to prepare calibration standards.

1. Calibration standards are prepared gravimetrically by mixing TE and TP, achieving TE levels of 0%, 1%, 2.5%, 5%, 10%, 15%, and 20% (as a w/w % of total fatty acids).

    It is best to use a heating block at 80 °C for 30 min to melt the TE and TP. Weigh TE in an amber vial and mix the TE in TP to a final weight of 1 gram. For example, if you are making a 1% *trans* fat, you will weigh 10 mg of the TE and take it to a final weight of 1 gram with TP. Make sure you record the weights precisely so you can calculate the final percentage of your TE in your mixture accurately.

    (*Note*: The melting point for TP and TE is 66 °C and 42 °C, respectively, so make sure the heating block is right next to the balance to prevent fast solidification of your prepared standards while completing weighting other standards.)

2. Warm the prepared standards in the heating block at 80 °C for 30 min (Fig. 12.4) and vortex them for 15 seconds prior to their spectral collections. Spectra measurement steps should follow Steps 3–6 in Sect. 12.2.6.1.

    (*Note*: Standards are measured in duplicate/triplicate by a FTIR spectrometer. Keep the standards in the heating block until applying them onto the ATR crystal to prevent solidification. A temperature control accessory (set at 70 °C) equipped on the spectrometer is highly recommended to use for preventing solidification.)

3. Record spectral data of standards and perform the second derivative of the spectra.

    After the second derivative transformation, record the band height of TE at around 966 cm⁻¹ and average the band heights of the duplicates/triplicates.

    (*Note:* The second derivative of the spectra can be transformed by using spectral process depending on the software applications. Make sure you allow the software to provide numbers with at least 5 decimal points. Use the "Fourier 2ⁿᵈ derivative" and select a gap size of 35 pts. The down TE band centered at around 966 cm⁻¹ is selected and its band height is displayed as shown in Fig. 12.5.)

4. Plot *trans*-fat contents (%) in standards versus band heights of *trans* fat on the second derivative transformed spectra and generate a regression fit (Fig. 12.6).

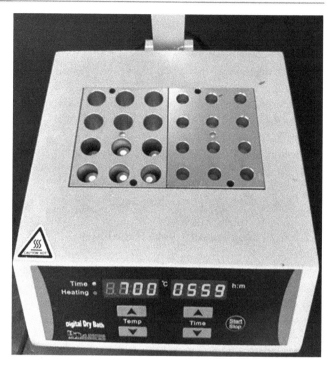

Fig. 12.4 Demonstration of using a heating block to warm the prepared standards

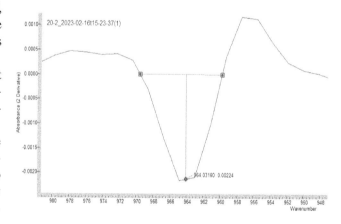

Fig. 12.5 Determination of TE band height (at ~966 cm⁻¹) on the second derivative transformed spectrum

Once the calibration curve is successfully generated, the *trans*-fat content in commercial butter and margarine products can be calculated.

5. Perform the second derivative on the sample spectra and record the heights of *trans* fat at around 966 cm⁻¹ following Step 3 in Sect. 12.2.6.2.

6. The recorded heights are used as the **Y** in the calibration curve equation of the line to calculate the corresponding **X** value (*trans*-fat content) in butter and margarine products.

    For example, if you have a recorded height of 0.00036 in a butter sample (Fig. 12.7), the *trans*-fat content would be calculated as follows, using the equation of the line shown in Fig. 12.6:

$$(0.00036 + 0.000004)/0.0001 = 3.64\%$$

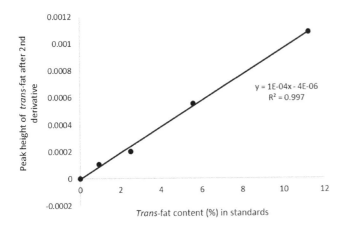

Fig. 12.6. A calibration curve of *trans*-fat content in standards versus band height of *trans* fat on the second derivative transformed spectra

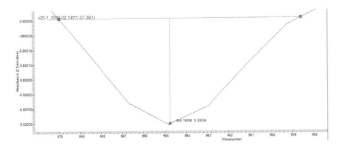

Fig. 12.7 Example of TE band height (at around 966 cm^{-1}) on the second derivative transformed spectrum of a commercial butter sample

7. Record the *trans*-fat values. Using the mean values, calculate the trans-fat content (gram) in your butter/margarine sample expressed in terms of per serving amount (14 grams) on its label.

12.2.6.3 Multivariate Data Analysis for Calculating Trans-Fat Contents

Partial least square regression (PLSR) is one of the widely used chemometric techniques to generate quantitative models. PLSR develops a regression model by correlating dependent (reference values) variables with a set of independent (spectra) variables. This calibrated model can be used for predicting the parameter of interest from independent variable(s) of a given test sample. The package used in the current R [4] code has been developed from spectral data acquired using an Agilent infrared spectrometer with a resolution of 4 cm^{-1}. Hence, the MIR spectra that will be used to predict *trans*-fat content using this code should be collected with 4 cm^{-1} resolution.

1. Installing R, RStudio, and Rtools
    R, **RStudio**, and **Rtools** are required to execute the code. The following steps will guide you through the installations.
    (a) Installation for *Windows Users.*
        (i) Installing **R** and **RStudio** for the first time: Install the latest version of R from https://cran.r--project.org/ depending on your operating system. Use https://www.rstudio.com/ to download R Studio and follow the installation steps to complete the setup.
        (ii) Upgrading R version if already installed: (If you have the latest version of R, you can skip this section.) To upgrade R to the latest version, search RGui in windows start menu. In the console paste the following code chunk (all 3 lines in the shaded box) and navigate through the installation steps.

```
install.packages("installr")
library("installr")
installr::updateR()
```

        (iii) Upgrading **Rtools**: (If you have the latest version of R tools, you can skip this section.) If you have Rtools installed, go to setting and uninstall prior to downloading the latest version (settings > apps > advanced app settings > Rtools > uninstall). You can find the latest version of Rtools here https://cran.r-project.org/bin/windows/Rtools/.
    (b) Installation for *macOS.*
        (i) **R** and **XQuartz**: Install the *latest release* of R from https://cran.r-project.org/bin/macosx/ depending on your macOS version. If you are unsure about your version, use any link (on left column) from the latest release section. From the same page, install and run "XQuartz."
        (ii) **Tools**: Download and run https://mac.r-project.org/tools/gfortran-8.2-Mojave.dmg
        (iii) **RStudio**: Install and run RStudio for macOS from https://posit.co/download/rstudio-desktop/
            (*Note:* Working with latest version of *Rtools* and *RStudio* is encouraged to avoid any errors. Leave the default options during installation steps. If you are installing the latest versions of applications, make sure to uninstall the previous files if you already have them.)

2. Predicting *trans*-fat content
    (a) Open "RStudio". Go to File > New file > R script to generate a new source file. Window as shown in Fig. 12.8 appears.

Fig. 12.8 RStudio interface highlighting important quadrants useful in executing the code

(*Note:* Most of the following steps involve copying and pasting the code chunks directly in the RStudio editor window (generally located in the top left quadrant of the screen, Fig. 12.8). A code chunk refers to *all* the lines (R code) given in the lightly shaded box. Code chunks should be entered in the editor quadrant (Fig. 12.8).)

Comments (lines beginning with '#') in the code explain the purpose of the code for user's reference. They can be pasted along with the code lines.

(b) Installing and loading packages:
  (i) Copy all R code lines given in the lightly shaded region below *Code 1,* paste it in RStudio editor window.

(ii) Highlight or select the code chunk you just pasted in RStudio

(iii) Run the code by clicking "run" on top of the quadrant's window (shown in Fig 12.8)

(*Note:* The following code installs and loads the required packages [5–8], if you are working with the code for the first time, it may take approximately a minute to download all packages. While downloading, it may ask if you want to update the package, press the Enter key in the console window to keep installing. The stop symbol in red [●] on top right of console window indicates that the code is still running. After this stop symbol disappears, you can run part 2.2 code.)

Code 1:

```
#Installs packages if not already installed
list.of.packages <- c("signal", "pls", "readxl","devtools")
new.packages <- list.of.packages[!(list.of.packages %in% installed.packages()[,"Package"])]
if(length(new.packages)) install.packages(new.packages)
install.packages(c('devtools','cachem'))
library("devtools")
devtools::install_github('nugurishreya/predicttransfat')

#loads the packages
library("predicttransfat")
library("signal")
library("pls")
library("readxl")
```

(c) Loading regression model and prediction file:

    (i) In continuation with the previous code lines, copy paste the following code chunk (***Code 2***) in editor window.

    (ii) Select or highlight the code chunk you just pasted and run it (click "run").

    (iii) After executing the code, a small window pops up allowing the user to select the excel file containing IR spectra whose *trans*-fat values must be predicted.

        (*Note:* If you cannot find this window, search for a blue-white squared icon in taskbar or minimize the RStudio as most of the time it gets hidden behind the RStudio window. Make sure the IR spectra is in excel format as the code is designed to read data only from an excel sheet. The first column of the excel sheet should contain the Sample ID and first row should have the wavenumbers. Spectral values of each sample should be entered along the rows. An excel sheet of prediction data having IR spectra of butter and margarine samples has been provided ("Spectra_for_Rprediction.xlsx") in a folder, you may consider this file as an example to execute the code.)

Code 2:

```
#obtaining the model under "cal_model" variable
cal_model <- transfat_model()

######predicting_model#######
#Selecting file
pred_data <- read_xlsx(file.choose()) #loads the file and names it as "pred_data"
colnames(pred_data) <- as.character(as.integer(colnames(pred_data))) #rounds off the wavenumbers
k <- pred_data[2:ncol(pred_data)]
pred <- data.frame(MIR = I(data.matrix(k)))

#pre-processing
der_pred <- apply(pred$MIR, 1, FUN=sgolayfilt, p=2, n=3, m=2, ts=1) #second derivative
smooth_pred <- apply(t(der_pred), 1, FUN=sgolayfilt, p=2, n=3, m=0, ts=1) #smoothing
```

(d) Finding the *trans*-fat wavenumber:

    (i) Referring to the first row of excel sheet, find a wavenumber closer to 966. Consider the *integer part* of it (mostly it should be either **965** or **966** or **967**). For example, the integer part to be considered if you have 967.6933 wavenumber will be 967, the integer part of 966.0125 will be 966.

    (ii) Enter this number in the first line of the following code chunk, ***Code 3*** (in the present code it is 966, highlighted in green font.

Enter the integer you found in the previous step in place of **966**).

    (iii) Copy the following code chunk (below ***Code 3***), paste it in the editor window.

    (iv) Select the pasted code in R editor and click "run" to execute.

        (*Note:* In this step, we are essentially trying to select the important region around 966 cm^{-1} corresponding to *trans*-fat characteristics. The output gives the *trans*-fat percentages in the console window (bottom left quadrant, Fig. 12.8).)

Code 3:

```
#selects the column number with important wavenumber, needs to be changed depending on
what wavenumber the user has in pred_data
x <- which(colnames(pred_data)=="966")
y <- x-86
z <- x+84
pred_region <- as.data.frame(t(smooth_pred)) %>% select(y:z) #selecting the important region
of trans fat
predict(cal_model, ncomp = 4, newdata = I(data.matrix(pred_region))) #predicting the values
```

12.3 QUESTIONS

1. Discuss the advantages and disadvantages of IR methods for quality assurance in contrast to other analysis methods.
2. Explain the advantages of performing second derivative in Sect. 12.2.6.2 and how it influences the sensitivity of the analysis.
3. Provide a reason why tripalmitin (TP) is chosen for generating the calibration curve with trielaidin (TE) in the *trans*-fat analysis.

REFERENCES

1. Tarrago-Trani MT, Phillips KM, Lemar LE, Holden JM (2006) New and existing oils and fats used in products with reduced trans-fatty acid content. Journal of the American Dietetic Association. 106:867–880, doi:https://doi.org/10.1016/j.jada.2006.03.010

2. Salas-Valerio W, Aykas DP, Hatta Sakoda BA, Ludeña-Urquizo FE, Ball C, Plans M, Rodriguez-Saona L (2022) In-field screening of trans-fat levels using mid- and near-infrared spectrometers for butters and margarines commercialized in the Peruvian market. Food Science & Technology 157:113074, doi:https://doi.org/10.1016/j.lwt.2022.113074

3. Mossoba M, Seiler A, Kramer JKG, Milosevic V, Milosevic M, Azizian H, Steinhart H (2009) Nutrition labeling: Rapid determination of total trans fats by using internal reflection infrared spectroscopy and a second derivative procedure. Journal of the American Oil Chemists Society 86:1037, doi:https://doi.org/10.1007/s11746-009-1444-x

4. R Core Team (2022). R: A language and environment for statistical computing. R Foundation for Statistical Computing, Vienna, Austria. URL https://www.R-project.org/.

5. Mevik B-H, Wehrens R (2007) The pls package: Principal component and partial least squares regression. In: RJ Stat. Softw. 18, doi:https://doi.org/10.18637/jss.v018.i02

6. Wickham, H, & Bryan J. (2019). readxl: Read excel files. *R package version*, *1*(1), 785.

7. Wickham, H, Hester, J, & Chang W. (2016). Devtools: Tools to make developing r packages easier. R package version, 1(0), 9000.

8. Developers S. (2013). Signal: signal processing. *URL http://r-forge.r-project. org/projects/signal.*

Moisture Content Determination

B. Pam Ismail and S. Suzanne Nielsen

Contents

B. P. Ismail (✉)
Department of Food Science and Nutrition, University of
Minnesota, St. Paul, MN, USA
e-mail: bismailm@umn.edu

S. S. Nielsen
Department of Food Science, Purdue University,
West Lafayette, IN, USA
e-mail: nielsens@purdue.edu

© The Author(s), under exclusive license to Springer Nature Switzerland AG 2024
B. P. Ismail, S. S. Nielsen (eds.), *Nielsen's Food Analysis Laboratory Manual*, Food Science Text Series,
https://doi.org/10.1007/978-3-031-44970-3_13

13.1 INTRODUCTION

13.1.1 Background

The moisture content of foods is important to food manufacturers for a variety of reasons. Moisture is an important factor in food quality, preservation, and resistance to deterioration. Determination of moisture content also is necessary to calculate the content of other food constituents on a uniform basis (i.e., dry weight basis). The dry matter that remains after moisture analysis is commonly referred to as total solids.

While moisture content is not given on a nutrition label, it must be determined to calculate total carbohydrate content. Moisture content of foods can be determined by a variety of methods, but obtaining accurate and precise data is commonly a challenge. The various methods of analysis have different applications, advantages, and disadvantages (see Sect. 13.1.2). In this experiment, several methods to determine the moisture content of foods will be used and the results compared. Note that in some cases, the method is not ideal for the sample type (e.g., moisture content of a product high in carbohydrate, such as baby cereal, using microwave oven), but the intent is to compare results to those obtained using more appropriate methods (i.e., vacuum oven for baby cereal).

The samples suggested for analysis (i.e., three cheese types, two cereal products, and raw almonds) are the same as those suggested for other components of proximate analysis (Chaps. 14, 15, and 16) and for the analysis of mineral analysis required on the nutrition label (Chap. 11). Each student or group of students will be assigned one product for the determination of moisture using different methods. It is recommended that the same product is assigned to the student or group of students for all proximate and mineral analyses. It is recommended that all analyses be performed in triplicate, as time permits.

13.1.2 Reading Assignment

Mauer LJ (2024) Moisture and total solids analysis, Ch. 15. In: Ismail BP and Nielsen SS (eds) Nielsen's food analysis, 6th edn. Springer, New York

13.1.3 Overall Objective

The objective of this experiment is to determine and compare the moisture contents of foods by various methods of analysis.

13.2 FORCED DRAFT OVEN

13.2.1 Objective

Determine the moisture content of the different food samples using a forced draft oven method.

13.2.2 Principle of Method

The sample is heated under specified conditions, and the loss of weight is used to calculate the moisture content of the sample.

13.2.3 Supplies

(*Note*: Samples for moisture analysis could be same products tested also for ash, fat, and protein analysis, by procedures in respective chapters. Content of ash, fat, and protein could then be expressed on both a wet weight basis (wwb) and dry weight basis (dwb).)

Food samples:
- Cheddar cheese
- Cheerios cereal
- Parmesan cheese
- Process cheese
- Raw almonds
- Rice baby cereal

Other supplies:
- Desiccators (with dried desiccant) (1 per food sample)
- Glass microfiber filters or glass wool
- Grater or grinder
- Petri dish
- Plastic gloves
- Spatulas (1 per food sample)
- Tongs (1 per food sample)
- Trays (to hold/transfer samples) (6)
- Weighing pans—disposable aluminum open pans (predried at 100 °C for 24 h) (18)

13.2.4 Equipment

- Analytical balance, 0.1 mg sensitivity
- Forced draft oven

13.2.5 Note

Glass microfiber filters (e.g., GF/A, Whatman, Newton, MA) or glass wool, pre-dried for 1 h at 100 °C, can be used to cover samples to prevent splattering in the forced draft oven.

13.2.6 Cautions and Hazards

Be sure to label all containers used with complete information, or record container information linked to each sample.

Use gloves or tongs when handling sample pans and crucibles. These pans have been dried and stored in desiccators prior to weighing. They will pick up moisture by sitting on the counter, so remove them from the desiccator only just before use. Open desiccators slowly to avoid damage and danger from broken glass.

13.2.7 Procedure

Instructions are given for analysis in triplicate.

1. Remove the pre-dried and pre-labeled pans from the desiccator with tongs. Make sure that you have the three correct pans for your sample (e.g., baby cereal pan Rep 1 to be placed in forced draft oven will be labeled as such (BF1)). Obtain three pieces of dried glass wool.
2. Place one sample pan and piece of glass wool on the balance and record the weight in your notebook and class data sheet. Take out the piece of glass wool with tongs and tare the sample pan weight.
3. Cheese samples must be grated, while Cheerios and almonds need to be ground. Avoid excessive grinding to prevent frictional heat. Place 2 g of your assigned sample in the pan and record the weight of the sample in your notebook and the class data spread sheet. Make sure the sample is spread in a thin layer on the bottom of the pan and place the piece of glass wool back on top of the sample.
4. Repeat Steps 2 and 3 to prepare two more replicates. You will end up with 3 replicates.
5. Place prepared samples in desiccator. Once all of the samples are collected, they will be placed into a forced draft oven at 98–100 °C for 24 h.
6. Teaching assistants will remove samples at 24 h, and place into desiccator to cool. Students will come back the following day to record the final weight of the samples.

13.2.8 Data and Calculations

Calculate percentage moisture (wt/wt) for each food sample ($n = 3$). Report the average % moisture of your product as well as the standard deviation, coefficient of variation, and % E_{Rel}.

$$\% \text{ moisture (wet weight basis, wwb)} = \frac{(\text{wt of water in sample})}{(\text{wt of wet sample})} \times 100$$

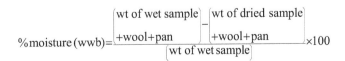

$$\% \text{ moisture (dry weight basis, dwb)}$$
$$= \frac{(\text{wt of water in sample})}{(\text{wt of sample} - \text{wt of water})} \times 100$$

(*Note:* Moisture on dwb is the ratio in percent of moisture to dry matter)

Sample	Rep	Pan + glass wool (g)	Wet sample (g)	Pan + wool + dry sample (g)	Dry sample (g)	Moisture content (wwb) (%)	Moisture content (dwb) (%)
Food Product # 1	1						
	2						
	3						

13.3 VACUUM OVEN

13.3.1 Objective

Determine the moisture content of the different food samples by the vacuum oven method.

13.3.2 Principle

The sample is heated under conditions of reduced pressure to remove water. Under vacuum, the boiling point of water will be reduced. The loss of weight is used to calculate the moisture content of the sample.

13.3.3 Supplies

Food samples:
- Cheddar cheese
- Cheerios cereal
- Parmesan cheese
- Process cheese
- Raw almonds
- Rice baby cereal

Other supplies:
- Cold trap for the vacuum oven
- Desiccator (with dried desiccant) (1 per food sample)
- Gloves
- Grater or grinder
- Liquid nitrogen
- Petri dish
- Spatulas for weighing (1 per food sample)
- Tongs (1 per food sample)
- Weighing pans—disposable aluminum open pans (pre-dried at 100 °C for 3 h) (18)

13.3.4 Equipment

- Analytical balance, 0.1 mg sensitivity
- Flow meter
- Vacuum oven (capable of pulling vacuum to <100 mm of mercury)
- Vacuum pump

13.3.5 Cautions and Hazards

See same information in Sect. 13.2.6.

13.3.6 Procedure

1. Remove the pre-dried and pre-labeled pans from desiccator with tongs. Make sure that you have the three correct pans for your sample.
2. Place one sample pan on the balance and record the weight in your lab notebook and data excel file, and then tare the sample pan weight.
3. Cheese samples must be grated, while Cheerios and almonds need to be ground. Avoid excessive grinding to prevent frictional heat. Place 2 g of your assigned sample in the pan and record the weight of the sample. Make sure the sample is spread in a thin layer on the bottom of the pan.
4. Repeat Steps 2 and 3 to prepare two more replicates. You will end up with 3 replicates.
5. Place prepared samples in desiccator. Once all of the samples are collected, they will be placed into a vacuum oven at 60 °C for 24 h at 27–29 inches of Hg, with a flow rate of 120 mL/min.
6. Teaching assistants will remove samples after 5 h, and place into desiccator to cool. Students will come back the following day to record the final weight of the samples.

13.3.7 Data and Calculations

Calculate percentage moisture (wt/wt) as in Sect. 13.2.8.

Sample	Rep	Pan (g)	Wet sample (g)	Pan + dry sample (g)	Dry sample (g)	Moisture content (wwb) (%)	Moisture content (dwb) (%)
Food Product # 1	1						
	2						
	3						

13.4   MICROWAVE DRYING OVEN

13.4.1 Objective

Determine the moisture content of the different food samples using a microwave drying oven.

13.4.2 Principle

The sample is heated using microwave energy, and the loss of weight is used to calculate the moisture content of the sample.

13.4.3 Supplies

Food samples:
- Cheddar cheese
- Cheerios cereal
- Parmesan cheese
- Process cheese
- Raw almonds
- Rice baby cereal

Other supplies:
- Gloves
- Grater or grinder
- Petri dish
- Spatulas for weighing (1 per food sample)

13.4.4 Equipment

- Microwave drying oven with internal balance (e.g., from CEM Corporation, Matthew, NC)

13.4.5 Procedure

For cheese samples:

1. Lift cover and place weighing pad onto balance in microwave. Tare balance.
2. Cheese samples must be grated.
3. Load approx. 4 g sample onto weighing pad. Spread product into a circular single layer.
4. Analyze moisture at 80% power for 2 min.
5. Microwave will report sample weight before and after as well as % moisture.
6. Repeat Steps 1–5 two more times. You will end up with 3 replicates.

For baby cereal, raw almonds, and Cheerios:

1. Lift cover and place weighing pad onto balance in the microwave. Tare balance.
2. Cheerios and almonds need to be ground. Avoid excessive grinding to prevent frictional heat.
3. Load approx. 2 g sample onto weighing pad. Spread product into a circular single layer.
4. Analyze moisture at 80% power for 3 min 30 s.
5. Microwave will report sample weight before and after as well as % moisture.
6. Repeat Steps 1–5 twice. You will end up with 3 replicates.

13.4.6 Data and Calculations

Sample	Rep	Wet sample (g)	Dry sample (g)	Moisture content (wwb) (%)	Moisture content (dwb) (%)
Food Product # 1	1				
	2				
	3				

13.5   RAPID MOISTURE ANALYZER

13.5.1 Objective

Determine the moisture content of the different food samples using a rapid moisture analyzer.

13.5.2 Principle

The sample placed on a digital balance is heated under controlled high-heat conditions, and the instrument automatically measures the loss of weight to calculate the percentage moisture or solids.

13.5.3 Supplies

Food samples:
- Cheddar cheese
- Cheerios cereal
- Parmesan cheese
- Process cheese
- Raw almonds
- Rice baby cereal

Other supplies:
- Gloves
- Grater or grinder
- Petri dish
- Spatula for weighing (1 per food sample)

13.5.4 Equipment

- Rapid moisture analyzer (e.g., from Computrac®, Arizona Instrument LLC., Chandler, AZ)

13.5.5 Procedure

Follow instructions from manufacturer for use of the rapid moisture analyzer, regarding the following:

- Obtaining results
- Select test material
- Taring instrument
- Testing sample
- Turning on instrument and warming up

13.5.6 Data and Calculations

Sample	Rep	Moisture content (wwb) (%)	Moisture content (dwb) (%)
Food Product # 1	1		
	2		
	3		

13.6 KARL FISCHER METHOD

13.6.1 Objective

Determine the moisture content of different food samples by the Karl Fischer (KF) method.

13.6.2 Principle

When the sample is titrated with the KF reagent, which contains iodine and sulfur dioxide, the iodine is reduced by sulfur dioxide in the presence of water from the sample. The water reacts stoichiometrically with the KF reagent. The volume of KF reagent required to reach the endpoint of the titration (visual, conductometric, or coulometric) is directly related to the amount of water in the sample.

13.6.3 Chemicals

	CAS No.	Hazards
Karl Fischer reagent		Toxic
2-Methoxyethanol	109-86-4	
Pyridine	110-86-1	
Sulfur dioxide	7446-09-5	
Iodine	7553-56-2	Harmful, dangerous to the environment
Methanol, anhydrous	67-56-1	Extremely flammable
Sodium tartrate dihydrate ($Na_2C_4H_4O_6 \cdot 2H_2O$)	868-18-8	

13.6.4 Reagents

- KF reagent
- Methanol, anhydrous
- Sodium tartrate dihydrate, 1 g, dried at 150 °C for 2 h

13.6.5 Hazards, Cautions, and Waste Disposal

Use the anhydrous methanol in an operating hood, since the vapors are harmful and it is toxic. Otherwise, adhere to normal laboratory safety procedures. Use appropriate eye and skin protection. The KF reagent and anhydrous methanol should be disposed of as hazardous wastes.

13.6.6 Supplies

Food samples:
- Cheddar cheese
- Cheerios cereal
- Parmesan cheese
- Process cheese
- Raw almonds
- Rice baby cereal

Other supplies:
- Gloves
- Grater or grinder
- Petri dish
- Spatula for weighing (1 per food sample)
- Weighing paper

13.6.7 Equipment

- Analytical balance, 0.1 g sensitivity
- KF titration unit, nonautomated unit (e.g., from Barnsted Themaline, Berkeley, CA, Aquametry Apparatus) or automated unit

13.6.8 Procedure

Instructions are given as for a nonautomated unit and for analysis in triplicate. If using an automated unit, follow instructions of the manufacturer.

13.6.8.1 Apparatus Setup

Assemble titration apparatus and follow instructions of manufacturer. The titration apparatus includes the following: buret, reservoir for reagent, magnetic stirring device, reaction/titration vessel, electrodes, and circuitry for dead stop endpoint determination. Note that the reaction/titration vessel of the KF apparatus (and the anhydrous methanol within the vessel) must be changed after analyzing several samples (exact number depends on type of sample). Remember that this entire apparatus is very fragile. To prevent contamination from atmospheric moisture, all openings must be closed and protected with drying tubes.

13.6.8.2 Standardizing Karl Fischer Reagent

The KF reagent is standardized to determine its water equivalence. Normally, this needs to be done only once a day, or when changing the KF reagent supply.

1. Add approximately 50 mL of anhydrous methanol to reaction vessel through the sample port.

2. Put the magnetic stir bar in the vessel and turn on the magnetic stirrer.
3. Remove the caps (if any) from drying tube. Turn the buret stopcock to the *filling position*. Hold one finger on the air-release hold in the rubber bulb and pump the bulb to fill the buret. Close the stopcock when the KF reagent reaches the desired level (at position 0.00 mL) in the buret.
4. Titrate the water in the solvent (anhydrous methanol) by adding enough KF reagent to just change the color of the solution from clear or yellow to dark brown. This is known as the KF endpoint. Note and record the conductance meter reading. (You may titrate to any point in the brown KF zone on the meter, but make sure that you always titrate to that same endpoint for all subsequent samples in the series.) Allow the solution to stabilize at the endpoint on the meter for at least 1 min before proceeding to the next step.
5. Weigh, to the nearest milligram, approximately 0.3 g of sodium tartrate dihydrate, previously dried at 150 °C for 2 h.
6. Fill the burette with the KF reagent, then titrate the water in the sodium tartrate dihydrate sample as in Sect. 13.6.8.2, Step 4. Record the volume (mL) of KF reagent used.
7. Calculate the KF reagent water (moisture) equivalence (KFR$_{eq}$) in mg H$_2$O/mL:

$$\text{KFR}_{eq} = \frac{36\,\text{g} / \text{mol} \times S \times 1000}{230.08\,\text{g} / \text{mol} \times A}$$

where:

S = weight of sodium tartrate dihydrate (g)
A = mL of KF reagent required for titration of sodium tartrate dihydrate

13.6.8.3 Titration of Sample

1. Prepare samples for analysis and place in reaction vessel as described below.

   If samples are in grated or powder form:
   - Use an analytical balance to weigh out approximately 0.3 g of sample, and record the exact sample weight (S) to the nearest milligram.
   - Remove the conductance meter from the reaction vessel, then transfer your sample to the reaction vessel through the sample port immediately. (Use an extra piece of weighing paper to form a cone-shaped funnel in the sample port, then pour your sample through the funnel into the reaction vessel.)
   - Put the conductance meter and stopper back in the reaction vessel. The color of the solution in the vessel should change to light yellow, and the meter will register below the KF zone on the meter.

If any samples analyzed are in liquid form:

- Use a 1-mL syringe to draw up about 0.1 mL of sample. Weigh the syringe with sample on an analytical balance and record the exact weight (S_1) to the nearest milligram.
- Inject 1–2 drops of sample into the reaction vessel through the sample port and then weigh the syringe again (S_0), to the nearest milligram.
- Sample weight (S) is the difference of S_1 and S_0.

$$S = S1 - S0$$

- Put the stopper back in the sample port of the reaction vessel. The color of the solution in the vessel should change to light yellow, and the meter will register below the KF zone on the meter.

2. Fill the burette, then titrate the water in the sample as in Sect. 13.6.8.2, Step 4. Record the volume (mL) of KF reagent used.
3. To titrate another sample, repeat Steps 5–7 above (Sect. 13.6.8.2) with the new sample. After titrating several samples (exact number depends on the nature of the sample), it is necessary to start with fresh methanol in a clean reaction vessel. Record the volume (mL) of KF reagent used for each titration.

13.6.9 Data and Calculations

Calculate the moisture content of the sample as follows:

$$\%H_2O = \frac{KFR_{eq} \times K_s}{S} \times 100$$

where:

KFR_{eq} = water equivalence of KF reagent (mg H_2O/mL)
K_s = mL of KF reagent required for titration of sample
S = weight of sample (mg)

Karl Fischer reagent water equivalence (KFR_{eq}):

Rep	Wt. sodium tartrate dihydrate (g)	Buret start (mL)	Buret end (mL)	Volume titrant (mL)	Calculated KFR_{eq}
1					
2					
3					
					$\bar{X} =$

Moisture content of samples by Karl Fischer method:

Sample	Rep	Wt. sample (g)	Buret start (mL)	Buret end (mL)	Volume titrant (mL)	Moisture content (%)

13.7 NEAR-INFRARED ANALYZER

13.7.1 Objective

Determine the moisture content of the different food samples using a near-infrared (NIR) analyzer.

13.7.2 Principle

Specific frequencies of infrared radiation are absorbed by the functional grc (i.e., the –OH stretch of the water molecule). The concentration of moisture in the sample is determined by measuring the energy that is reflected or transmitted by the sample, which is inversely proportional to the energy absorbed.

13.7.3 Supplies

Food samples:
- Cheddar cheese
- Cheerios cereal
- Parmesan cheese
- Process cheese
- Raw almonds
- Rice baby cereal

Other supplies:
- Gloves
- Grater or grinder
- Pans and sample preparation tools for near-infrared analyzer
- Petri dish
- Spatula for weighing (1 per food sample)

13.7.4 Equipment

- Near-infrared analyzer

13.7.5 Procedure

Follow instructions from manufacturer for use of the near-infrared analyzer, regarding the following:

- Calibrating instrument
- Obtaining results
- Turning on instrument and warming up
- Testing sample

13.7.6 Data and Calculations

Sample	Rep	Moisture content (wwb) (%)	Moisture content (dwb) (%)
Food Product # 1	1		
	2		
	3		

13.8 QUESTIONS

1. In a table, summarize the moisture results from all products for all methods (include the average, CV and $\%E_{Rel}$).
2. What are the advantages and disadvantages of each method used to determine moisture content?
3. If you need to determine the moisture content of a milk sample, which method would be best to use, and would a pre-drying step be necessary? Explain.
4. Compare your results to reported values and indicate which method is most accurate and precise for each type of food. Justify your answer and compare and contrast the different methods used.
5. When you need to determine food perishability, which would give you a better indication, moisture content or water activity? Explain your answer.
6. Explain the theory/principles involved in predicting the concentrations of various constituents in a food sample by NIR analysis. Why do we say "predict" and not "measure"? What assumptions are being made?
7. Your quality control lab has been using a forced draft oven method to make moisture determinations on various products produced in your plant. You have been asked to evaluate the feasibility of switching to new methods (the specific one would depend on the product) for measuring moisture content.
   (a) Describe how you would evaluate the accuracy and precision of any new method.
   (b) What common problems or disadvantages with the forced draft oven method would you seek to reduce or eliminate using any new method?

Acknowledgments Arizona Instrument Corp., Tempe, AZ, is acknowledged for its partial contribution of a Computrac moisture analyzer for use in developing a section of this laboratory exercise.

RESOURCE MATERIALS

Cereal & Grains Association (formerly AACC International) (2010) Approved methods of analysis, 11th edn. (online). Cereals & Grains Assoc. St. Paul, MN

AOAC International (2023) Official methods of analysis, 22nd edn. (online). AOAC International, Rockville, MD

Mauer LJ (2024) Moisture and total solids analysis, Ch. 15. In: Ismail BP and Nielsen SS (eds) Nielsen's food analysis, 6th edn. Springer, New York

Wehr HM, Frank JF (eds.) (2004) Standard methods for the examination of dairy products, 17th edn. American Public Health Association, Washington, DC

Ash Content Determination

14

B. Pam Ismail

Contents

14.1 INTRODUCTION

14.1.1 Background

Ash refers to the inorganic residue remaining after either ignition or complete oxidation of organic matter in a food sample. The inorganic residue consists mainly of the minerals present in the food sample. Determining the ash content is part of the proximate analysis for nutritional evaluation. Also, ashing is the first step in the preparation of a sample for specific elemental analysis (Chap. 11). Two major types of ashing procedures are commonly used, dry ashing and wet ashing. Dry ashing involves heating the sample at elevated temperatures (500–600 °C) in a muffle furnace. Water and volatiles will evaporate, and organic matter will burn in the presence of oxygen and convert to CO_2 and oxides of N_2. In contrast, wet ashing is based on oxidizing organic matter

using acids and oxidizing agents or their combination. Minerals are thus solubilized without oxidation. However, for proximate analysis and for the determination of total ash content, dry ashing is used.

Foods with high-moisture content, such as Cheddar cheese, are often dried prior to ashing. Foods with high-fat content, such as full fat Cheddar cheese, also may need to be dried and their fat extracted prior to ashing. The ash content can be expressed on a wet basis or a dry basis.

The samples suggested for ash analysis (i.e., three cheese types, two cereal products, and raw almonds) are the same as those suggested for other proximate analysis components (Chaps. 13, 15, and 16) and for the analysis of minerals required on a nutrition label (Chap. 11). Each student or group of students will be assigned one product for the determination of ash content using dry ashing. It is recommended that the same product is assigned to the student or group of students for all proximate and mineral analyses. It is recommended that all analyses be performed in triplicate, as time permits.

B. P. Ismail (✉)
Department of Food Science and Nutrition, University of Minnesota, St. Paul, MN, USA
e-mail: bismailm@umn.edu

© The Author(s), under exclusive license to Springer Nature Switzerland AG 2024
B. P. Ismail, S. S. Nielsen (eds.), *Nielsen's Food Analysis Laboratory Manual*, Food Science Text Series,
https://doi.org/10.1007/978-3-031-44970-3_14

14.1.2 Reading Assignment

Harris GK (2024) Ash analysis, Ch. 16. In: Ismail BP, Nielsen SS (eds) Nielsen's food analysis, 6th edn. Springer, New York

14.1.3 Objective

Determine the ash content of a variety of food products by the dry ashing technique and express on a wet weight basis (wwb) and dry weight basis (dwb).

14.1.4 Principle of Method

Organic materials are incinerated at elevated temperatures (550 °C) in a muffle furnace, and inorganic matter (ash) remains. Ash content is measured by weight of inorganic matter remaining.

14.1.5 Chemical

	CAS no.	Hazards
Hydrochloric acid (HCl)	7647-01-0	Corrosive

14.1.6 Hazards, Cautions, and Waste Disposal

Concentrated hydrochloric acid is corrosive; avoid breathing vapors and contact with skin and clothes. The muffle furnace is extremely hot. Use gloves and tongs when handling crucibles. The crucibles have been dried and stored in desiccators prior to weighing. They will pick up moisture by sitting on the counter, so remove them from the desiccators only just before use. Open desiccators slowly to avoid damage and danger from broken glass.

14.1.7 Supplies

Food samples:
- Cheddar cheese
- Cheerios cereal
- Parmesan cheese
- Process cheese
- Raw almonds
- Rice baby cereal

Other supplies:
- Ashed crucibles (numbered) (prewashed with 0.2 N HCl, heated in a muffle furnace at 550 °C for 24 h, and stored in a desiccator prior to use)
- Desiccators (with dried desiccant)
- Plastic gloves
- Spatulas (1 per food sample)
- Tongs (1 per food sample)

14.1.8 Equipment

- Analytical balance
- Electric muffle furnace

14.2 PROCEDURE

(*Note:* Food products high in moisture, such as cheese, will need be dried (before ashing) by the teaching assistant, using the vacuum oven method in Chap. 13. The moisture content obtained will be used to calculate the ash content on a wet weight basis. For dry food products such as those listed above, drying is not needed before ashing. However, moisture content obtained in Chap. 13 will be used to calculate ash content on a dwb.)

1. Remove ashed crucibles from the desiccator and record weight and number of crucible in the table.
2. Accurately weigh ca. 2 g of sample (note that cheese samples are pre-dried and placed in desiccators) into the crucible and record weight on the spreadsheet. Prepare triplicate samples for each type of food product analyzed.
3. Place crucibles in muffle oven at 550 °C for 24 h.
4. Turn off the muffle furnace and allow it to cool (might take a few hours).
5. Remove crucibles from the muffle furnace and place into a desiccator to cool (note that this may need to be done by a teaching assistant). Return the following day to weigh the ashed sample and record weight of crucible plus ashed sample in the table.

14.3   DATA AND CALCULATIONS

Weight of ash $=($ weight of crucible and ash $)$ $-$ weight of crucible

$\%\text{Ash} = ($ weight of ash $/$ original sample weight $) \times 100$

Report the average ash %, standard deviation, and coefficient of variation for the food product analyzed. Also calculate average ash % on a dry basis, using the average % moisture value determined in the moisture analysis experiment.

(*Note:* For cheese samples, the % ash obtained is on a dwb, since samples to be ashed need to be pre-dried. Also, calculate % ash on a wwb, using the moisture % obtained in the moisture analysis experiment.)

Converting wwb to dwb:

$\%\text{ash on dwb} = \%\text{ash on wwb}$
$\times 100 / (100 - \%\text{moisture content})$

Converting dwb to wwb:

$\%\text{ash on wwb} = \%\text{ash on dwb}$
$\times (100 - \%\text{moisture content}) / 100$

Rep	Crucible number	Crucible wt. (g)	Crucible + un-ashed sample (g)	Un-ashed sample (g)	Crucible + ashed sample (g)	Ash (g)	Moisture (%) (previously determined)	Ash (% wwb)	Ash (% dwb)
Sample A									
1									
2									
3									
								$\bar{X} =$	$\bar{X} =$
								SD =	SD =
								CV =	CV =
								$\%E_{Rel} =$	$\%E_{Rel} =$
Sample B									
1									
2									
3									
$\bar{X} =$	$\bar{X} =$								
								SD =	SD =
								CV =	CV =
								$\%E_{Rel} =$	$\%E_{Rel} =$

14.4   QUESTIONS

1. For this laboratory, we used a dry ashing technique. In what instance would you want to use a wet ashing technique?
2. What are the advantages and disadvantages to using a dry ashing technique?
3. What are the disadvantages of using a wet ashing technique?
4. Why was it necessary to pre-acid wash, pre-ash, and dry in a desiccator the crucibles prior to use?
5. Were your results comparable to USDA reported values? Explain any discrepancies.

RESOURCES

Harris GK (2024) Ash analysis, Ch. 16. In: Ismail BP, Nielsen SS (eds) Nielsen's food analysis, 6th edn. Springer, New York

Fat Content Determination

15

B. Pam Ismail and S. Suzanne Nielsen

Contents

B. P. Ismail (✉)
Department of Food Science and Nutrition,
University of Minnesota, St. Paul, MN, USA
e-mail: bismailm@umn.edu

S. S. Nielsen
Department of Food Science, Purdue University,
West Lafayette, IN, USA
e-mail: nielesns@purdue.edu

© The Author(s), under exclusive license to Springer Nature Switzerland AG 2024
B. P. Ismail, S. S. Nielsen (eds.), *Nielsen's Food Analysis Laboratory Manual*, Food Science Text Series,
https://doi.org/10.1007/978-3-031-44970-3_15

15.1 INTRODUCTION

15.1.1 Background

The term "lipid" refers to a group of compounds that are sparingly soluble in water but show variable solubility in a number of organic solvents (e.g., ethyl ether, petroleum ether, acetone, ethanol, methanol, benzene). The lipid content of a food determined by extraction with one solvent may be quite different from the lipid content as determined with another solvent of different polarity. Fat content is determined often by solvent extraction methods (e.g., Soxhlet, Goldfisch, Mojonnier), but it also can be determined by non-solvent wet extraction methods (e.g., Babcock, Gerber) and by instrumental methods that rely on the physical and chemical properties of lipids (e.g., infrared, density, X-ray absorption). The method of choice depends on a variety of factors, including the nature of the sample (e.g., dry versus moist), the purpose of the analysis (e.g., official nutrition labeling or rapid quality control), and instrumentation available (e.g., Babcock uses simple glassware and equipment; infrared requires an expensive instrument).

This experiment includes the Soxhlet (and Goldfisch, if Soxhlet unit is not available), Mojonnier, and Babcock methods. If samples analyzed by these methods can be tested by an instrumental method for which equipment is available in your laboratory, data from the analyses can be compared. Due to solvent hindrance, it is best that high-moisture foods (such as cheeses) are predried prior to analysis by the Soxhlet and Goldfisch methods. Foods high in carbohydrates and/or protein are best analyzed by the Mojonnier method due to acid hydrolysis or the alkaline solubilization step prior to solvent extraction. High-moisture foods do not need to be predried prior to fat analysis by Mojonneir, due to the shaking steps involved in the method. The Babcock method is used only for dairy foods (i.e., milk, cheese, and cream).

The experiment specifies the use of petroleum ether as the solvent for the Soxhlet and Goldfisch methods, for reasons of cost and safety. Anhydrous ethyl ether could be used for both extraction methods, but appropriate precautions must be taken in handling the solvent.

The samples suggested for fat analysis (i.e., three cheese types, two cereal products, and raw almonds) are the same as those suggested for other proximate analysis components (Chap. 13, 14 and 16) and for the analysis of minerals required on a nutrition label (Chap. 11). Each student or group of students will be assigned one product for the determination of fat content using the different methods, and results will be compared. It is recommended that the same product be assigned to the student or group of students for all proximate and mineral analyses. It is recommended that all analyses be performed in triplicate, as time permits. In this laboratory exercise, the moisture content data are used to express the fat content on a wet and dry weight basis (dwb). Food products high in moisture, such as cheese samples, will need to be dried (before determination of fat content by organic solvent extraction methods) by the teaching assistant, following the vacuum oven method for drying (Chap. 13). The moisture content obtained using the vacuum oven method will be used to calculate the fat content in cheese samples on a wet weight basis (wwb). The moisture content data will also be used to express the fat content on a dwb for the other samples.

15.1.2 Reading Assignment

Chen B, Ellefson WC (2024) Fat analysis. Ch. 17. In: Ismail BP, Nielsen SS (eds) *Nielsen's food analysis*, 6th edn. Springer, New York.

15.1.3 Objective

Determine the fat content of various foods by the Soxhlet, (or Goldfisch) and Mojonnier methods, and determine the fat content of whole milk by the Babcock method.

15.2 SOXHLET METHOD

15.2.1 Principle of Method

Fat is extracted, semicontinuously, with an organic solvent. Solvent is heated and volatilized, then condensed above the sample. Solvent drips onto the sample and soaks it to extract the fat. At 15–20 min intervals, the solvent is siphoned to the heating flask, to start the process again. Fat content is measured by weight loss of sample or weight of fat removed.

15.2.2 Chemicals

	CAS no.	Hazards
Petroleum ether	8032-32-4	Harmful, highly flammable, dangerous for the environment
(or ethyl ether)	60–29-7	Harmful, extremely flammable

15.2.3 Hazards, Precautions, and Waste Disposal

Petroleum ether and ethyl ether are fire hazards; avoid open flames, breathing vapors, and contact with skin. Ether is extremely flammable, is hygroscopic, and may form explosive peroxides. Otherwise, adhere to normal laboratory safety procedures. Wear gloves and safety glasses at all times. Petroleum ether and ether liquid wastes must be disposed of in designated hazardous waste receptacles.

15.2.4 Supplies

Food samples:

- Cheddar cheese
- Cheerios cereal
- Parmesan cheese
- Process cheese
- Rice baby cereal
- Raw almonds

Other supplies:

- Beaker, 250 mL
- Cellulose extraction thimbles, predried in a vacuum oven at 70 °C for 24 h
- Desiccator
- Glass boiling beads
- Glass wool, predried in a vacuum oven at 70 °C for 24 h
- Graduated cylinder, 500 mL
- Mortar and pestle
- Plastic gloves
- Plastic gloves

- Spatulas (1 per food sample)
- Tongs (1 per food sample)
- Tape (to label beaker)
- Weighing pan (to hold 30-g snack food)

15.2.5 Equipment

- Analytical balance
- Coffee grinder
- Soxhlet extractor, with glassware
- Vacuum oven

15.2.6 Procedure

(Instructions are given for analysis in triplicate.)

1. Record the fat content of your food product as reported on the package label. Also record the serving size so you can calculate g fat/100-g product.
2. For predried cheese samples, slightly grind ~30-g sample with a mortar and pestle (excessive grinding will lead to greater loss of fat in mortar). Cheese samples are predried by your teaching assistant following the vacuum oven method (Chap. 13). Cheerios and raw almonds need to be ground in a coffee grinder.
3. Wearing plastic gloves, remove three predried cellulose extraction thimbles from the desiccator. Label the thimbles on the outside with your initials and a number (use a lead pencil), then weigh accurately on an analytical balance.
4. Place ~2–3 g of sample in the thimble. Reweigh. Place a small plug of dried glass wool in each thimble. Reweigh.
5. Place the three samples in a Soxhlet extractor. Put ~350-mL petroleum ether in the flask, add several glass boiling beads, and extract for 6 h or longer. Place a 250-mL beaker labeled with your name below your samples on the Soxhlet extraction unit. Samples in thimbles will be placed in the beaker after extraction and before drying.
6. Remove thimbles from the Soxhlet extractor using tongs and air-dry overnight in a hood. Place thimbles in a desiccator then reweigh.

15.2.7 Data and Calculations

Using the weights recorded in the tables below, calculate the percent fat (wt/wt) on a wwb and dwb determined by the Soxhlet extraction method. If the fat content of the food you analyzed was given on the label, report this theoretical value.

Name of food:
Label g fat/serving:
Label serving size (g):
Label g fat/100-g product:

Data from Soxhlet extraction:

Sample	Rep	Thimble (g)	Thimble + Sample (g)	Thimble + Sample + Glass wool (g)	Sample (g)	Extracted Sample + Thimble + Glass wool (g)	Fat (g)	% Fat (wwb)	% Fat (dwb)
	1								
	2								
	3								
								Av	Av
								SD	SD
								CV	CV
								%E_{Rel}	%E_{Rel}

$$\text{Weight of Fat} = \left(\text{weight sample} + \text{thimble} + \text{glass wool}\right) - \left(\begin{array}{l}\text{weight of thimble} + \text{glass wool} \\ + \text{defatted sample}\end{array}\right)$$

$$\%\text{Fat} = \left(\text{weight of fat} / \text{weight of sample}\right) \times 100$$

(*Note:* For all cheese samples, the % fat obtained is on a dwb, since samples are predried. For all low moisture foods, the % fat you obtain is on a wwb, and you need to use the moisture % (on a wwb) obtained from Chap. 13 results to calculate the fat % on a dwb. Use the equations shown in Chap. 14, Sect. 14.3 to covert dwb to wwb, and wwb to dwb.)

15.2.8 Questions

1. The Soxhlet extraction procedure utilized petroleum ether. What were the advantages of using it rather than ethyl ether?
2. What were the advantages of using the Soxhlet extraction method rather than the Goldfisch extraction method?
3. If the fat content measured here differed from that reported on the nutrition label, how might this be explained?

15.3 GOLDFISCH METHOD

(Could be used if Soxhlet unit not available)

15.3.1 Principle

Fat is extracted, continuously, with an organic solvent. Solvent is heated and volatilized, then is condensed above the sample. Solvent continuously drips through the sample to extract the fat. Fat content is measured by weight loss of sample or weight of fat removed.

15.3.2 Chemicals

Same as for Sect. 15.2.2.

15.3.3 Hazards, Precautions, and Waste Disposal

Same as for Sect. 15.2.2.

15.3.4 Supplies

Same as for Sect. 15.2.2.

15.3.5 Equipment

- Analytical balance.
- Goldfisch extraction apparatus.
- Vacuum oven.

15.3.6 Procedure

(Instructions are given for analysis in triplicate.)
   (*Note:* Analyze samples in triplicate.)

1. Follow Steps 1–4 in Sect. 15.2.6.
2. Place the thimble in the Goldfisch condenser bracket. Push the thimble up so that only about 1–2 cm is below the bracket. Fill the reclaiming beaker with petroleum ether (50 mL) and transfer to beaker. Seal beaker to apparatus using gasket and metal ring. Start the water flow through the condenser. Raise the hot plate up to the beaker, turn on, and start the ether boiling. Extract for 4 h at a condensation rate of 5–6 drops per second.
3. Follow Step 6 in Sect. 15.2.6.

15.3.7 Data and Calculations

Record the data and do the calculations for the Goldfisch method using the sample tables and equations given for the Soxhlet method in Sect. 15.2.7. Calculate the percent fat (wt/wt) on a wwb and dwb. If the fat content of the food you analyzed was given on the label, report this theoretical value.

Name of food:
Label g fat/serving:
Label serving size (g):
Label g fat/100-g product:

15.3.8 Questions

1. What would be the advantages and disadvantages of using ethyl ether rather than petroleum ether in a solvent extraction method, such as the Goldfisch method?
2. What were the advantages of using the Goldfisch extraction method rather than the Soxhlet extraction method?
3. If the fat content measured here differed from that reported on the nutrition label, how might this be explained?

15.4 Mojonnier Method

15.4.1 Principle

Fat is extracted with a mixture of ethyl ether and petroleum ether. The extract containing the fat is dried and expressed as percent fat by weight.

The assay uses not only ethyl ether and petroleum ether, but also ammonia, hydrochloric acid (HCl), and ethanol. Ammonia dissolves the casein in cheese and neutralizes the acidity of the product to reduce its viscosity. Hydrochloric acid is used to digest the macronutrients in high carbohydrate products to limit interference and enhance the extraction of fat components. Ethanol prevents gelation of the protein and ether and aids in the separation of the ether-water phase. Ethyl ether and petroleum ether serve as lipid solvents, and petroleum ether decreases the solubility of water in the ether phase.

15.4.2 Chemicals

	CAS no.	Hazards
Ammonium hydroxide	1336-21-6	Corrosive, dangerous for the environment
Hydrochloric acid (HCl)	7647-01-1	Corrosive
Ethanol	64–17-5	Highly flammable
Petroleum ether	8032-32-4	Harmful, highly flammable, dangerous for the environment
(or Ethyl ether)	60–29-7	Harmful, extremely Flammable

15.4.3 Reagents

(**It is recommended that these solutions be prepared by the laboratory assistant before class.)

- Deionized distilled (dd) water.
- 8.5 M Hydrochloric acid (25 parts of 37% HCl + 11 parts of water, v/v)
- Phenolphthalein indicator.

15.4.4 Hazards, Precautions, and Waste Disposal

Ethanol, ethyl ether, and petroleum ether are fire hazards; avoid open flames, breathing vapors, and contact with skin. Ether is extremely flammable, is hygroscopic, and may form explosive peroxides. Ammonia is corrosive; avoid contact and breathing vapors. Otherwise, adhere to normal laboratory safety procedures. Wear gloves and safety glasses at all times. Use acids in a hood. Petroleum ether and ether liquid wastes must be disposed of in designated hazardous waste receptacles. The aqueous waste can go down the drain with a water rinse.

15.4.5 Supplies

Food samples:

- Cheddar cheese
- Cheerios cereal
- Parmesan cheese
- Process cheese
- Raw almonds
- Rice baby cereal

Other supplies:

- Burettes (50 mL) for dispensing ether and petroleum ether
- Mojonnier extraction flasks, with stoppers
- Mojonnier fat dishes with lids
- Mojonnier rack for the extraction flasks
- Plastic gloves
- Tongs

15.4.6 Equipment

- Analytical balance
- Hot plate
- Mojonnier apparatus (with centrifuge, vacuum oven, and cooling desiccator)
- Water bath

15.4.7 Notes

Reagents must be added to the extraction flask in the following order: water, ammonia (or HCl), alcohol, ethyl ether, and petroleum ether. The burets on the dispensing cans or tilting pipets are graduated for measuring the proper amount. Make duplicate or triplicate determinations.

15.4.8 Procedure for Cheese Samples (Adapted from AOAC 933.05)

1. Weigh 1 g, in duplicate samples, finely grated cheese, directly into Mojonnier extraction flasks. Record the weight on the spread sheet.
2. Add 9 mL dd water and 1 mL ammonium hydroxide.
3. Mix until smooth and heat in water bath at 70–80 °C until casein is well softened (overnight; Steps 1–3 will be carried out by the teaching assistant).
4. Cool solution, add 3 drops phenolphthalein indicator, and then begin reagent addition.
    (a) While the solution is cooling, remove your labeled weighing dishes from the desiccator. Use tongs to handle the dishes at all times to prevent adding any additional weight to the dishes from moisture or oils on your hands.
    (b) Record the weight of the dish on the spread sheet and then place back into the desiccator until use.
5. Add 10 mL ethanol from the burette set-up. Stopper the flask and gently mix by rocking the flask back and forth for 20 sec.
6. Add 25 mL ether from the burette set-up. Stopper the flask and gently mix by rocking the flask back and forth for 20 sec.
7. Finally, add 25 mL petroleum ether. Again, stopper the flask and gently mix by rocking the flask back and forth for 20 sec.
8. Place the flasks into the centrifuge. Operate the centrifuge to run at around 600 rpm. Turn the crank 15 times (approximately one turn per second), then let go of the handle and allow the machine to stop on its own. There should be a clean separation of aqueous and ether phases.
9. Decant ether solution into your preweighed weighing dish.
10. When ether solution is decanted into dishes, be careful not to pour over any suspended solids or aqueous phase into the weighing dish.
11. Evaporate ether from dishes on a hot plate of Mojonnier unit while conducting the second extraction.
12. For the second extraction, add reagents in the following order with 20 sec of gentle rocking in between each addition: 5 mL ethanol, 25 mL diethyl ether, and 25 mL petroleum ether.

13. Centrifuge the flasks as in Step 8 and pour ether extract into the respective fat dish. Be careful to remove all the ether but none of the other liquid in the flask.
14. Completely evaporate ether. When all ether is gone, place the dishes in the vacuum oven connected to the Mojonnier apparatus at 70–75 °C for 10 min with a vacuum of at least 20 in.
15. Cool dishes in desiccators for 7 min.
16. Accurately weigh each dish with fat. Record weight on the spread sheet.

15.4.9 Procedure for Cereal, Cheerios, Raw Almonds (Adapted from AOAC 922.06)

1. Place 2 g of sample into extraction flask, record the weight, add 2 mL ethanol, and swirl to moisten all particles to prevent lumping on the addition of acid.
2. Add 10 mL hydrochloric acid, mix well, and set in 70–80 °C water bath for 30 min. Stir at frequent intervals.
3. Add 10 mL ethanol and cool.
    (a) While the solution is cooling, remove your prelabeled weighing dishes from the desiccator.
    (b) Record the weight of the dishes on the spread sheet and then place back into the desiccator until use.
4. Add 25 mL ether to flask, stopper flask, and gently mix by rocking the flask back and forth for 1 min.
5. Add 25 mL petroleum ether, stopper, and gently mix by rocking the flask back and forth for 1 min.
6. Place the flasks into the centrifuge. Operate the centrifuge to run at around 600 rpm. Turn the crank 15 times (approximately one turn per second), then let go of the handle and allow the machine to stop on its own. There should be a clean separation of aqueous and ether phases.
7. Decant ether solution into a preweighed weighing dish.
8. When ether solution is decanted into dishes, be careful not to pour over any suspended solids or aqueous phase into the weighting dish.
9. Evaporate ether from dishes on a hot plate of the instrument while conducting the second extraction.
10. For the second extraction, add reagents in the following order: 5 mL ethanol, 25 mL diethyl ether, and 25 mL petroleum ether. Mix the flask between every addition by rocking the flask back and forth for 1 minute.
11. Centrifuge the flasks as in Step 7 and pour ether extract into the respective fat dish. Be careful to remove all the ether but none of the other liquid in the flask.
12. Completely evaporate ether. When all is gone, place the dishes in the vacuum oven connected to the Mojonnier centrifuge at 70–75 °C for 10 min with a vacuum of at least 20 in.
13. Cool dishes in desiccators for 7 min.
14. Accurately weigh each dish with fat. Record weight on the spread sheet.

15.4.10 Data and Calculations

Calculate the fat content of each sample.

Sample	Rep	Sample (g)	Aluminum Dish (g)	Fat + Aluminum Dish (g)	Fat (g)	% Fat (wwb)	% Fat (dwb)
	1						
	2						
	3						
						Av	Av
						SD	SD
						CV	CV
						%E$_{Rel}$	%E$_{Rel}$

$$\%Fat = 100 \times \left[\left(wt\ dish + fat\right) - \left(wt\ dish\right)\right] / wt\ sample$$

(*Note:* All samples are analyzed on a wwb. You need to use the moisture % (on wwb) obtained from Chap. 13 results to calculate the fat % on a dwb. Use the equations shown in Chap. 14, Sect. 14.3, to covert wwb to dwb)

15.4.11 Questions

1. List possible causes for high and low results in a Mojonnier fat test.
2. How would you expect the elimination of alcohol from the Mojonnier procedure to affect the results? Why?

15.5 BABCOCK METHOD

15.5.1 Principle

Sulfuric acid is added to a known amount of dairy sample in a Babcock bottle specific for the dairy product (in this case, dairy milk and cheese). The acid digests the protein, generates heat, and releases the fat. Centrifugation and hot water addition isolate the fat into the graduated neck of the bottle. The Babcock fat test uses a volumetric measurement to express the percent of fat in milk or meat by weight.

15.5.2 Note

The fat column in the Babcock test should be at 57–60 °C when read. The specific gravity of liquid fat at that temperature is approximately 0.90 g per mL. The calibration on the graduated column of the test bottle reflects this fact and enables one to make a volumetric measurement, which expresses the fat content as percent by weight.

15.5.3 Chemicals

	CAS no.	Hazards
Glymol (red reader)	8042-47-5	Toxic, irritant
Sulfuric acid	7664-93-9	Corrosive

15.5.4 Hazards, Precautions, and Waste Disposal

Concentrated sulfuric acid is extremely corrosive; avoid contact with skin and clothes and breathing vapors. Wear gloves and safety glasses at all times. Otherwise, adhere to normal laboratory safety procedures. Sulfuric acid and glymol wastes must be disposed of in a designated hazardous waste receptacle.

For safety and accuracy reasons, dispense the concentrated sulfuric acid from a bottle fitted with a repipettor (i.e., automatic bottle dispenser). Fit the dispenser with a thin, semirigid tube to dispense directly and deep into the Babcock bottle while mixing contents. Set the bottle with dispenser on a tray to collect spills. Wear corrosive- and heat-resistant gloves when mixing the sulfuric acid with samples.

15.5.5 Supplies

- Babcock bottles (3)
- Babcock caliber (or shrimp divider)
- Measuring pipette, 10 mL
- Pipette bulb or pump
- Plastic gloves
- Standard milk pipette (17.6 mL)
- Thermometer
- Whole milk

15.5.6 Equipment

- Analytical balance
- Babcock centrifuge
- Water bath

15.5.7 Procedure

(Instructions are given for analysis in triplicate.)

1. Adjust milk sample to ca. 38 °C and mix until homogenous. Using a standard milk pipette, pipette 17.6 mL of milk into each of three Babcock bottles. After the pipette

has emptied, blow out the last drops of milk from the pipette tip into the bottle. Allow milk samples to adjust to ca. 22 °C.

2. Dispense ca. 17.5 mL of sulfuric acid (specific gravity 1.82–1.83) and carefully add into the test bottle, with mixing during and between additions, taking care to wash all traces of milk into the bulb of the bottle. Time for complete acid addition should not exceed 20 s. Mix the milk and acid thoroughly. Be careful not to get any of the mixture into the column of the bottle while shaking. Heat generated behind any such lodged mixture may cause a violent expulsion from the bottle.

3. Place bottles in a centrifuge heated to 60 °C. Be sure bottles are counterbalanced. Position bottles so that bottlenecks will not be broken in horizontal configuration. Be sure that the heater of the centrifuge is on.

4. Centrifuge the bottles for 5 min after reaching the proper speed (speed will vary depending upon the diameter of the centrifuge head).

5. Stop the centrifuge and add hot dd water (60 °C) until the liquid level is within 0.6 cm of the neck of the bottle. Carefully permit the water to flow down the side of the bottle. Again, centrifuge the bottles for 2 min.

6. Stop the centrifuge and add enough hot dd water (60 °C) to bring the liquid column near the top graduation of the scale. Again, centrifuge the bottles for 1 min.

7. Remove the bottles from the centrifuge and place in a heated (55–60 °C, preferably 57 °C) water bath deep enough to permit the fat column to be below the water level of the water bath. Allow bottles to remain at least 5 min before reading.

8. Remove the samples from the water bath one at a time, and quickly dry the outside of the bottle. Add glymol (red reader) to the top of the fat layer. Immediately use a divider or caliper to measure the fat column to the nearest 0.05%, holding the bottle in a vertical position at eye level. Measure from the highest point of the upper meniscus to the bottom of the lower meniscus.

9. Reject all tests in which the fat column is milky or shows the presence of curd or charred matter, or in which the reading is indistinct or uncertain. The fat should be clear and sparkling, the upper and lower meniscus clearly

defined, and the water below the fat column should be clear.

10. Record the readings of each test, and determine the mean % fat and the standard deviation.

15.5.8 Data and Calculations

Rep	Measured % fat
1	
2	
3	
$\bar{X} =$	
SD =	
CV =	
%E_{Rel} =	

15.5.9 Questions

1. What are the possible causes of charred particles in the fat column of the Babcock bottle?
2. What are the possible causes of undigested curd in the Babcock fat test?
3. Why is sulfuric acid preferred over other acids for use in the Babcock fat test?

Acknowledgments The authors of this chapter thank Charles E. Carpenter (Department of Nutrition, Dietetics and Food Sciences, Utah State University, Logan UT, USA) for contributing material for the chapter when it was originally developed.

RESOURCE MATERIALS

AOAC International (2023) Official methods of analysis, 22th edn, (On-line). AOAC International, Rockville, MD

Chen B, Ellefson WC (2024) Fat analysis. Ch. 17. In: Ismail BP and Nielsen SS (eds) Nielsen's food analysis, 6th edn. Springer, New York

Wehr HM, and Frank JF (eds) (2004) Standard methods for the examination of dairy products. 17th edn. American Public Health Administration, Washington, DC

Protein Nitrogen Determination

16

B. Pam Ismail and S. Suzanne Nielsen

Contents

16.1 INTRODUCTION

16.1.1 Background

Protein analysis is important for a variety of reasons, including nutritional labeling, functional property investigation, and biological activity determination. The protein content of foods can be determined by numerous methods. The Kjeldahl method and the nitrogen combustion (Dumas) method for protein analysis are based on nitrogen determination. Both methods are official and widely used. While the Kjeldahl method has been the method of choice for over 100 years, the

B. P. Ismail (✉)
Department of Food Science and Nutrition, University of Minnesota, St. Paul, MN, USA
e-mail: bismailm@umn.edu

S. S. Nielsen
Department of Food Science, Purdue University, West Lafayette, IN, USA
e-mail: nielesns@purdue.edu

© The Author(s), under exclusive license to Springer Nature Switzerland AG 2024
B. P. Ismail, S. S. Nielsen (eds.), *Nielsen's Food Analysis Laboratory Manual*, Food Science Text Series,
https://doi.org/10.1007/978-3-031-44970-3_16

availability of automated instrumentation for the Dumas method is contributing to the substitution of the Kjeldahl method.

Food proteins on average have 16% nitrogen but this percentage will vary depending on the amino acid composition of the protein. Although nitrogen is largely unique to protein, other nitrogenous compounds (such as chlorophyll, nucleic acids, amino sugars, vitamins, among others) will contribute to the total nitrogen. Therefore, both the Kjeldahl and the Dumas methods provide the crude protein content of food.

The samples suggested for protein analysis (i.e., three cheese types, two cereal products, and raw almonds) are the same as those suggested for other proximate analysis components (Chap. 13, 14 and 15) and for the analysis of minerals required on a nutrition label (Chap. 11). Each student or group of students will be assigned one product for the determination of protein content using the two methods and results are compared. It is recommended that the same product is assigned to the student or group of students for all proximate and mineral analyses. It is also recommended that all analyses be performed in triplicate, as time permits. In this laboratory exercise, the moisture content data are used to express the protein content on a dry weight basis (dwb).

16.1.2 Reading Assignment

Dee DR and Chang SKC (2024) Protein analysis. Ch. 18. In: Ismail BP, Nielsen SS (eds) *Nielsen's food analysis*, 6th edn. Springer, New York.

16.1.3 Notes

Both the Kjeldahl and nitrogen combustion methods are commonly done with automated instruments. The descriptions below are based on the availability of such automated instrumentation. If protein content of samples analyzed by the Kjeldahl method and/or nitrogen combustion has been estimated in a previous experiment by near-infrared analysis, values can be compared among methods. After collecting your data, you will compare your calculated results to the values declared on the label and reported by the USDA nutrient database.

16.2   Kjeldahl Nitrogen Method

16.2.1 Objective

Determine the protein content of various food products using the Kjeldahl method.

16.2.2 Principle of Method

The Kjeldahl procedure measures the nitrogen content of a sample. The protein content then can be calculated assuming a ratio of protein to nitrogen for the specific food being analyzed. The Kjeldahl procedure can be basically divided into three parts: (1) digestion, (2) distillation, (3) titration. During digestion, the sample is mixed with concentrated sulfuric acid (H_2SO_4) and a catalyst, then heated to approx. 420 °C in order to break down the organic matter and release the nitrogen in an ammonium form. In the distillation step, the digested sample is diluted with water, made alkaline with sodium hydroxide (NaOH), and the nitrogen is distilled off as ammonia (NH_3). The NH_3 is distilled into a solution of boric acid and is converted to ammonium and borate ion. The amount of NH_3 nitrogen in this solution is quantified by titration with a standardized hydrochloric acid (HCl) solution. The borate ion that is generated (proportional to the amount of nitrogen) is quantified by titrating with a 0.1 *N* HCl. The endpoint is detected via a methyl red indicator. A reagent blank is carried through the analysis and the volume of HCl titrant required for this blank is subtracted from each determination.

16.2.3 Chemicals

	CAS no.	Hazards
Boric acid (H_3BO_3)	10,043-35-3	
Bromocresol green	76-60-8	
Ethanol, 95%	64-17-5	Highly flammable
Hydrochloric acid, conc. (HCl)	7647-01-0	Corrosive
Methyl red	493-52-7	
Kjeldahl digestion tablets		Irritant
Potassium sulfate (K_2SO_4)	7778-80-5	
Cupric sulfate	7758-98-7	
Titanium dioxide (TiO_2)	13,463-67-7	
Sodium hydroxide (NaOH)	1310-73-2	Corrosive
Sulfuric acid, conc. (H_2SO_4)	7664-93-9	Corrosive
Tris (hydroxymethyl) aminomethane (THAM)	77-86-1	Irritant

16.2.4 Reagents

(** It is recommended that these solutions be prepared by a laboratory assistant before class.)

- Sulfuric acid (concentrated, N-free)
- Catalyst/salt mixture (Kjeldahl digestion tablets)
  - Contains potassium sulfate, cupric sulfate, and titanium dioxide

- (*Note:* There are several types of Kjeldahl digestion tablets that contain somewhat different chemicals.)
- Sodium hydroxide solution, 32%, w/v, NaOH in deionized distilled (dd) water **
  - Dissolve 1280 g sodium hydroxide (NaOH) pellets in ~3.5 L dd water. Cool. Add dd water to make up to 4.0 L.
- Boric acid solution **
  - In a 4-L flask, dissolve 160 g boric acid in ca. 2 L boiled, and still very hot, dd water. Mix and then add an additional 1.5 L of boiled, hot dd water. Cool to room temperature under tap water (caution: glassware may break due to sudden cooling) or leave overnight. When using the rapid procedure, the flask must be shaken occasionally to prevent recrystallization of the boric acid. Add 40 mL of bromocresol green solution (100 mg bromocresol green/100 mL ethanol) and 28 mL of methyl red solution (100 mg methyl red/100-mL ethanol). Dilute to 4 L with water and mix carefully. Transfer 25 mL of the boric acid solution to a receiver flask and distill a digested blank (a digested catalyst/salt/acid mixture). The contents of the flask should then be a neutral gray. If not, titrate with 0.1 N NaOH solution until this color is obtained. Calculate the amount of NaOH solution necessary to adjust the boric acid solution in the 4 L flask with the formula:

$$mL\ 0.1 N\ NaOH = \frac{(mL\ titer) \times (4000\ mL)}{(25\ mL)}$$

Add the calculated amount of 0.1 N NaOH solution to the boric acid solution. Mix well. Verify the adjustment results by distilling a new blank sample. Place adjusted solution into a bottle equipped with a 50 mL repipettor.

- Standardized HCl solution**
  - Dilute 3.33 mL conc. HCl to 4 L with dd water. Empty old HCl solution from the titrator reservoir and rinse three times with a small portion of the new HCl solution. Fill the titrator with the new HCl solution to be standardized. Using a volumetric pipet, dispense 10-mL aliquots of the THAM solution prepared as described below into 3 Erlenmeyer flasks (50 mL). Add 3–5 drops indicator (3 parts 0.1% bromocresol green in ethanol to 1 part of 0.2% methyl red in ethanol) to each flask and swirl. Titrate each solution with the HCl solution to a light pink endpoint. Note the acid

volume used and calculate the normality as described below.
Calculation to standardize HCl solution:

$$Normality = \frac{mL\ THAM \times THAM\ Normality}{average\ acid\ volume(AAV)} = \frac{20\ mL \times 0.01 N}{AAV}$$

Write the normality of the standardized HCl solution on the stock container.

- Tris (hydroxymethyl) aminomethane (THAM) Solution - (0.01 N).

  Place 2 g of THAM in a crucible. Leave in a drying oven (95 °C) overnight. Let cool in a desiccator. In a 1-L volumetric flask, dissolve 1.2114 g of oven-dried THAM in distilled water. Dilute to volume.

(*Note:* Alternatively, both standardized HCl and 4% boric acid can be purchased, and in this case, methyl red indicator (1% prepared in ethanol) is used.)

16.2.5 Hazards, Cautions, and Waste Disposal

Concentrated sulfuric acid is extremely corrosive; avoid breathing vapors and contact with skin and clothes. Concentrated sodium hydroxide is corrosive. Wear corrosion-resistant gloves and safety glasses at all times. Perform the digestion in an operating hood with an aspirating fume trap attached to the digestion unit. Allow samples to cool in the hood before removing the aspirating fume trap from the digestion unit. Otherwise, adhere to normal laboratory safety procedures. The waste of combined sulfuric acid and sodium hydroxide has been largely neutralized (check pH to ensure it is pH 3-9), so it can be discarded down the drain with a water rinse. However, for disposing any chemical wastes, follow good laboratory practices outlined by environmental health and safety protocols at your institution.

For safety and accuracy reasons, dispense the concentrated sulfuric acid from a bottle fitted with a repipettor (i.e., automatic dispenser). Fit the dispenser with a thin, semirigid tube to dispense directly into the Kjeldahl tube. Set the bottle with dispenser on a tray to collect spills.

16.2.6 Supplies

Food samples:

- Cheddar cheese
- Cheerios cereal
- Parmesan cheese
- Process cheese
- Rice baby cereal
- Raw almonds

Other supplies:

- Burettes (50 mL) if automatic titrator is not available
- Digestion tubes (2 for blank and 3 for each food sample)
- Erlenmeyer flasks, 250 mL (2 for blank and 3 for each food sample)
- Spatulas
- Weighing paper (Nitrogen free if available)

16.2.7 Equipment

- Analytical balance
- Automatic titrator, if available
- Kjeldahl digestion and distillation systems

16.2.8 Procedure

(Instructions are given for analysis in triplicate. Follow manufacturer's instructions for specific Kjeldahl digestion and distillation system used. Some instructions given here may be specific for one type of Kjeldahl system.)

16.2.8.1 Digestion

1. Weigh samples on weighing paper and record sample weight. Add the paper and sample to Kjeldahl tube. Prepare samples in duplicate.
   (a) 0.25 g for cheese samples and raw almonds
   (b) 1 g for baby cereal and Cheerios
2. Prepare two blank tubes for each food product. For the blank, add only the weighing paper to each tube.
3. Add 2 catalyst tablets and approx. 30 mL concentrated sulfuric acid to each tube (under the hood).
4. Place tubes into digestion block, cover, and allow to digest under the hood (~2 hr. at 420 °C).
5. Allow samples to digest completely; samples should be clear (but neon green/blue).
6. Take samples off digestion block and allow to cool with exhaust system turned on.

7. Carefully dilute digest with an appropriate volume of distilled water (50-100 mL). Swirl each tube.

16.2.8.2 Distillation

1. Follow appropriate procedure to start up distillation system.
2. Dispense appropriate volume of boric acid solution (with 2-3 drops methyl red, if boric acid solution does not already contain an indicator) into the receiving flask. Place receiving flask on distillation system. Make sure the tube coming from the distillation of the sample is submerged in the boric acid solution.
3. Put sample tube from Sect. 16.2.8.1 in place, making sure it is seated securely, and proceed with the distillation until completed. In this distillation process, a set volume of NaOH solution will be delivered to the tube, and a steam generator will distill the sample for a set period of time. Follow directions as instructed by the teaching assistant (32% NaOH should be added at 3:1 ratio, base to acid; in this case, ~ 90–120 mL of NaOH should be added to the glass tube). This should be tested beforehand by the teaching assistant to ensure adequate volume.
4. Upon completing distillation of one sample, proceed with a new sample tube and receiving flask.
5. After completing distillation of all samples, follow manufacturer's instructions to shut down the distillation unit.

16.2.8.3 Titration

1. Record the normality of the standardized HCl solution, as determined by the teaching assistant.
2. If using an automated pH meter titration system, follow manufacturer's instructions to calibrate the instrument. Put a magnetic stir bar in the receiver flask from Sect. 16.2.8.2 and place it on a stir plate. Keep the solution stirring briskly while titrating, but do not let the stir bar hit the electrode. Titrate each sample and blank to an endpoint pH of 4.2. Record volume of HCl titrant used.
3. If using a colorimetric endpoint, put a magnetic stir bar in the receiver flask, place it on a stir plate, and keep the solution stirring briskly while titrating. Titrate each sample and blank with the standardized HCl solution to the first faint gray color (or pink color if methyl red indicator is used). Record volume of HCl titrant used for the blank and the sample.

16.2.9 Data and Calculations

Calculate the percent nitrogen and the percent protein for each of your duplicate or triplicate food samples, then determine average values. The teaching assistant will provide the nitrogen conversion factor for each food sample based on literature values. The food sample you analyzed was not a dried sample. You need to use the moisture % (on wet weight basis, wwb) obtained from Chap. 13 results to calculate the protein % on a dry weight basis (dwb). Use the equations shown in Chap. 14, Sect. 14.3 to convert wwb to dwb.

$$\%N = \text{Normality HCl}^* \times \frac{\text{corrected acid vol.} (mL)^{**}}{\text{g of sample}}$$
$$\times \frac{14\,g\,N}{mol} \times 100$$

* Normality is in $mol/1000\,mL$

** Corrected acid vol. $= (mL\,std.acid\,for\,sample)$
$$- (mL\,std.for\,blank)$$

$$\%\text{Protein} = \%N \times \text{Protein Factor}$$

	Rep	Sample (g)	HCl (ml)	Nitrogen (%)	Nitrogen Conversion Factor	% Protein (wwb)	% Protein (dwb)
Blank	1	0		–	–	–	–
	2	0		–	–	–	–
	Av	0	Av				
Sample	1						
	2						
	3						
						Av	Av
						SD	SD
						CV	CV
						$\%E_{Rel}$	$\%E_{Rel}$

16.2.10 Questions

1. If the alkali pump timer on the distillation system was set to deliver 25 mL of 50% NaOH and 7 mL of concentrated H_2SO_4 was used to digest the sample, how many milliliters of the 50% NaOH is actually required to neutralize the amount of sulfuric acid used in the digestion? How would your results have been changed if the alkali pump timer malfunctioned and delivered only 15 mL of the 50% NaOH? (Molarity of conc. $H_2SO_4 = 18$).
2. Could phenolphthalein be used as an indicator in the Kjeldahl titration? Why or why not?
3. Describe the function of the following chemicals used in this determination:
   (a) Catalyst pellet.
   (b) Sulfuric acid.
   (c) Sodium hydroxide.
   (d) Boric acid.
   (e) Hydrochloric acid.
4. Why was it not necessary to standardize the boric acid solution?
5. Explain why different conversion factors were used for converting nitrogen to protein for the different products.
6. For each of the disadvantages of the Kjeldahl method, give another protein analysis method that overcomes (at least partially) that disadvantage.

16.3 NITROGEN COMBUSTION METHOD

16.3.1 Objective

Determine the protein content of the different food samples using the Dumas (Nitrogen combustion) method.

16.3.2 Principle of Method

The Dumas method quantifies nitrogen by combustion of the sample. Samples are weighed and combusted at high temperatures (700–1000 C, generally 900–950 C) which releases nitrogen (as a gas) and other combustion products (water, carbon dioxide, etc.). The combustion products are removed through various adsorbents and the nitrogen gas released is quantified using a thermal conductivity detector. Nitrogen content is determined using a calibration curve and then converted to % protein using a conversion factor that is based on the ratio of protein to nitrogen for that particular type of food.

16.3.3 Chemicals

	CAS no.	Hazards
Ethylenediaminetetraacetic acid (EDTA), disodium salt ($Na_2EDTA \cdot 2H_2O$)	60-00-4	Irritant

16.3.4 Hazards, Cautions, and Waste Disposal

Check instructions of manufacturer for any hazards, especially those associated with maintenance of instrument.

16.3.5 Supplies

Food samples:

- Cheddar cheese
- Cheerios cereal
- Parmesan cheese
- Process cheese
- Rice baby cereal
- Raw almonds

Other supplies:

- Gelatin sample capsules
- Spatulas

16.3.6 Equipment

- Analytical balance
- Nitrogen combustion unit

16.3.7 Procedure

1. Weigh appropriate sample amount into foil sample capsule. Record sample weight.
   (a) For cheese samples and raw almonds weigh approximately 0.12–0.13 g sample (in duplicate).
   (b) For baby cereal and Cheerios cereal, weigh 0.16–0.17 g sample (in duplicate).
2. Follow manufacturer instructions for analyzing the sample with the help of the teaching assistant. Before the lab, the teaching assistant will ensure adequate calibration. After completion of the analysis, also follow manufacturer's instruction for shutdown of the unit.
3. Data generated from Dumas will be % nitrogen.

16.3.8 Data and Calculations

Record the percent nitrogen content for each of your duplicate or triplicate samples. Calculate protein content from percent nitrogen data, and determine average percent protein. The teaching assistant will provide the nitrogen conversion factor for each food sample based on literature values. The food sample you analyzed was not a dried sample. You need to use the moisture % (on wet basis) obtained from Chap. 13 results to calculate the protein % on dry basis. Use the equations shown in Chap. 14, Sect. 14.3, to convert wwb to dwb.

	Rep	Sample (g)	Nitrogen (%)	Nitrogen Conversion Factor	% Protein (wwb)	% Protein (dwb)
Sample	1					
	2					
	3					
					Av	Av
					SD	SD
					CV	CV
					$\%E_{Rel}$	$\%E_{Rel}$

16.3.9 Questions

1. What are the advantages of the nitrogen combustion method compared to the Kjeldahl method?
2. Explain why EDTA or glycine can be used as standards to calibrate the nitrogen analyzer.
3. If you analyzed the food sample by both the Kjeldahl and nitrogen combustion methods, compare the results. What might explain any differences?
4. Using all the most accurate proximate values you obtained from Chaps.13, 14, 15, and 16, determine the % total carbohydrate in your sample on both dry and wet weight bases.

RESOURCE MATERIALS

AOAC International (2023) Official methods of analysis, 22nd edn, (On-line). Method 960.52 (Micro-Kjeldahl method) and Method 992.23 (Generic combustion method). AOAC International, Rockville, MD

Dee DR and Chang SKC (2024) Protein analysis. Ch. 18. In: Ismail BP, Nielsen SS (eds) Nielsen's food analysis, 6th edn. Springer, New York

Total Carbohydrate by Phenol-Sulfuric Acid Method

17

S. Suzanne Nielsen

Contents

17.1 INTRODUCTION

17.1.1 Background

The phenol-sulfuric acid method (Dubois et al. 1956) is a simple and rapid colorimetric method to determine total carbohydrates in a sample. The method detects virtually all classes of carbohydrates, including mono-, di-, oligo-, and polysaccharides. Although the method detects almost all carbohydrates, the absorptivity of the different carbohydrates varies. Thus, unless a sample is known to contain only one carbohydrate, the results must be expressed arbitrarily in terms of one carbohydrate.

In this method, the concentrated sulfuric acid breaks down any polysaccharides, oligosaccharides, and disaccharides to monosaccharides. Pentoses (5-carbon compounds) then are dehydrated to furfural, and hexoses (6-carbon compounds) to hydroxymethyl furfural. These compounds then react with phenol to produce a yellow-gold color. For products that are very high in xylose (a pentose), such as wheat bran or corn bran, xylose should be used to construct the standard curve for the assay and measure the absorption at 480 nm. For products that are high in hexose sugars, glucose is commonly used to create the standard curve, and the absorption is measured at 490 nm. The color for this reaction is stable for several hours, and the accuracy of the method is within ±2% under proper conditions.

Carbohydrates are the major source of calories in soft drinks, beer, and fruit juices, supplying 4 Cal/g carbohydrate. In this experiment, you will create a standard curve with a glucose standard solution, use it to determine the carbohydrate concentration of soft drinks and beer, then calculate the caloric content of those beverages.

S. S. Nielsen (✉)
Department of Food Science, Purdue University,
West Lafayette, IN, USA
e-mail: nielsens@purdue.edu

© The Author(s), under exclusive license to Springer Nature Switzerland AG 2024
B. P. Ismail, S. S. Nielsen (eds.), *Nielsen's Food Analysis Laboratory Manual*, Food Science Text Series,
https://doi.org/10.1007/978-3-031-44970-3_17

17.1.2 Reading Assignment

Huber KC, BeMiller JN (2024) Carbohydrate analysis, Ch. 19. In: Ismail BP, Nielsen SS (eds) Nielsen's food analysis, 6th edn. Springer, New York

17.1.3 Objective

Determine the total carbohydrate content of soft drinks and beers.

17.1.4 Principle of Method

Carbohydrates (simple sugars, oligosaccharides, polysaccharides, and their derivatives) react in the presence of strong acid and heat to generate furan derivatives that condense with phenol to form stable yellow-gold compounds that can be measured spectrophotometrically.

17.1.5 Chemicals

	CAS no.	Hazards
D-Glucose ($C_6H_{12}O_6$)	50-99-7	
Phenol (C_6H_6O)	108-95-2	Toxic
Sulfuric acid (H_2SO_4)	7664-93-9	Corrosive

17.1.6 Reagents

(**It is recommended that these solutions be prepared by the laboratory assistant before class.)

- Glucose std. solution, 100 mg/L**
- Phenol, 80% wt/wt in H_2O, 1 mL**
  Prepare by adding 20 g deionized distilled (dd) water to 80 g of redistilled reagent grade phenol (crystals).
- Sulfuric acid, concentrated

17.1.7 Hazards, Cautions, and Waste Disposal

Use concentrated H_2SO_4 and the 80% phenol solution with caution. Wear gloves and safety glasses at all times, and use good lab technique. The concentrated H_2SO_4 is very corrosive (e.g., to clothes, shoes, skin). The phenol is toxic and must be discarded as hazardous waste. Other waste not containing phenol likely may be put down the drain using a water rinse, but follow good laboratory practices outlined by environmental health and safety protocols at your institution.

17.1.8 Supplies

- Beer (lite and regular, of same brand)
- Bottle to collect waste
- Cuvettes (tubes) for spectrophotometer
- Erlenmeyer flask, 100 mL, for dd water
- Erlenmeyer flasks, 500 mL, for beverages (2)
- Gloves
- Mechanical, adjustable volume pipettors, 1000 μL and 100 μL (or 200 μL), with plastic tips
- Pasteur pipettes and bulb
- Parafilm®
- Pipette bulb or pump
- Repipettor (for fast delivery of 5-mL conc. H_2SO_4)
- Soft drinks (clear-colored, diet and regular, of same brand)
- Test tubes, 16–20-mm internal diameter (20)
- Test tube rack
- Volumetric flasks, 100 mL *or* 2 volumetric flasks, 1000 mL (4)
- Volumetric pipette, 5 mL
- Volumetric pipettes, 10 mL (2)

17.1.9 Equipment

- Spectrophotometer
- Vortex mixer
- Water bath, maintained at 25 °C

17.2 PROCEDURE

(Instructions are given for analysis in duplicate.)

1. Standard curve tubes: Using the glucose standard solution (100 mg glucose/L) and dd water as indicated in the table below, pipette aliquots of the glucose standard into clean test tubes (duplicates for each concentration) such that the tubes contain 0–100 μL of glucose (use 1000-μL mechanical pipettor to pipette samples), in a total volume of 2 mL. These tubes will be used to create a standard curve, with values of 0–100-μg glucose/2 mL. The 0-μg glucose/2 mL sample will be used to prepare the reagent blank.

	μg Glucose/2 mL					
	0	20	40	60	80	100
mL glucose stock solution	0	0.2	0.4	0.6	0.8	1.0
mL dd water	2.0	1.8	1.6	1.4	1.2	1.0

2. Record caloric content from label: You will analyze for total carbohydrate content: (1) a regular and diet soft drink of the same brand, and/or (2) a regular and lite beer

of the same brand. Before you proceed with the sample preparation and analysis, record the caloric content on the nutrition label of the samples you will analyze.

3. Decarbonate the beverages: With the beverages at room temperature, pour approximately 100 mL into a 500-mL Erlenmeyer flask. Shake gently at first (try not to foam the sample if it is beer) and continue gentle shaking until no observable carbon dioxide bubbles appear. If there is any noticeable suspended material in the beverage, filter the sample before analysis.

4. Sample tubes: So the sample tested will contain 20–100-µg glucose/2 mL, the dilution procedure and volumes to be assayed are given below.

		Dilution	Volume assayed (mL)
Soft drink	Regular	1:2000	1
	Diet	0	1
Beer	Regular	1:2000	1
	Lite	1:1000	1

Recommended dilution scheme for 1:2000 dilution:

(a) Pipette 5 mL of beverage into a 100-mL volumetric flask, and dilute to volume with dd water. Seal flask with Parafilm® and mix well (this is a 1:20 dilution). Then, pipette 1.0 mL of this 1:20 diluted beverage into another 100-mL volumetric flask. Dilute to volume with dd water. Seal flask with Parafilm® and mix well.

OR

(b) Pipette 1.0 mL of beverage into a 1000-mL volumetric flask, and dilute to volume with dd water. Seal flask with Parafilm® and mix well. Then, in a test tube, combine 1 mL of the 1:1000 diluted beverage and 1 mL dd water. Mix well.

Recommended dilution scheme for 1:1000 dilution:

(a) Pipette 10 mL of beverage into a 100-mL volumetric flask, and dilute to volume with dd water. Seal flask with Parafilm® and mix well (this is a 1:10 dilution). Then, pipette 1.0 mL of this 1:10 diluted beverage into another 100-mL volumetric flask. Dilute to volume with dd water. Seal flask with Parafilm® and mix well.

OR

(b) Pipette 1.0 mL of beverage into a 1000-mL volumetric flask, and dilute to volume with dd water. Seal flask with Parafilm® and mix well.

5. After dilution as indicated, pipette 1.0 mL of standard (Part 1) or sample (Part 4) into a test tube and add 1.0 mL of dd water. Repeat this for each standard and sample, preparing duplicates of each for analysis. Each tube will now have a volume of 2.0 mL.

6. Phenol addition: To each tube from Part 5, add 0.05 mL 80% phenol (use 100 or 200-µL mechanical pipettor). Mix on a Vortex test tube mixer.

7. H_2SO_4 addition: To each tube from Part 6, add 5.0 mL H_2SO_4. The sulfuric acid reagent should be added rapidly to the test tube. Direct the stream of acid against the liquid surface rather than against the side of the test tube in order to obtain good mixing. (These reactions are driven by the heat produced upon the addition of H_2SO_4 to an aqueous sample. Thus, the rate of addition of sulfuric acid must be standardized.) Mix on a Vortex test tube mixer. Let tubes stand for 10 min and then place in a 25 °C bath for 10 min (i.e., to cool them to room temperature). Vortex the test tubes again before reading the absorbance.

8. Reading absorbance: Wear gloves to pour samples from test tubes into cuvettes. Do not rinse cuvettes with water between samples. Zero the spectrophotometer with the standard curve sample that contains 0-µg glucose/2 mL (i.e., blank). Retain this blank sample in one cuvette for later use. Read absorbances of all other samples at 490 nm. Read your standard curve tubes from low to high concentration (i.e., 20 µg/2 mL up to 100 µg/2 mL), and then read your beverage samples. To be sure that the outside of the cuvettes is free of moisture and smudges, wipe the outside of the cuvette with a clean paper wipe prior to inserting it into the spectrophotometer for a reading.

9. Absorbance spectra: Use one of the duplicate tubes from a standard curve sample with an absorbance reading of 0.5–0.8. Determine the absorbance spectra from 450–550 nm by reading the tube at 10 nm intervals. Zero the spectrophotometer with the blank at each 10-nm interval.

17.3 DATA AND CALCULATIONS

1. Summarize your procedures and results for all standards and samples in the tables immediately below. Use the data for the standard curve samples in the first table to calculate the equation for the line, which is used to calculate the concentrations in the original samples reported in the second table.

   Standard Curve:

Sample identity	A_{490} msmt 1	A_{490} msmt 2	Avg.
Blank			
Std. 20 µg			
Std. 40 µg			
Std. 60 µg			
Std. 80 µg			
Std. 100 µg			

Samples:

Sample identity	A_{490}	μg glucose/2 mL	Dilution scheme	Glucose equivalent	
				$\mu g/mL$, original sample	g/L, original sample
Soft drink, reg.					
Soft drink, reg.					
Soft drink, diet					
Soft drink, diet					
Beer, reg.					
Beer, reg.					
Beer, lite					
Beer, lite					

Sample calculation for soft drink, regular:

Equation of the line : $y = 0.011x + 0.1027$

$$y = 0.648$$

$$x = 49.57\,\mu g\,/\,2\,mL$$

$$C_i = C_f\left(V_2\,/\,V_1\right)\left(V_4\,/\,V_3\right)$$

(See Chap. 3 in this laboratory manual, C_i = initial concentration; C_f = final concentration)

$$C_i = \left(49.57\ \mu g\,glucose\,/\,2\ mL\right)\times\left(2000\ mL\,/\,1\ mL\right)$$
$$\times\left(2\ mL\,/\,1\ mL\right) = 99140\ \mu g\,/\,mL$$
$$= 99.14\ mg\,/\,mL$$
$$= 99.14\ g\,/\,L$$

2. Construct a standard curve for your total carbohydrate determinations, expressed in terms of glucose (A_{490} versus μg glucose/2 mL). Determine the equation of the line for the standard curve.
3. Calculate the concentration of glucose in your soft drink samples and beer samples, in terms of (a) grams/liter and (b) g/12 fl. oz. (*Note:* 29.56 mL/fl. oz.)
4. Calculate the caloric content (based only on carbohydrate content) of your soft drink samples and beer samples in terms of Cal/12 fl. oz.

Sample		g Glucose/12 fl. oz.	Measured Cal/12 fl. oz.	Nutrition label Cal/12 fl. oz.
Soft drink	Regular			
	Diet			
Beer	Regular			
	Lite			

5. Plot the absorbance spectra obtained by measuring the absorbance between 450 and 550 nm.

Wavelength	450	460	470	480	490	500	510	520	530	540	550
Abs.											

17.4   QUESTIONS

1. What are the advantages, disadvantages, and sources of error for this method to determine total carbohydrates?
2. Your lab technician performed the phenol-H_2SO_4 analysis on food samples for total carbohydrates but the results showed low precision, and the values seemed a little high. The technician had used new test tubes (they had never been used, and were taken right from the cardboard box). What most likely caused these results? Why? Describe what happened.
3. If you started with a glucose standard solution of 10-g glucose/L, what dilution of this solution would be necessary such that you could pipette 0.20, 0.40, 0.60, 0.80, 1.0 mL of the diluted glucose standard solution into test tubes and add water to 2 mL for the standard curve tubes (20–100 μg/2 mL)? Show all calculations.
4. If you had not been told to do a 2000-fold dilution of a soft drink sample, and if you know the approximate carbohydrate content of regular soft drinks (US Department of Agriculture Nutrient Database for Standard Reference indicates ca. 3 g carbohydrate/fl. oz.), how could you have calculated the 2000-fold dilution was appropriate if you wanted to use 1 mL of diluted soft drink in the assay. Show all calculations.
5. How does your calculated value compare to the caloric content on the food label? Do the rounding rules for Calories explain any differences (See Nielsen 2024, Table 3.3)? Does the alcohol content (assume 4–5% alcohol at 7 Cal/g) of beer explain any differences?
6. Was it best to have read the absorbance for the standard curve and other samples at 490 nm? Explain why a wavelength in this region is appropriate for this reaction.

RESOURCE MATERIALS

Huber KC, BeMiller JN (2024) Carbohydrate analysis, Ch. 19. In: Ismail BP, Nielsen SS (eds) Nielsen's food analysis, 6th edn. Springer, New York

Dubois M, Gilles KA, Hamilton JK, Rebers PA, Smith F (1956) Colorimetric method for determination of sugars and related substances. Anal Chem 28: 350–356

Nielsen SS (2024) Nutrition labeling. Ch. 3. In: Ismail BP, Nielsen SS (eds). Nielsen's food analysis, 6th edn. Springer, New York

Vitamin C Determination by Indophenol Method

18

S. Suzanne Nielsen

Contents

18.1 INTRODUCTION

18.1.1 Background

Vitamin C is an essential nutrient in the diet, but it is easily reduced or destroyed by exposure to heat and oxygen during processing, packaging, and storage of food. The instability of vitamin C makes it more difficult to ensure an accurate listing of vitamin C content on the nutrition label.

The official method of analysis for vitamin C determination of juices is the 2,6-dichloroindophenol titrimetric method (AOAC Method 967.21). While this method is not official for other types of food products, it is sometimes used as a rapid, quality control test for a variety of food products, rather than the more time-consuming microfluorometric method (AOAC Method 984.26). The procedure outlined below is from AOAC Method 967.21.

18.1.2 Reading Assignment

AOAC International (2023) Official methods of analysis, 22th edn. (On-line). AOAC International, Rockville, MD

Pegg RB (2024) Vitamin analysis. Ch. 20. In: Ismail BP, Nielsen SS (eds) Nielsen's food analysis, 6th edn. Springer, New York

S. S. Nielsen (✉)
Department of Food Science, Purdue University,
West Lafayette, IN, USA
e-mail: nielsens@purdue.edu

18.1.3 Objective

Determine the vitamin C content of various orange juice products using the indicator dye 2,6-dichloroindophenol in a titration method.

18.1.4 Principle of Method

Ascorbic acid reduces the indicator dye to a colorless solution. At the endpoint of titrating an ascorbic acid-containing sample with dye, excess unreduced dye is a rose-pink color in the acid solution. The titer of the dye can be determined using a standard ascorbic acid solution. Food samples in solution then can be titrated with the dye and the volume for the titration used to calculate the ascorbic acid content.

18.1.5 Chemicals

	CAS no.	Hazards
Acetic acid (CH$_3$COOH)	64-19-7	Corrosive
Ascorbic acid	50-81-7	
2,6-Dichloroindophenol (DCIP) (sodium salt)	620-45-1	
Metaphosphoric acid (HPO$_3$)	37267-86-0	Corrosive
Sodium bicarbonate (NaHCO$_3$)	144-55-8	

18.1.6 Reagents

(**It is recommended that reagents be prepared by the laboratory assistant before class.)

- Ascorbic acid standard solution (prepare only at the time of use)**
    - Accurately weigh (on an analytical balance) approximately 50 mg ascorbic acid [preferably US Pharmacopeia (USP) Ascorbic Acid Reference Standard]. Record this weight. Transfer to a 50-mL volumetric flask. Dilute to volume *immediately before use* with the metaphosphoric acid-acetic acid solution (see below for preparation of this solution).
- Indophenol solution – dye**
    - To 50 mL deionized distilled (dd) water in a 150-mL beaker, add and stir to dissolve 42 mg sodium bicarbonate, then add and stir to dissolve 50 mg 2,6-dichloroindophenol sodium salt. Dilute mixture to 200 mL with dd water. Filter through fluted filter paper into an amber bottle. Close the bottle with a stopper or lid and store refrigerated until used.

- Metaphosphoric acid-acetic acid solution**
    - To a 250-mL beaker, add 100 mL dd water then 20 mL acetic acid. Add and stir to dissolve 7.5 g metaphosphoric acid. Dilute mixture to 250 mL with distilled water. Filter through fluted filter paper into a bottle. Close the bottle with a stopper or lid and store refrigerated until used.

18.1.7 Hazards, Precautions, and Waste Disposal

Preparation of reagents involves corrosives. Use appropriate eye and skin protection. Otherwise, adhere to normal laboratory safety procedures. Waste likely may be put down the drain using a water rinse, but follow good laboratory practices outlined by environmental health and safety protocols at your institution.

18.1.8 Supplies

(Used by students)

- Beaker, 250 mL
- Bottles, glass, 200–250 mL, one amber and one clear, both with lids or stoppers (2)
- Buret, 50 or 25 mL
- Erlenmeyer flasks, 50 mL (or 125 mL) (9)
- Fluted filter paper, two pieces
- Funnel, approx. 6–9 cm diameter (to hold filter paper)
- Funnel, approx. 2–3 cm diameter (to fill buret)
- Glass stirring rods (2)
- Graduated cylinder, 25 mL
- Graduated cylinder, 100 mL
- Orange juice samples
    - Use products processed and packaged in various ways (e.g., canned, reconstituted frozen concentrate, fresh squeezed, not-from-concentrate). Filter juices through cheesecloth to avoid problems with pulp when pipetting. Record from the nutrition label for each product the vitamin C content and percent of the Daily Value (if available).
- Pipette bulb or pump
- Ring stand
- Spatulas (3)
- Volumetric flask, 50 mL
- Volumetric flask, 200 mL
- Volumetric flask, 250 mL
- Volumetric pipettes, 2 mL (2)
- Volumetric pipette, 5 mL
- Volumetric pipette, 7 mL
- Volumetric pipette, 10 or 20 mL
- Weighing boats or paper

18.1.9 Equipment

- Analytical balance

18.1.10 Notes

The instructor may want to assign one or two types of orange juice samples to each student (or lab group) for analysis, rather than having all students analyze all types of orange juice samples. Quantities of supplies and reagents specified are adequate for each student (or lab group) to standardize the dye and analyze one type of orange juice sample in triplicate.

18.2 PROCEDURE

(Instructions are given for analysis in triplicate.)

18.2.1 Standardization of Dye

1. Pipette 5 mL metaphosphoric acid-acetic acid solution into each of three 50-mL Erlenmeyer flasks.
2. Add 2.0 mL ascorbic acid standard solution to each flask.
3. Using a funnel, fill the buret with the indophenol solution (dye) and record the initial buret reading.
4. Place the Erlenmeyer flask under the tip of the buret. Slowly add indophenol solution to standard ascorbic acid solution until a light but distinct rose-pink color persists for >5 s (takes about 15–17 mL). Swirl the flask as you add the indophenol solution.
5. Note the final buret reading and calculate the volume of dye used.
6. Repeat Steps 3–5 for the other two standard samples. Record the initial and final buret readings and calculate the volume of dye used for each sample.
7. Prepare blanks: Pipette 7.0 mL metaphosphoric acid-acetic acid solution into each of three 50-mL Erlenmeyer flasks. Add to each flask a volume of distilled water approximately equal to the volume of dye used above (i.e., average volume of dye used to titrate three standard samples).
8. Titrate the blanks in the same way as Steps 3–5 above. Record initial and final buret readings for each titration of the blank, and then calculate the volume of dye used.

18.2.2 Analysis of Juice Samples

1. Pipet into each of three 50-mL Erlenmeyer flasks 5 mL metaphosphoric acid-acetic acid solution and 2 mL orange juice.

2. Titrate each sample with the indophenol dye solution (as you did in Sect. 18.2.1, Steps 3–5 above) until a light but distinct rose-pink color persists for >5 s.
3. Record the initial and final readings, and calculate the difference to determine the amount of dye used for each titration.

18.3 DATA AND CALCULATIONS

18.3.1 Data

	Rep	Buret start (mL)	Buret end (mL)	Vol. titrant (mL)
Ascorbic acid standards	1			
	2			
	3			
				$\bar{X} =$
Blank	1			
	2			
	3			
				$\bar{X} =$
Sample	1			
	2			
	3			
				$\bar{X} =$

18.3.2 Calculations

1. Using the data obtained in standardization of the dye, calculate the titer using the following formula:

$$\text{Titer} = F = \frac{\text{mg ascorbic acid in volume of standard solution titrated}^{**}}{(\text{average mL dye used to titrate standards}) - (\text{average mL dye used to titrate blank})}$$

** mg ascorbic acid in volume of standard solution titrated
$= (\text{mg of ascorbic acid} / 50 \, \text{mL}) \times 2 \, \text{mL}$

2. Calculate the ascorbic acid content of the juice sample in mg/mL, using the equation that follows and the volume of titrant for each of your replicates. Calculate the mean and standard deviation of the ascorbic acid content for your juice (in mg/mL). Obtain from other lab members the mean ascorbic acid content (in mg/mL) for other types of juice. Use these mean values for each type of juice to express the vitamin C content of the juice samples as milligrams ascorbic acid/100 mL and as milligrams ascorbic acid/8 fl. oz. (29.56 mL/fl. oz.).

$$\text{mg ascorbic acid} / \text{mL} = (X - B) \times (F / E) \times (V / Y)$$

where:

X = mL for sample titration
B = average mL for sample blank titration
F = titer of dye (= mg ascorbic acid equivalent to 1.0 mL indophenol standard solution)
E = mL assayed (= 2 mL)
V = volume of initial assay solution (= 7 mL)
Y = volume of sample aliquot titrated (= 7 mL)

Ascorbic acid (AA) content for replicates of orange juice sample:

Replicate	mg AA/mL
1	
2	
3	
	$\bar{X}$ =
	SD =

Summary of ascorbic acid (AA) content of orange juice samples:

Sample Identity	mg AA/mL	mg AA/100 mL	mg AA/8 fl. oz.
1			
2			
3			
4			

Example calculation:

Weight of ascorbic acid used = 50.2 mg

Average volume of titrant used:

Ascorbic acid standards = 15.5 mL

Blanks = 0.10 mL

Volume of titrant used for orange juice sample = 7.1 mL

$$\text{Titer} = F = \left[(50.2\,\text{mg}/50\,\text{mL}) \times 2\,\text{mL}\right]/(15.5\,\text{mL} - 0.10\,\text{mL})$$
$$= 0.130\,\text{mg}/\text{mL}$$

$$\text{mg ascorbic acid}/\text{mL} = (7.1\,\text{mL} - 0.10\,\text{mL})$$
$$\times (0.130\,\text{mg}/2\,\text{mL}) \times (7\,\text{mL}/7\,\text{mL})$$
$$= 0.455\,\text{mg}/\text{mL}$$

$$0.455\,\text{mg}/\text{mL} = 45.4\,\text{mg}/100\,\text{mL}$$

$$0.455\,\text{mg ascorbic acid}/\text{mL juice} \times 29.56\,\text{mL}/\text{fl.oz.} \times 8\,\text{fl.oz.}$$
$$= 107.6\,\text{mL ascorbic acid}/8\,\text{fl.oz.}$$

18.4 QUESTIONS

1. By comparing results obtained for various orange juice products, did heat and/or oxygen exposure during processing and storage of the samples analyzed seem to affect the vitamin C content?
2. How do the results you have available for the juice samples analyzed compare to (1) values listed on the nutrition label (if available) for the same juice product and (2) values in the US Department of Agriculture Nutrient Database for Standard Reference (web address: http://ndb.nal.usda.gov/)? For the nutrition label values, convert percent of Daily Value to mg/8 fl. oz., given that the Daily Value for vitamin C is 90 mg. Why might the vitamin C content determined for a specific orange juice product not match the value as calculated from percent of Daily Value on the nutrition label?

Ascorbic acid content of orange juices (mg AA/8 fl. oz.):

Sample identity	Lab values	USDA database	Nutrition label
1			
2			
3			
4			

3. Why was it necessary to standardize the indophenol solution?
4. Why was it necessary to titrate blank samples?
5. Why might the vitamin C content as determined by this method be underestimated in the case of the heat-processed juice samples?

RESOURCE MATERIALS

AOAC International (2023) Official methods of analysis, 22th edn. (Online). AOAC International, Rockville, MD
Pegg RB (2024) Vitamin analysis. Ch. 20. In: Ismail BP, Nielsen SS (eds) Nielsen's food analysis, 6th edn. Springer, New York

Water Hardness Testing by Complexometric Determination of Calcium

S. Suzanne Nielsen

Contents

19.1 INTRODUCTION

19.1.1 Background

Ethylenediaminetetraacetate (EDTA) complexes with numerous mineral ions, including calcium and magnesium. This reaction can be used to determine the amount of these minerals in a sample by a complexometric titration.

Endpoints in the titration are detected using indicators that change color when they complex with mineral ions. Calmagite and eriochrome black T (EBT) are such indicators that change from blue to pink when they complex with calcium and magnesium. In the titration of a mineral-containing solution with EDTA, the solution turns from pink to blue at the endpoint with either indicator. The pH affects a complexometric EDTA titration in several ways and must be carefully controlled. A major application of EDTA titration is testing the hardness of water, for which the method described is an official one (Standard Methods for the Examination of Water and Wastewater, Method 2340C; AOAC Method 920.196).

S. S. Nielsen (✉)
Department of Food Science, Purdue University,
West Lafayette, IN, USA
e-mail: nielsens@purdue.edu

Hardness of water also can be tested by a more rapid test strip method. Such test strips are available from various companies. The strips contain EDTA and an indicator chemical to cause a color change when the calcium and magnesium in water react with the EDTA.

19.1.2 Reading Assignment

Ward RE (2024) Traditional methods for mineral analysis. Ch. 21. In: Ismail BP Nielsen SS (eds) *Nielsen's food analysis*, 6th edn. Springer, New York.

19.1.3 Objective

Determine the hardness of water by EDTA titration and with Quantab® test strips.

19.2 EDTA TITRIMETRIC METHOD FOR TESTING HARDNESS OF WATER

19.2.1 Principle of Method

Ethylenediaminetetraacetic acid (EDTA) forms a stable 1:1 complex with calcium or magnesium at pH 10. The metal ion indicators, calmagite and eriochrome black T (EBT), are pink when complexed to metal ions but blue when no metal ions are complexed to them. The indicators bind to metal ions less strongly than does EDTA. When the indicator is added to a solution containing metal ions, the solution becomes pink. When EDTA is added as titrant to the mineral-containing sample, metal ions preferentially complex with the EDTA, leaving the indicator without a metal ion to complex. When enough EDTA has been titrated to complex with all the metal ions present, the indicator appears blue. This blue color is the endpoint of the titration. The volume and concentration of the EDTA in the titration are used to calculate the concentration of calcium in the sample, which is expressed as mg calcium carbonate/liter. Stoichiometry of the reaction is 1 mol of calcium complexing with 1 mol of EDTA.

19.2.2 Chemicals

	CAS no.	Hazards
Ammonium chloride (NH_4Cl)	12125-02-9	Harmful
Ammonium hydroxide (NH_4OH)	1336-21-6	Corrosive, dangerous for the environment
Calcium carbonate ($CaCO_3$)	471-34-1	

	CAS no.	Hazards
Calmagite [3-Hydroxy-4-(6-hydroxy-*m*-tolylazo) Naphthalene-1-sulfonic acid]	3147-14-6	
Ethylenediaminetetraacetic acid, disodium salt ($Na_2EDTA \cdot 2H_2O$)	60-00-4	Irritant
Hydrochloric acid, concentrated (HCl)	7647-01-0	Corrosive
Magnesium chloride, hexahydrate ($MgCl_2 \cdot 6H_2O$)	7791-18-6	
Magnesium sulfate, heptahydrate ($MgSO_4 \cdot 7H_2O$)	10034-99-8	

19.2.3 Reagents

(**It is recommended that these solutions be prepared by the laboratory assistant before class.)

- Buffer Solution**

    Dissolve 16.9 g NH_4Cl in 143 mL concentrated NH_4OH. In 50 mL deionized distilled (dd) water, dissolve 1.179 g $Na_2EDTA \cdot 2H_2O$ (analytical reagent grade) and either 780 mg $MgSO_4$. $7H_2O$ *or* 644 mg $MgCl_2$. $6H_2O$. Combine these two solutions with mixing and dilute to 250 mL with dd water. Store in tightly stoppered Pyrex or plastic bottle to prevent loss of ammonia (NH_3) or pickup of carbon dioxide (CO_2). Dispense this buffer solution with a repipette system. Discard buffer when 1–2 mL added to a sample fails to give pH 10.0 ± 0.1 at the endpoint of the titration.

- Calcium standard solution, 1.00 mg $CaCO_3$/mL** (modified from official method; omit use of methyl red indicator)

    Use primary standard or special reagent that is low in heavy metals, alkalis, and magnesium. Dry $CaCO_3$ at 100 °C for 24 hr. Accurately weigh ca. 1.0 g $CaCO_3$. Transfer to a 500-mL Erlenmeyer flask. Place a funnel in the neck of the flask, and add HCl (1:1, conc. HCl:H_2O) a little at a time, until all the $CaCO_3$ has dissolved (make sure all the $CaCO_3$ in the neck of the flask has been washed down with HCl). Add 200 mL dd water and boil a few minutes to expel CO_2. Cool. Adjust to pH 3.8 with 3 *M* NH_4OH or HCl (1: 1, conc. HCl: H_2O), as required. Transfer quantitatively to a 1-L volumetric flask, and dilute to volume with dd water (1 mL = 1.00 mg $CaCO_3$).

- EDTA Standard Solution, 0.01 *M*

    Weigh 3.723 g $Na_2EDTA \cdot 2H_2O$. Dilute to 1 L with dd water. Store in polyethylene (preferable) or borosilicate glass bottles. Standardize this solution using the calcium standard solution as described in the Procedure.

- Hydrochloric acid, 1: 1 with water**

  To 10 mL of dd water, add 10 mL concentrated HCl. Mix carefully.
- Calmagite**

  Dissolve 0.10 g calmagite in 100 mL dd water. Use 1 mL per 30 mL solution to be titrated. Put in a bottle with eye dropper.

19.2.4 Notes

In this experiment, calmagite will be used as the indicator dye rather than EBT. Unlike EBT, calmagite is stable in aqueous solution. Calmagite gives the same color change as EBT, but with a sharper endpoint.

To give a satisfactory endpoint, magnesium ions must be present. To ensure this, a small amount of neutral magnesium salt is added to the buffer.

The specified pH of 10.0 + 0.1 is a compromise situation. With increasing pH, the sharpness of the endpoint increases. However, at high pH, the indicator dye changes color and there is the risk of precipitating calcium carbonate ($CaCO_3$) or magnesium hydroxide. The tendency toward $CaCO_3$ precipitation is the reason for the titration duration time limit of 5 min.

Fading or indistinct endpoints can be caused by interference from some metal ions. Certain inhibitors can be added before titration to reduce this interference, but the inhibitors specified are toxic (i.e. sodium cyanide) or malodorous. Magnesium salt of 1,2-cyclohexanediaminetetraacetic acid (MgCDTA), which selectively complexes heavy metals, may be substituted for these inhibitors. However, for samples with high concentrations of heavy metals, a non-EDTA method is recommended. In this experiment, inhibitors or MgCDTA will not be used.

19.2.5 Hazards, Precautions, and Waste Disposal

Adhere to normal laboratory safety procedures. Wear gloves and safety glasses at all times. The buffer solution, which contains ammonium hydroxide, should be disposed of as hazardous waste. Other wastes likely may be put down the drain using a water rinse, but follow good laboratory practices outlined by environmental health and safety protocols at your institution.

19.2.6 Supplies

- Buret, 25 or 50 mL
- Erlenmeyer flasks, 125 mL (9)

- Funnel (to fill buret)
- Graduated cylinders, 25 mL (3)
- Graduated cylinder, 50 mL
- Graduated cylinder of larger volumes may be necessary; for example, 100 mL or larger; size to be determined by trial in Sect. 19.2.8.2
- Mechanical pipettor, 1000 μL, with plastic tips
- Pasteur pipette and bulb
- Spatula
- Volumetric flask, 1000 mL
- Volumetric pipette, 10 mL
- Weighing paper/boat

19.2.7 Equipment

- Analytical balance
- Drying oven, 100 °C
- Hot plate
- pH meter

19.2.8 Procedure

(Modified from Method 2340 Hardness, "Standard Methods for the Examination of Water and Wastewater", 24th ed.) (Instructions are given for analysis in triplicate.)

19.2.8.1 Standardization of EDTA Solution

1. Pipette 10 mL of calcium standard solution into each of three 125-mL Erlenmeyer flasks.
2. Adjust to pH 10.0 ± 0.05 with buffer solution. (If possible, do this pH adjustment with the buffer in an operating hood, due to its odor.) As necessary, use the HCl solution (1: 1) in pH adjustment.
3. Add 1 mL of calmagite to each flask, then titrate each flask with EDTA solution slowly, with continuous stirring, until last reddish tinge disappears, adding last few drops at 3–5 s intervals. Color at endpoint is blue in daylight and under daylight fluorescent lamp. Color may first appear lavender or purple, but will then turn to blue. Complete titration within 5 min from time of buffer addition.
4. Record the volume of EDTA solution used for each titration.

19.2.8.2 Titration of Water Sample

1. Dilute 25 mL tap water sample (or such volume as to require <15 mL titrant) to ca. 50 mL with dd water in 125-mL Erlenmeyer flask. For tap distilled water, test 50 mL, without dilution. Prepare samples in triplicate. [Official method recommends the following: For water of

low hardness (<5 mg/L), use 100–1000 mL specimen, proportionately larger amounts of reagents, micro-buret, and blank of distilled water equal to specimen volume.]

2. Adjust pH to 10 ± 0.05 as described in Sect. 19.2.8.1, Step 2.
3. Titrate each sample with EDTA standard solution slowly, as described in Sect. 16.2.8.1, Step 3, for standardization of EDTA solution.
4. Record the volume of EDTA solution used for each titration.

19.2.9 Data and Calculations

Calculate molarity of calcium standard solution:

$$\text{Molarity of calcium solution} = \frac{\text{g}\,CaCO_3}{(100.09\,\text{g}/\text{mol})(\text{liter solution})}$$

$$= \text{mol calcium}/\text{L}$$

Standardization of EDTA solution:

Rep	Buret start (mL)	Buret start (mL)	Volume titrant (mL)	Molarity
1				
2				
3				
				$\bar{X} =$
				SD =

Calculate molarity of EDTA solution:

$$\text{mol calcium} = \text{mol EDTA}$$

$$M_1 V_1 = M_2 V_2$$

$$\left(M_{\text{Ca solution}}\right)\left(V_{\text{Ca solution,L}}\right) = \left(M_{\text{EDTA solution}}\right)\left(V_{\text{EDTA solution,L}}\right)$$

Solve for $M_{\text{EDTA solution}}$

Titration of water sample with EDTA solution:

Rep	Buret start (mL)	Buret start (mL)	Volume titrant (mL)	g Ca/L	mg CaCO₃/L
1					
2					
3					
				$\bar{X} =$	$\bar{X} =$
				SD =	SD =

Calcium content of water sample (g Ca/L and g CaCO₃/L):

$$\text{mol calcium} = \text{mol EDTA}$$

$$M_1 V_1 = M_2 V_2$$

$$\left(M_{\text{Ca in sample}}\right)\left(V_{\text{sample,liter}}\right) = \left(M_{\text{EDTA solution}}\right)$$
$$\left(V_{\text{EDTA solution used in titration,L}}\right)$$

Solve for $M_{\text{Ca in sample}}$:

$$M_{\text{Ca in sample}}\,x\,40.085\,\text{g Ca}/\text{mol} = \text{g Ca}/\text{L}$$

$$(\text{g Ca}/\text{L})(100.09\,\text{g}\,CaCO_3/40.085\,\text{g Ca})$$
$$x(1000\,\text{mg}/\text{g}) = \text{mg}\,CaCO_3/\text{L}$$

19.2.10 Questions

1. If a sample of water is thought to have a hardness of approximately 250 mg/L CaCO₃, what size sample (i.e., how many mL) would you use so that you would use approximately 10 mL of your EDTA solution?
2. Why were you asked to prepare the CaCl₂ solution by using CaCO₃ and HCl rather than just weighing out CaCl₂?
3. In this EDTA titration method, would overshooting the endpoint in the titration cause an over-or underestimation of calcium in the sample? Explain your answer.

19.3   TEST STRIPS FOR WATER HARDNESS

19.3.1 Note

All information given is for AquaChek test strips, from Environmental Test Systems, Inc., a HACH Company, Elkhart, IN. Other similar test strips could be used. Any anion (e.g., magnesium, iron, copper) that will bind the EDTA may interfere with the AquaChek test. Very strong bases and acids also may interfere.

19.3.2 Principle of Method

The test strips have a paper, impregnated with chemicals, that is adhered to polystyrene for ease of handling. The major chemicals in the paper matrix are calmagite and EDTA, and minor chemicals are added to minimize reaction time, give long-term stability, and maximize color distinction between levels of water hardness. The strips are dipped into the water to test for total hardness caused by calcium and magnesium. The calcium displaces the magnesium bound to EDTA, and the released magnesium binds to calmagite, causing the test strip to change color.

19.3.3 Chemicals

	CAS no.	Hazards
Calcium carbonate (CaCO₃)	471-34-1	Harmful
Calmagite	3147-14-7	

	CAS no.	Hazards
Ethylenediaminetetraacetic acid, disodium salt ($Na_2EDTA \cdot 2H_2O$)	60-00-4	Irritant
Hydrochloric acid, concentrated (HCl)	7647-01-0	Corrosive
Other proprietary chemicals in test strip		

19.3.4 Reagents

(**It is recommended that this solution be prepared by the laboratory assistant before class.)

• Calcium standard solution, 1.000 mg $CaCO_3$/mL**
   Prepare as described in Sect. 19.2.3, using $CaCO_3$ and concentrated HCl.

19.3.5 Hazards, Precautions, and Waste Disposal

No precautions are needed in use of the test strip. Adhere to normal laboratory safety procedures. Wastes likely may be put down the drain using a water rinse, but follow good laboratory practices outlined by environmental health and safety protocols at your institution.

19.3.6 Supplies

• AquaChek® Test Strips (Environmental Test Systems, Inc., a HACH Company, Elkhart, IN)
• Beakers, 100 mL (2)

19.3.7 Procedure

(*Note:* Test the same standard calcium solution as used in Sect. 19.2.8.1 and the same tap water and distilled was as used in Sect. 19.2.8.2.)

1. Dip the test strip into a beaker filled with water or the standard calcium solution. Follow instructions on strip about how to read it, relating color to ppm $CaCO_3$.
2. Convert ppm $CaCO_3$ as determined with the test strips to mg $CaCO_3$/L and g Ca/L.

19.3.8 Data and Calculations

Sample	Rep (ppm $CaCO_3$)			Rep (mg $CaCO_3$/L)			Rep (g Ca/L)		
	1	2	3	1	2	3	1	2	3
Tap water									
Tap distilled water									
Standard Ca solution									

19.3.9 Question

1. Compare and discuss the accuracy and precision of the EDTA titration and test strip methods to measure calcium carbonate contents of the water samples and the calcium standard solution.

RESOURCE MATERIALS

American Public Health Association (APHA), American Water Works Association, Water Environment Federation (2023). Standard methods for the examination of water and wastewater, 24th edn. Lipps WC, Braun-Howard EB, Baxter TE (eds) APHA Press, Washington, DC.

Ward RE (2024) Traditional methods for mineral analysis. Ch. 21. In: Ismail BP, Nielsen SS (eds) Nielsen's food analysis, 6th edn. Springer, New York.

Phosphorus Determination by Murphy-Riley Method

20

Young-Hee Cho and S. Suzanne Nielsen

Contents

20.1 INTRODUCTION

20.1.1 Background

Phosphorus is one of the important minerals found in foods. Murphy-Riley method, a dry ashing colorimetric method, has been widely used to measure the phosphorus content in natural waters as well as in foods. In this procedure, it is necessary to ash the sample prior to analysis. The method described here is applicable to most foods following ashing. If the food sample has a low magnesium content, several milliliters of saturated $Mg(NO_3)_2 \cdot 6H_2O$ in ethanol should be added to the sample prior to ashing to prevent volatilization and loss of phosphorus at the high temperatures used in ashing.

20.1.2 Reading Assignment

Ward RE (2024) Traditional methods for mineral analysis. Ch. 21, In: Ismail BP, Nielsen SS (ed) *Nielsen's food analysis*, 6th edn. Springer, New York

20.1.3 Objective

Determine the phosphorus content of milk using a colorimetric method, with the Murphy and Riley reagent.

20.1.4 Principle of Method

Ammonium molybdate reacts with phosphorus to form phosphomolybdate. This complex is then reduced by ascorbic acid with antimony serving as a catalyst. The reduced phosphomolybdate complex has an intense blue color. The maximum absorptivity of this complex is at approximately 880 nm. This

Y.-H. Cho (✉) · S. S. Nielsen
Department of Food Science, Purdue University,
West Lafayette, IN, USA
e-mail: cho173@purdue.edu; nielsens@purdue.edu

wavelength is above the operating range of most spectrophotometers. The absorptivity of the molybdate complex is great enough in the region of 600–700 nm, however, to allow determination of the phosphorus content of most foods at any wavelength within this region. Most spectrophotometers have a maximum usable wavelength of 600–700 nm.

20.1.5 Chemicals

	CAS no.	Hazard(s)
Ammonium molybdate	13106-76-8	Irritant
Potassium antimonyl tartrate	28300-74-5	Harmful
Ascorbic acid	50-81-7	
Monopotassium phosphate	7778-77-0	
Sulfuric acid	7664-93-9	Corrosive

20.1.6 Reagents

(**It is recommended that these solutions be prepared by the laboratory assistant before class. Glassware cleaned by phosphate-free detergents and double-distilled or distilled deionized water must be used in all dilutions and preparation of all reagents.)

- Ammonium molybdate**
  Dissolve 48 g of ammonium molybdate to 1000 mL in a volumetric flask.
- Potassium antimonyl tartrate**
  Dissolve 1.10 g of potassium antimonyl tartrate to 1000 mL in a volumetric flask.
- Ascorbic acid**
  Prepare fresh daily. Dissolve 2.117 g to 50 mL in a volumetric flask.
- Phosphorus stock solution and working standard solution**
  Dissolved 0.6590 g of dried (105 °C, 2h) monopotassium phosphate (KH_2PO_4) to 1000 mL in a volumetric flask. This solution contains 0.150 mg phosphorus/mL (i.e., 150 µg/mL). To prepare a working standard solution, dilute 10 mL of this solution to 100 mL in a volumetric flask. The solution now contains 15 µg phosphorus/mL.
- 2.88 N Sulfuric acid (H_2SO_4)**
  Prepare 200 mL of 2.88 N sulfuric acid (H_2SO_4) using concentrated sulfuric acid (H_2SO_4) and double-distilled or distilled deionized water. Always add concentrated acid to water, not water to concentrated acid. Do not use a mechanical pipette to pipette concentrated sulfuric acid and 2.88 N sulfuric acid, since this would corrode the pipettor.
- Murphy and Riley reagent**
  Combine 14 mL of 2.88 N H_2SO_4, 2 mL of ammonium molybdate solution, and 2 mL of potassium antimonyl tartrate solution in a 50 mL Erlenmeyer flask.

20.1.7 Hazards, Precautions, and Waste Disposal

Adhere to normal laboratory safety procedures. Wear gloves, lab coat, and safety glasses at all times. Care should be exercised in pipetting any solution that contains antimony since it is a toxic compound. Waste containing antimony, sulfuric acid, and molybdate must be discarded as hazardous waste. Other wastes likely may be put down the drain using a water rinse, but follow good laboratory practices outlined by environmental health and safety protocols at your attention.

20.1.8 Supplies

- Crucible (preheated at 550 °C for 24 hr)
- Glass funnel
- Glass stirring rods
- Kimwipes
- Mechanical adjustable volume pipettes, 1000 µL with pipette tips
- Nonfat liquid milk, 5 g
- Repipettor (for fast delivery of 2 mL H_2SO_4)
- Test tubes (13 × 100 mm)
- Volumetric flask, 250 mL
- Whatman No. 41 ashless filter paper

20.1.9 Equipment

- Analytical balance, 0.1 mg sensitivity
- Forced draft oven
- Hot plate
- Muffle furnace
- Spectrophotometer
- Vortex mixer

20.2   PROCEDURE

(Instructions are given for analysis in duplicate.)

20.2.1 Ashing

1. Predry crucible at 550°C for 24 h and weigh accurately.
2. Accurately weigh ca. 5 g of sample in the crucible.
3. Heat on the hot place until a majority of water has been evaporated, to avoid splattering and sample loss if placed directly in the oven or muffle furnace.
4. Dry in a forced draft oven at 100°C for 3 hr.
5. Ash in a muffle furnace at 550°C for 18–24 hr.

20.2.2 Phosphorus Measurement

1. Preparation of Murphy-Riley reagent: Add 2 mL of ascorbic acid solution to preprepared 18 mL of Murphy-Riley reagent. Swirl to mix.

2. Standards: Prepare phosphorus standards using working standard solution (15 µg phosphorus/mL) and water as indicated in the table below. Pipet aliquots of the phosphorus standard into clean test tubes (duplicated for each concentration), and add water so each test tube contains 4 mL. Add 1 mL of Murphy-Riley (M&R) reagent to each tube. The final volume of each tube should be 5 mL. Mix with a vortex mixer. Allow the color to develop at room temperature for 10–20 minutes. Read the absorbance at 700 nm.

µg P/5 mL	15 µg P/mL	Vol. water (mL)	Vol. M&R (mL)
0 (blank)	0	4.0	1.0
1.5	0.1	3.9	1.0
3.0	0.2	3.8	1.0
4.5	0.3	3.7	1.0
6.0	0.4	3.6	1.0

3. Sample analysis: (*Note*: *Do not* use an automatic pipettor to obtain 2 mL of the sulfuric acid. This would corrode the pipettor.) Analyze one ashed sample. Carefully moisten ash in crucible with H_2O and then add 2 mL of 2.88 N H_2SO_4. Filter through ashless filter paper into a 250 mL volumetric flask. Thoroughly rinse crucible, ash, and filter paper. Dilute to volume with distilled water and mix. Analyze in duplicate by combining 0.2 mL of sample, 3.8 mL of distilled water, and 1 mL of Murphy and Riley reagent. Mix and allow to react for 10–20 minutes. Absorbance is read at 700 nm as per the phosphorus standard solutions.

20.3 DATA AND CALCULATIONS

1. Report data obtained, giving A_{700} of the duplicate standards and sample and the average A_{700}.

µg P/tube (5mL)	Absorbance (700 nm)		
	1	2	Average
1.5			
3.0			
4.5			
6.0			
Sample			

2. Construct a standard curve for your phosphorus determination, expressed in terms of phosphorus (A_{700} vs. µg phosphorus/5 mL). Determine the equation of the line for the phosphorus standard curve.

3. Calculate the concentration of phosphorus in your milk sample expressed in terms of mg phosphorus/100 g sample. Show all calculations.

Sample calculation for milk sample:

Weight of milk sample: 5.0150 g

$A_{700} = 0.394$

Equation of the line: $y = 0.12x + 0.0066$

$$y = 0.394$$

$$x = 3.2\,\mu g\,P\,/\,5\,mL$$

P content of milk sample

$$= \frac{3.2\,\mu g\,P}{5\,mL} \times \frac{5\,mL}{0.2\,mL} \times \frac{250\,mL}{5.0150g}$$

$$= \frac{820\,\mu g\,P}{g} = \frac{82\,mg\,P}{100g}$$

20.4 QUESTIONS

1. How does your value compare to the literature value (US Department of Agriculture Nutrient Database for Standard Reference) for the phosphorus content of milk?

2. If you had not been told to use a 250 mL volumetric flask to prepare your ashed milk sample, how could you have calculated the dilution scheme was appropriate if you wanted to use 0.2 mL ashed milk sample in the assay. The US Department of Agriculture Nutrient Database for Standard Reference indicates ca. 101 mg phosphorus/100g. Show all calculations.

3. What are the possible sources of error using this method to determine the phosphorus content of milk?

4. What was the function of ascorbic acid in the assay?

RESOURCE MATERIALS

Murphy J, Riley JP (1962) A modified single solution method for the determination of phosphate in natural waters. Anal. Chim. Acta 27:31–36

Ward RE (2024) Traditional methods for mineral analysis. Ch. 21, In: Ismail BP, Nielsen SS (ed) Nielsen's food analysis, 6th edn. Springer, New York

Sodium Determination Using Ion-Selective Electrodes, Mohr Titration, and Test Strips

21

S. Suzanne Nielsen

Contents

S. S. Nielsen (✉)
Department of Food Science, Purdue University,
West Lafayette, IN, USA
e-mail: nielsens@purdue.edu

© The Author(s), under exclusive license to Springer Nature Switzerland AG 2024
B. P. Ismail, S. S. Nielsen (eds.), *Nielsen's Food Analysis Laboratory Manual*, Food Science Text Series,
https://doi.org/10.1007/978-3-031-44970-3_21

21.1 INTRODUCTION

21.1.1 Background

Sodium content of foods can be determined by various methods, including an ion-selective electrode (ISE), the Mohr or Volhard titration procedure, or indicator test strips. These methods are official methods of analysis for numerous specific products. All these methods are faster and less expensive procedures than analysis by atomic absorption spectroscopy or inductively coupled plasma-optical emission spectroscopy. This experiment allows one to compare sodium analysis of several food products by ISE, Mohr titration, and Quantab® Chloride Titrators.

21.1.2 Reading Assignment

Ward RE (2024) Traditional methods for mineral analysis. Ch. 21. In: Ismail BP, Nielsen SS (ed) *Nielsen's food analysis*, 6th edn. Springer, New York.

21.2 ION-SELECTIVE ELECTRODES

21.2.1 Objective

Determine the sodium content of various foods with sodium and/or chloride ion-selective electrodes.

21.2.2 Principle of Method

The principle of ISE is the same as for measuring pH, but by varying the composition of the glass in the sensing electrode, the electrode can be made sensitive to sodium or chloride ions. Sensing and reference electrodes are immersed in a solution that contains the element of interest. The electrical potential that develops at the surface of the sensing electrode is measured by comparing the reference electrode with a fixed potential. The voltage between the sensing and reference electrodes relates to the activity of the reactive species. Activity (A) is related to concentration (C) by $A = \gamma C$, where γ is the activity coefficient, which is a function of ionic strength. By adjusting the ionic strength of all test samples and standards to a nearly constant (high) level, the Nernst equation can be used to relate electrode response to concentration of the species being measured.

21.2.3 Chemicals

	CAS No.	Hazards
Ammonium chloride (NH$_4$Cl)	12,125-02-9	Harmful
Ammonium hydroxide (NH$_4$OH)	1336-21-6	Corrosive, dangerous for the environment
Nitric acid (HNO$_3$)	7697-37-2	Corrosive
Potassium nitrate (KNO$_3$)	7757-79-1	
Sodium chloride (NaCl)	7647-14-5	Irritant
Sodium nitrate (NaNO$_3$)	7631-99-4	Harmful, oxidizing

21.2.4 Reagents

(**If these solutions are not purchased, it is recommended that these solutions be prepared by the laboratory assistant before class.)

(*Note:* You can use a chloride and/or sodium ion-selective electrode, with the appropriate associated solutions (commercially available from companies that sell the electrodes): electrode rinse solution, ionic strength adjuster, reference electrode fill solution, standard solution, electrode storage solution)

- Electrode rinse solution**
  For sodium electrode, dilute 20-mL ionic strength adjuster to 1 L with deionized distilled (dd) water. For chloride electrode, use deionized distilled water.
- Ionic strength adjuster (ISA)**
  For sodium electrode, 4 M NH$_4$Cl and 4 M NH$_4$OH. For chloride electrode, 5 M NaNO$_3$.
- Nitric acid, 0.1 N**
  Dilute 6.3-mL conc. HNO$_3$ to 1 L with dd water.
- Reference Electrode Fill Solution**
  For sodium electrode, 0.1 M NH$_4$Cl. For chloride electrode, 10% KNO$_3$.
- Standard solutions**: 1000 ppm, sodium and/or chloride
  Use the 1000 ppm sodium or chloride solution to prepare 50 mL each of the following concentrations: 10, 20, 100, 500, and 1000 ppm sodium or chloride.

21.2.5 Hazards, Precautions, and Waste Disposal

Adhere to normal laboratory safety procedures. Wear gloves and safety glasses at all times. Ammonium hydroxide waste should be discarded as hazardous waste. Other waste likely can be put down the drain using a water rinse, but follow good laboratory practices outlined by environmental health and safety protocols at your institution.

21.2.6 Supplies

- Beakers, 250 mL (or sample cups to hold 100 mL) (16–18)
- Food products: catsup, cottage cheese, potato chips, sports drink (e.g., Gatorade, white or clear)
- Graduated cylinder, 100 mL
- Magnetic stir bars
- Pipette bulb or pump
- Spatulas (3)
- Volumetric flasks, 100 mL (16–18)
- Volumetric flasks, 50 mL (2)
- Volumetric pipette, 2 mL
- Volumetric pipettes, 5 mL (9)
- Watch glass
- Weighing paper

21.2.7 Equipment

- Analytical balance
- Chloride electrode (e.g., Van London-pHoenix Company, Houston, TX, Chloride Ion Electrode, Cat. # CL01502)
- Direct concentration readout ISE meter (i.e., suitable meter with millivolt accuracy to 0.1 mV)
- Heating plate with stirrer
- Magnetic stirrer
- Sodium electrode (e.g., Van London-pHoenix Company, Houston, TX, Sodium Ion Electrode, Cat. # NA71502)

21.2.8 Procedure

(Replicate the preparation and analysis of standards and samples as specified by an instructor.)

21.2.9 Sample Preparation (General Instructions)

1. Treat specific samples as described below (i.e., pre-homogenized and/or diluted if necessary, as per Technical Services of ISE manufacturer), and then add 5 g or 5 mL of prepared sample to a 100-mL volumetric flask, then add 2-mL ISA, and dilute to volume with dd water. (See instructions specific for each type of food product below. Samples with high fat levels may require fat removal.

Consult technical services of the company that manufactures the ISE.)

Specific Samples

Sports drink: No dilution is required before use.

Catsup: Accurately weigh ca. 1 g catsup into 50-mL volumetric flask, and dilute to volume with dd water. Mix well.

Cottage cheese: Accurately weigh ca. 1 g of finely grated cheese into a 250-mL beaker containing a stir bar. Add 100 mL 0.1 N HNO$_3$. Cover beaker with a watch glass and boil gently for 20 min on stirrer/hot plate in a hood. Remove from the hot plate and cool to room temperature in the hood.

Potato chips: Accurately weigh ca. 5 g of potato chips into a 250-mL beaker. Crush chips with a glass stirring rod. Add 95 mL boiling dd water and stir. Filter water extract into a 100-mL volumetric flask, using a funnel with glass wool. Let cool to room temperature and dilute to volume.

2. Prepare standards by adding 5 mL standard of proper dilution (e.g., 10, 20, 100, 500, 1000 ppm sodium or chloride) to a 100-mL volumetric flask. Add 2 mL ISA, then dilute to volume with dd water.

(*Note:* Sample/standard preparation calls for identical 1: 20 dilution of each (i.e., 5 mL diluted to 100 mL). Therefore, since samples and standards are treated the same, no correction for this dilution needs to be made in calibration or calculation of results).

21.2.10 Sample Analysis by ISE

1. Condition sodium electrode as specified by the manufacturer.
2. Assemble, prepare, and check sodium and reference electrodes as described in electrode instruction manuals.
3. Connect electrodes to meter according to meter instruction manual.
4. For instruments with direct concentration readout capability, consult meter manual for correct direct measurement procedures.
5. Using the pH meter set on mV scale, determine the potential (mV) of each standard solution (10, 20, 100, 500, 1000 ppm), starting with the most dilute standard. Use a uniform stirring rate, with a magnetic stir bar in each solution, placed on a magnetic stir plate.
6. Rinse electrodes with electrode rinse solution between standards.

7. Measure samples and record the mV reading. As you rinse electrodes with electrode rinse solution between measurements, be careful not to get rinse solutions into the hole for outer fill solution in the reference electrode (or ensure that the hole is covered).

8. After use, store sodium electrode and reference electrode as specified by manufacturer.

21.2.11 Data and Calculations

1. Prepare a standard curve, with electrode response plotted against concentration on a log scale. (Plot actual concentration values on the log scale, not log values.) Concentrations may be determined by reading directly off the standard curve or using a calculated equation of the line.

2. Use the standard curve and the mV readings for the samples to determine the sodium and/or chloride concentrations in ppm for the food samples as analyzed.

3. Convert the ppm sodium and/or chloride values for the food samples to mg/mL for the sports drink, catsup, cheese, and potato chips.

4. Taking into account the dilution of the samples, calculate the sodium and/or chloride content for catsup, cheese, and potato chips (in mg/g) (on a wet weight basis). Summarize the data and calculated results in one table. Show all sample calculations below each table.

5. Calculate sodium chloride content of each food, based on the (a) chloride content and/or (b) sodium content.

6. Calculate the sodium content of each food, based on the sodium chloride content.

7. Compare the sodium/sodium chloride contents of the foods you analyzed to those reported in the US Department of Agriculture (USDA) Nutrient Database for Standard Reference (http://ndb.nal.usda.gov).

21.2.12 Question

1. If you used both a sodium and chloride ISE, which electrode worked better, concerning accuracy, precision, and time to response? Explain your answer, with appropriate justification.

21.3 MOHR TITRATION

21.3.1 Objective

Determine the sodium content of various foods using the Mohr titration method to measure chloride content.

21.3.2 Principle of Method

The Mohr titration is a direct titration method to quantitate chloride ions and then to calculate sodium ions. The chloride-containing sample solution is titrated with a standard solution of silver nitrate. After the silver from silver nitrate has complexed with all the available chloride in the sample, the silver reacts with chromate that has been added to the sample, to form an orange-colored solid, silver chromate. The volume of silver used to react with the chloride is used to calculate the sodium content of the sample.

21.3.3 Chemicals

	CAS no.	Hazards
Potassium chloride (KCl)	7447-40-7	Irritant
Potassium chromate (K_2CrO_4)	7789-00-6	Toxic, dangerous for environment
Silver nitrate ($AgNO_3$)	7761-88-8	Corrosive, dangerous for environment

21.3.4 Reagents

(** It is recommended that these solutions be prepared by laboratory assistant before class.)

- Potassium chloride
- Potassium chromate, 10% solution**
- Silver nitrate solution, ca. 0.1 M**
  Prepare approximately 400 mL of the ca. 0.1 M $AgNO_3$ [molecular weight (MW) 169.89] for each student or lab group. Students should accurately standardize the solution, as described in the Sect. 21.3.8.1.

21.3.5 Hazards, Precautions, and Waste Disposal

Wear gloves and safety glasses at all times, and use good lab technique. Potassium chromate may cause serious skin sensitivity reactions. The use of crystalline $AgNO_3$ or solutions of the silver salt can result in dark brown stains caused by photodecomposition of the salt to metallic silver. These stains are the result of poor technique on the part of the analyst, with spilled $AgNO_3$ causing discoloration of the floor. If you do spill this solution, immediately sponge up the excess solution and thoroughly rinse out the sponge at a sink. Then come back with the clean, rinsed sponge and mop up the area at least 3–4 times to remove all of the silver nitrate. Also, be sure to rinse all pipettes, burets, beakers, flasks, etc. to remove residual $AgNO_3$ when you are finished with this experiment. Otherwise, these items also will stain, and drip stains are likely to appear on the floor. Potassium chromate and silver nitrate must be disposed of as a hazardous waste. Other waste likely can be put down the drain using a water rinse, but follow good laboratory practices outlined by environmental health and safety protocols at your institution.

21.3.6 Supplies

- Beakers, 250 mL (9)
- Brown bottle, 500 mL
- Buret, 25 mL
- Erlenmeyer flasks, 125 mL (3)
- Erlenmeyer flasks, 250 mL (4)
- Food products: cottage cheese (30 g), potato chips (15 g), sports drink (15 mL) (e.g., Gatorade, white or clear)
- Funnel
- Glass wool
- Graduated cylinder, 25 mL
- Magnetic stir bars (to fit 125 or 250-mL flasks)
- Pipette bulb or pump
- Spatulas
- Weighing paper and boats
- Volumetric flask, 100 mL (2)
- Volumetric pipette, 1 mL

21.3.7 Equipment

- Analytical balance.
- Hot plate.
- Magnetic stir plate.

21.3.8 Procedure

(Instructions are given for analysis in triplicate.)

21.3.8.1 Standardization of ca. 0.1 M AgNO₃

1. Transfer 400 mL of the 0.1 M $AgNO_3$ solution to a brown bottle. This solution will be standardized and then used to titrate the food samples. Fill a buret with this $AgNO_3$ solution.
2. Prepare the primary standard (KCl, MW = 74.55) solution in triplicate. Accurately weigh to four decimal places about 100 mg KCl into three 125-mL Erlenmeyer flasks. Dissolve in dd water (about 25 mL), add 2–3 drops of K_2CrO_4 solution. (Caution: potassium chromate may cause serious skin sensitivity reactions!)

3. Put a magnetic stir bar in each flask with the KCl solution, and place the beaker on a magnetic stir plate below the buret for titration. Using the $AgNO_3$ solution in the buret, titrate the KCl solutions to the appearance of the first permanent, pale, pink-orange color. (*Note*: you will first get a white precipitate, then green color, and then the pink-orange color.) This endpoint is due to the formation of Ag_2CrO_4. The solution must be vigorously stirred during the addition of the $AgNO_3$ solution to avoid erroneous results.
4. Record volume of $AgNO_3$.
5. Calculate and record molarity of $AgNO_3$.

$$\frac{g\,KCl}{\left(mL\,AgNO_3\right)} \times \frac{1\,mol\,KCl}{74.555\,g} \times \frac{1000\,mL}{1\,L}$$
$$= mol\ of\ AgNO_3\,/\,L = M\,AgNO_3$$

6. Label the bottle of $AgNO_3$ with your name and the molarity of the solution.

21.3.8.2 Sample Analysis by Mohr Titration

Cottage Cheese
1. Accurately weigh 10 g of cottage cheese in triplicate into 250-mL beakers.
2. Add about 15 mL of warm dd water (50-55 °C) to each beaker. Mix to a thin paste using a glass stirring rod or spatula. Add another ca. 25 mL dd water to each beaker until the sample is dispersed.
3. Quantitatively transfer each solution to a 100-mL volumetric flask, rinsing beaker, and magnetic stir bar with dd water several times. Dilute to volume with dd water.
4. Filter each solution through glass wool. Transfer 50 mL of each solution to 250-mL Erlenmeyer flasks.
5. Add 1 mL of potassium chromate indicator to each 50 mL of filtrate.
6. Titrate each solution with standardized ca. 0.1 M $AgNO_3$, to the first visible pale red-brown color that persists for 30 s. Record the volume of titrant used.

Potato Chips
1. Weigh accurately approximately 5 g of potato chips in duplicate into 250 mL beakers, then add 95 mL boiling dd water to each beaker.
2. Stir the mixture vigorously for 30 s, wait for 1 min, stir again for 30 s, then let cool to room temperature.
3. Filter each solution through glass wool in a 100-mL volumetric flask and dilute to volume with dd water. Transfer 50 mL of each solution to 250-mL Erlenmeyer flasks.
4. Add 1 mL of potassium chromate indicator to each 50 mL of filtrate.

5. Titrate each solution with standardized ca. 0.1 M AgNO₃, to the first visible pale red-brown color that persists for 30 s. Record the volume of titrant used.

Sports Drink (Clear or White)

1. Pipette accurately 10 mL of sports drink in duplicate into 250-mL beakers, then add 95 mL boiling dd water to each beaker.
2. Stir the mixture vigorously for 30 s, wait 1 min, stir again for 30 s.
3. Transfer 50 mL of each solution to 250-mL Erlenmeyer flasks.
4. Add 1 mL of potassium chromate indicator to each 50 mL of prepared sample.
5. Titrate each solution with standardized ca. 0.1 M AgNO₃, to the first visible pale red-brown color that persists for 30 s. Record the volume of titrant used.

21.3.9 Data and Calculations

1. Calculate the chloride content and the sodium chloride content of each replicated sample, then calculate the mean and standard deviation for each type of sample. Express the values in terms of percent, wt/vol, for the cottage cheese and potato chips, and percent, vol/vol, for the sports drink. Note that answers must be multiplied by the dilution factor.

$$\%\text{chloride} = \frac{\text{mL of AgNO3}}{\text{g}\left(\text{or mL}\right)\text{sample}} \times \frac{\text{mol AgNO}_3}{\text{L}} \times \frac{35.5\,\text{g Cl}}{\text{mol NaCl}}$$
$$\times \frac{1\,\text{L}}{1000\,\text{mL}} 100 \times \text{dilution factor}$$

$$\%\text{sodium chloride}\left(\text{salt}\right) = \frac{\text{mL of AgNO}_3}{\text{g}\left(\text{or mL}\right)\text{sample}} \times \frac{\text{mol AgNO}_3}{\text{liter}}$$
$$\times \frac{58.5\,\text{g}}{\text{mol NaCl}} \times \frac{1\,\text{L}}{1000\,\text{mL}} \times 100$$
$$\times \text{dilution factor}$$

Sample	Rep	Buret start (mL)	Buret end (mL)	Vol. AgNO₃ (mL)	% Cl	% NaCl
Cottage cheese	1					
	2					
	3					
					$\overline{X} =$ SD	
Potato chips	1					
	2					
	3					

Sample	Rep	Buret start (mL)	Buret end (mL)	Vol. AgNO₃ (mL)	% Cl	% NaCl
						$\overline{X} =$ SD
Sports drink	1					
	2					
	3					
						$\overline{X} =$ SD

21.3.10 Questions

1. Show the calculations of how to prepare 400 mL of an approximately 0.1 M solution of AgNO₃ (MW = 169.89).
2. Would this Mohr titration procedure as described above work well to determine the salt content of grape juice or catsup? Why or why not?
3. How did this method differ from what would be done using a Volhard titration procedure? Include in your answer what additional reagents would be needed.
4. Would overshooting the endpoint result in an over- or underestimation of the salt content using the (a) Mohr titration or (b) Volhard titration?

21.4 QUANTAB® TEST STRIPS

21.4.1 Objective

To measure the chloride content of foods using Quantab® Chloride Titrators, then calculate the sodium chloride content.

21.4.2 Principle of Method

Quantab® Chloride Titrators are thin, chemically inert plastic strips. These strips are laminated with an absorbent paper impregnated with silver nitrate and potassium dichromate, which together form brown-silver dichromate. When the strip is placed in an aqueous solution that contains chlorine, the liquid rises up the strip by capillary action. The reaction of silver dichromate with chloride ions produces a white column of silver chloride in the strip. When the strip is completely saturated with the liquid, a moisture-sensitive signal across the top of the titrator turns dark blue to indicate the completion of the titration. The length of the white color change is proportional to the chloride concentration of the liquid being tested. The value on the numbered scale is read at the tip of the color change and then is converted to percent salt using a calibration table.

21.4.3 Chemicals

	CAS no.	Hazards
Sodium chloride (NaCl)	7647-14-5	Irritant

21.4.4 Reagents

(**It is recommended that these solutions be prepared by laboratory assistant before class.)

- Sodium chloride stock solution**
  Accurately weigh 5.00 g of dried sodium chloride and quantitatively transfer to a 100-mL volumetric flask. Dilute to volume with dd water and mix thoroughly.
- Sodium chloride standard solutions**
  Dilute 2 mL of the stock solution to 1000 mL with dd water in a volumetric flask to create a 0.010% sodium chloride solution to use as a standard solution with the low range Quantab® Chloride Titrators.
  Dilute 5 mL of the stock solution to 100 mL with dd water in a volumetric flask to create a 0.25% sodium chloride solution to use as a standard solution with the high range Quantab® Chloride Titrators.

21.4.5 Supplies

- Beakers, 200 mL (5)
- Filter paper (when folded as a cone, should fit into a 200-mL beaker)
- Funnels
- Glass wool
- Glass stirring rod
- Graduated cylinder, 100 mL
- Quantab® Chloride Titrators, range: 0.05% to 1.0% Na Cl; 300–6000 ppm Cl (High Range, HR) and 0.005% to 0.1%, 30–600 ppm Cl (Low Range, LR) (Environmental Test Systems/Hach Company, Elkhart, IN, 1–800–227-4224)
- Spatulas
- Sports drink, 10 mL (i.e., same one used in Sects. 19.2 and 19.3)
- Volumetric flasks, 100 mL (2)

21.4.6 Equipment

- Hot plate
- Top loading balance

21.4.7 Procedure

(Instructions are given for analysis in triplicate.)

21.4.7.1 Standard Solutions of Sodium Chloride

1. Transfer 50 mL of the 0.25% standard sodium chloride solution to a 200-mL beaker.
2. Fold a piece of filter paper into a cone-shaped cup and place it point end down into the beaker. This will allow liquid from the beaker to seep through the filter paper at the pointed end.
3. Using the 0.25% sodium chloride standard solution, place the lower end of the High Range Quantab® Strip (0.05% to 1.0%) into the filtrate within the pointed end of the filter paper cone, being sure not to submerge the titrator more than 1.0 in.
4. Thirty seconds after the moisture-sensitive signal string at the top of the titrator turns dark blue or a light brown, record the Quantab® reading at the tip of the yellow-white peak, to the nearest 0.1 units on the titrator scale.
5. Using the calibration chart included with the Quantab® package, convert the Quantab® reading to percent sodium chloride (NaCl) and to ppm chloride (Cl⁻). Note that each lot of Quantab® has been individually calibrated. Be sure to use the correct calibration chart (i.e., the control number on the product being used must match the control number on the bottle).
6. Repeat Steps 1–5 given above (Sect. 21.4.7.1) using the 0.01% sodium chloride standard solution with the Low Range Quantab® Strip.

21.4.7.2 Sample Analysis with Quantab® Test Strips

Cottage Cheese
1. Weigh accurately approximately 5 g of cottage cheese into a 200-mL beaker, then add 95 mL boiling dd water.
2. Stir mixture vigorously for 30 s, wait for 1 min, stir again for 30 s, then let cool to room temperature.
3. Fold a piece of filter paper into a cone-shaped cup and place it point end down the beaker. This will allow liquid from the beaker to seep through the filter paper at the pointed end.
4. Testing with both the Low Range and the High Range Quantab® Test Strips, place the lower end of the Quantab® into the filtrate within the pointed end of the filter paper cone, being sure not to submerge the titrator more than 2.5 cm.
5. Thirty seconds after the moisture-sensitive signal string at the top of the titrator turns dark blue or a light brown, record the Quantab® reading at the tip of the yellow-white peak, to the nearest 0.1 units on the titrator scale.
6. Using the calibration chart included with the Quantab® package, convert the Quantab® reading to percent sodium chloride (NaCl) and to ppm chloride (Cl⁻). Note that each lot of Quantab® has been individually calibrated. Be sure

to use the correct calibration chart (i.e., the control number on the product being used must match the control number on the bottle).

7. Multiply the result by the dilution factor 20 to obtain the actual salt concentration in the sample.

Potato Chips

1. Weigh accurately approximately 5 g of potato chips into a 200-mL beaker. Crush chips with a glass stirring rod. Add 95-mL boiling dd water and stir.
2. Filter water extract into a 100-mL volumetric flask, using a funnel with glass wool. Let cool to room temperature and dilute to volume. Transfer to a 200-mL beaker.
3. Follow Steps 3–7 from the procedure for cottage cheese, Sect. 21.4.7.2.

Catsup

1. Weigh accurately approximately 5 g of catsup into a 200-mL beaker. Add 95-mL boiling dd water and stir.
2. Filter water extract into a 100-mL volumetric flask. Let cool to room temperature and dilute to volume. Transfer to a 200-mL beaker.
3. Follow Steps 3–7 from the procedure for cottage cheese, Sect. 21.4.7.2.

Sports Drink

1. Weigh accurately approximately 10 mL of sports drink into a 200-mL beaker. Add 95-mL boiling dd water and stir.
2. Follow Steps 3–7 from the procedure for cottage cheese, Sect. 21.4.7.2.

21.4.8 Data and Calculations

	From calibration chart				Corrected for dilution factor			
	% NaCl		ppm Cl		% NaCl		ppm Cl	
Rep.	LR	HR	LR	HR	LR	HR	LR	HR
Catsup 1 2 3								
					$\bar{X} =$ SD =	$\bar{X} =$ SD =	$\bar{X} =$ SD =	$\bar{X} =$ SD =
Cottage cheese 1 2 3								
					$\bar{X} =$ SD =	$\bar{X} =$ SD =	$\bar{X} =$ SD =	$\bar{X} =$ SD =

	From calibration chart				Corrected for dilution factor			
	% NaCl		ppm Cl		% NaCl		ppm Cl	
Rep.	LR	HR	LR	HR	LR	HR	LR	HR
Potato chips 1 2 3								
					$\bar{X} =$ SD =	$\bar{X} =$ SD =	$\bar{X} =$ SD =	$\bar{X} =$ SD =
Sports drink 1 2 3								
					$\bar{X} =$ SD =	$\bar{X} =$ SD =	$\bar{X} =$ SD =	$\bar{X} =$ SD =

21.5 SUMMARY OF RESULTS

Summarize in a table the sodium chloride content (mean and standard deviation) of the various food products as determined by the three methods described in this experiment. Include in the table the sodium chloride contents of the foods from the nutrition label and those published in the USDA Nutrient Database for Standard Reference (web address: http://ndb.nal.usda.gov/).

Sodium chloride content (%) of foods by various methods:

Food product		Ion-selective electrode	Mohr titration	Quantab® titrator	Nutrition label	USDA database
Catsup	$\bar{X} =$ SD =					
Cottage cheese	$\bar{X} =$ SD =					
Potato chips	$\bar{X} =$ SD =					
Sports drink	$\bar{X} =$ SD =					

21.6 QUESTIONS

1. Based on the results and characteristics of the methods, discuss the relative advantages and disadvantages of each method of analysis for these applications.
2. Comparing your results to data from the nutrition label and USDA Nutrient Database, what factors might explain any differences observed?

Acknowledgments Van London-pHoenix Company, Houston, TX, is acknowledged for its contribution of the sodium and chloride ion-selective electrodes, and related supplies, for use in developing a section of this laboratory exercise. Environmental Test Systems/HACH Company, Elkhart, IN, is acknowledged for contributing the Quantab® Chloride Titrators for use in developing a section of this laboratory exercise.

RESOURCE MATERIALS

AOAC International (2023) Official methods of analysis, 22nd edn. (On-line). Method 941.18, Standard solution of silver nitrate; Method 983.14, Chloride (total) in cheese. AOAC International, Rockville, MD

AOAC International (2023) Official methods of analysis, 22nd edn. (On-line). Method 976.25, Sodium in foods for special dietary use, ion selective electrode method. AOAC International, Rockville, MD

AOAC International (2023) Official methods of analysis, 22nd edn. (On-line). Method 971.19, Salt (chlorine as sodium chloride) in meat, fish, and cheese; Indicating strip method. AOAC International, Rockville, MD

Ward RE (2024) Traditional methods for mineral analysis. Ch. 21. In: Ismail BP, Nielsen SS (ed) Nielsen's food analysis, 6th edn. Springer, New York

Environmental Test Systems (2022) Quantab® Technical Bulletin. Quantab salt (cheese). Online. Environmental Test Systems, Elkhart, IN Website: Does Hach have information on applications that the Quantab Chloride Strips can be used in? (see link for "Salt in cheese" for general instructions)

Van London-pHoenix Company, Houston, TX. Product literature. http://www.vl-pc.com/default/index.cfm/ion-selective

Wehr HM, Frank JF (eds) (2004) Standard methods for the examination of dairy products, 17th edn., Part 15.053 Chloride (Salt). American Public Health Association, Washington, DC

Standard Solutions and Titratable Acidity

22

S. Suzanne Nielsen

Contents

22.1 INTRODUCTION

22.1.1 Background

Many types of chemical analyses are made using a method in which a constituent is titrated with a solution of known strength to an indicator endpoint. Such a solution is referred to as a standard solution. From the volume and concentration of standard solution used in the titration, and the sample size, the concentration of the constituent in the sample can be calculated.

The assay for titratable acidity is a volumetric method that uses a standard solution and, most commonly, the indicator phenolphthalein. In the titration, a standard solution of sodium hydroxide reacts with the organic acids present in the sample. The normality of the sodium hydroxide solution, the volume used, and the volume of the test sample are used to calculate titratable acidity, expressing it in terms of the predominant acid present in the sample. A standard acid such

S. S. Nielsen (✉)
Department of Food Science, Purdue University,
West Lafayette, IN, USA
e-mail: nielsens@purdue.edu

© The Author(s), under exclusive license to Springer Nature Switzerland AG 2024
B. P. Ismail, S. S. Nielsen (eds.), *Nielsen's Food Analysis Laboratory Manual*, Food Science Text Series,
https://doi.org/10.1007/978-3-031-44970-3_22

as potassium acid phthalate can be used to determine the exact normality of the standard sodium hydroxide used in the titration.

The phenolphthalein endpoint in the assay for titratable acidity is pH 8.2, where there is a significant color change from clear to pink. When colored solutions obscure the pink endpoint, a potentiometric method is commonly used. A pH meter is used to titrate such a sample to pH 8.2.

22.1.2 Reading Assignment

Tyl C (2024) pH and titratable acidity. Ch. 22. In: Ismail BP, Nielsen SS (eds) *Nielsen's food analysis*, 6th edn. Springer, New York

22.1.3 Notes

1. Carbon dioxide (CO_2) acts as an interfering substance in determining titratable acidity, by the following reactions:

$$H_2O + CO_2 \leftrightarrow H_2CO_3 \text{ (carbonic acid)}$$

$$H_2CO_3 \leftrightarrow H^+ + HCO_3^- \text{ (bicarbonate)}$$

$$HCO_3 \leftrightarrow H^+ + CO_3^{-2} \text{ (carbonate)}$$

In these reactions, buffering compounds and hydrogen ions are generated. Therefore, CO_2-free water is prepared and used for standardizing acids and bases and for determining titratable acidity. An Ascarite® trap is attached to bottles of CO_2-free water, so that as air enters the bottle when water is siphoned out, the CO_2 is removed from the air.

2. Ascarite® is a silica base coated with NaOH, and it removes CO_2 from the air by the following reaction:

$$2NaOH + CO_2 \rightarrow Na_2CO_3 + H_2O$$

22.2 PREPARATION AND STANDARDIZATION OF BASE AND ACID SOLUTIONS

22.2.1 Objective

Prepare and standardize solutions of sodium hydroxide and hydrochloric acid.

22.2.2 Principle of Method

A standard acid can be used to determine the exact normality of a standard base and vice versa.

22.2.3 Chemicals

	CAS no.	Hazards
Ascarite®	81133-20-2	Corrosive
Ethanol (CH_3CH_2OH)	64-17-5	Highly flammable
Hydrochloric acid (HCl)	7647-01-0	Corrosive
Phenolphthalein	77-09-8	Irritant
Potassium acid phthalate ($HOOCC_6H_4COOK$)	877-24-7	Irritant
Sodium hydroxide (NaOH)	1310-73-2	Corrosive

22.2.4 Reagents

(** It is recommended that these solutions be prepared by the laboratory assistant before class.)

(*Note:* Preparation of NaOH and HCl solutions is described under Procedure.)

- Ascarite® trap**

    Put the Ascarite® in a syringe that is attached to the flask of CO_2-free water (see note about CO_2-free water).
- Carbon dioxide-free water**

    Prepare 1.5 L of CO_2-free water (per person or group) by boiling deionized, distilled (dd) water for 15 min in a 2-L Erlenmeyer flask. After boiling, stopper the flask with a rubber stopper through which is inserted a tube attached to an Ascarite® trap. Allow the water to cool with Ascarite® protection.
- Ethanol, 100 mL
- Hydrochloric acid, concentrated
- Phenolphthalein indicator solution, 1% **
- Dissolve 1.0 g in 100 mL ethanol. Put in a bottle with eyedropper.
- Potassium acid phthalate (KHP)**

    3–4 g, dried in an oven at 120°C for 2 h cooled and stored in a closed bottle inside a desiccator until use
- Sodium hydroxide, pellets

22.2.5 Hazards, Precautions, and Waste Disposal

Use appropriate precautions in handling concentrated acid and base. Otherwise, adhere to normal laboratory safety procedures. Wear gloves and safety glasses at all times.

Waste likely may be put down the drain using a water rinse, but follow good laboratory practices outlined by environmental health and safety protocols at your institution.

22.2.6 Supplies

- Beaker, 50 mL (for waste NaOH from buret)
- Beaker, 100 mL
- Buret, 25 or 50 mL
- Erlenmeyer flasks, 250 mL (5)
- Erlenmeyer flask, 1 L
- Funnel, small, to fit top of 25 or 50 mL buret
- Glass stirring rod
- Glass storage bottle, 100 mL
- Graduated cylinder, 50 mL
- Graduated cylinder, 1 L
- Graduated pipette, 1 mL
- Graduated pipette, 10 mL
- Parafilm®
- Pipette bulb or pump
- Plastic bottle, with lid, 50 or 100 mL
- Plastic bottle, with lid, 1 L
- Spatula
- Squirt bottle, with dd water
- Volumetric flask, 50 mL
- Volumetric flask, 100 mL
- Weighing paper/boat
- White piece of paper

22.2.7 Equipment

- Analytical balance
- Forced draft oven (heated to 120°C)
- Hot plate

22.2.8 Calculations Required Before Lab

1. Calculate how much NaOH to use to prepare 50 mL of 25% NaOH (wt/vol) in water (see Table 2.1 in Chap. 2 for definition of wt%).
2. Calculate how much concentrated HCl to use to prepare 100 mL of ca. 0.1 N HCl in water (concentrated HCl = 12.1 N).

22.2.9 Procedure

1. Prepare 25% (wt/vol) NaOH solution: Prepare 50 mL of 25% NaOH (wt/vol) in dd water. To do this, weigh out the appropriate amount of NaOH and place it in a 100-mL beaker. While adding about 40 mL of dd water, stir the NaOH pellets with a glass stirring rod. Continue stirring until all pellets are dissolved. Quantitatively transfer the NaOH solution into a 50-mL volumetric flask. Dilute to volume with dd water. The solution must be cooled to room temperature before final preparation. Store this solution in a plastic bottle and label appropriately.

2. Prepare ca. 0.1 N HCl solution: Prepare 100 mL of ca. 0.1 N HCl using concentrated HCl (12.1 N) and dd water. (*Note*: Do not use a mechanical pipettor to prepare this, since the acid can easily get into the shaft of the pipettor and cause damage.) To prepare this solution, place a small amount of dd water in a 100-mL volumetric flask, pipette in the appropriate amount of concentrated HCl, then dilute to volume with dd water. Mix well, transfer into a glass bottle, seal bottle, and label appropriately.

3. Prepare ca. 0.1 N NaOH solution: Transfer 750 mL CO_2-free water to a 1-L plastic storage bottle. Add ca. 12.0 mL of well-mixed 25% (wt/vol) NaOH solution prepared in Step 1. Mix thoroughly. This will give an approximately 0.1 N solution. Fill the buret with this solution using a funnel. Discard the first volume of the buret and then refill the buret with the NaOH solution.

4. Standardize ca. 0.1 N NaOH solution: Accurately weigh about 0.8 g of dried potassium acid phthalate (KHP) into each of three 250-mL Erlenmeyer flasks. Record the exact weights. Add ca. 50 mL of cool CO_2-free water to each flask. Seal the flasks with Parafilm® and swirl gently until the sample is dissolved. Add 3 drops of phenolphthalein indicator and titrate, against a white background, with the NaOH solution being standardized. Record the beginning and ending volume on the buret. Titration should proceed to the faintest tinge of pink that persists for 15 s. after swirling. The color will fade with time. Record the total volume of NaOH used to titrate each sample. Data from this part will be used to calculate the mean normality of the diluted NaOH solution.

5. Standardize ca. 0.1 N HCl solution: Devise a scheme to standardize (i.e., determine the exact N) the ca. 0.1 N HCl solution that you prepared in Step 2. Remember that you have your standardized NaOH to use. Do analyses in at least duplicate. Record the volumes used.

22.2.10 Data and Calculations

Using the weight of KHP and the volume of NaOH titrated in Sect. 22.2.9, Step 4, calculate the normality of the diluted NaOH solution as determined by each titration and then calculate the mean normality (molecular weight (MW) potassium acid phthalate = 204.228). The range of triplicate determinations for normality should be less than 0.2% with good technique.

Rep	Weight of KHP (g)	Buret start (mL)	Buret end (mL)	Vol. NaOH titrated (mL)		N NaOH
1						
2						
3						
					$\bar{X} =$	
					SD =	

Sample calculation:

$$Weight\ of\ KHP = 0.8115\,g$$

$$MW\ of\ KHP = 204.228\,g\,/\,mol$$

$$Vol.of\ ca.0.10\,N\ NaOH\ used\ in\ titration = 39\,mL$$

$$Mol\,KHP = 0.8115\,g\,/\,204.228\,g\,/\,mol$$
$$= 0.003974\,mol$$

$$Mol\,KHP = Mol\,NaOH$$
$$= 0.003974\,mol = N\,NaOH\,x\,L\,NaOH$$

$$0.003978\,mol\,NaOH\,/\,0.039\,L\,NaOH = 0.1019\,N$$

With the volumes of HCl and NaOH used in Sect. 22.2.9, Step 5, calculate the exact normality of the HCl solution as determined by each titration and then calculate the mean normality.

Rep	Vol. HCl (mL)	Vol. NaOH (mL)	N HCl
1			
2			
			$\bar{X} =$
			SD =

22.2.11 Questions

1. What does 25% NaOH (wt/vol) mean? How would you prepare 500 mL of a 25% NaOH (wt/vol) solution?

2. Describe how you prepared the 100 mL of ca. 0.1 N HCl. Show your calculations.

3. If you had not been told to use 12 mL of 25% NaOH (wt/vol) to make 0.75 L of ca. 0.1 N NaOH, how could you have determined this was the appropriate amount? Show all calculations.

4. Describe in detail how you standardized your ca. 0.1 N HCl solution.

22.3   TITRATABLE ACIDITY AND PH

22.3.1 Objective

Determine the titratable acidity and pH of food samples.

22.3.2 Principle of Method

The volume of a standard base used to titrate the organic acids in foods to a phenolphthalein endpoint can be used to determine the titratable acidity.

22.3.3 Chemicals

	CAS no.	Hazards
Ascarite	81133-20-2	Corrosive
Ethanol (CH₃CH₂OH)	64-17-5	Highly flammable
Hydrochloric acid (HCl)	7647-01-0	Corrosive
Phenolphthalein	77-09-8	Irritant
Sodium hydroxide (NaOH)	1310-73-2	Corrosive

22.3.4 Reagents

(** It is recommended that these items/solutions be prepared by the laboratory assistant before class.)

- Ascarite® trap**
  Put the Ascarite® in a syringe that is attached to the flask of CO_2-free water.
- Carbon dioxide-free water**
  Prepared and stored as described in Sect. 22.2.4
- Phenolphthalein indicator solution, 1%**
  Prepared as described in Sect. 22.2.4
- Sodium hydroxide, ca. 0.1 N
  From Sect. 22.2.9, Step 4; exact N calculated
- Standard buffers, pH 4.0 and 7.0

22.3.5 Hazards, Precautions, and Waste Disposal

Adhere to normal laboratory safety procedures. Wear safety glasses at all times. Waste likely may be put down the drain using a water rinse, but follow good laboratory practices outlined by environmental health and safety protocols at your institution.

22.3.6 Supplies

- Apple juice, 60 mL
- Beakers, 250 mL (3)
- Burets, 25 or 50 mL (2)
- Erlenmeyer flasks, 250 mL (4)
- Funnel, small, to fit top of 25 or 50 mL buret
- Graduated cylinder, 50 mL
- Soda, clear, 80 mL
- Volumetric pipettes, 10 or 20 mL

22.3.7 Equipment

- Hot plate
- pH meter

22.3.8 Procedure

22.3.8.1 Soda

Do at least duplicate determinations for unboiled soda and for boiled soda sample; open soda well before use to allow escape of carbon dioxide, so that the sample can be pipetted.

1. Unboiled soda: Pipette 20 mL of soda into a 250-mL Erlenmeyer flask. Add ca. 50 mL CO_2-free dd water. Add 3 drops of a 1% phenolphthalein solution and titrate with standardized NaOH (ca. 0.1 N) to a faint pink color (NaOH in buret, from Method A). Record the beginning and ending volumes on the buret to determine the total volume of NaOH solution used in each titration. Observe the endpoint. Note whether the color fades.
2. Boiled soda: Pipette 20 mL of soda into a 250-mL Erlenmeyer flask. Bring the sample to boiling on a hot plate, swirling the flask often. Boil the sample only 30–60 s. Cool to room temperature. Add ca. 50 mL CO_2-free dd water. Add 3 drops of the phenolphthalein solution and titrate as described above. Record the beginning and ending volumes on the buret to determine the total volume of NaOH solution used in each titration. Observe the endpoint. Note whether the color fades.

22.3.8.2 Apple Juice

1. Standardize the pH meter with pH 7.0 and 4.0 buffers, using instructions for the pH meter available.
2. Prepare as described below three apple juice samples that will be compared:
   A – Apple juice (set aside, to recall original color)
   B – Apple juice; titrate with std. NaOH
   C – Apple juice; add phenolphthalein; titrate with std. NaOH; follow pH during titration
   Procedure: Into each of three (A, B, C) 250-mL beakers, pipette 20 mL of apple juice. Add ca. 50 mL CO_2-free water to each. To Sample C, add 3 drops of a 1% phenolphthalein solution. Using two burets filled with the standardized NaOH solution (ca. 0.1 N), titrate Samples B and C simultaneously. Follow the pH during titration of Sample C containing phenolphthalein. (*Note*: If only one buret is available, titrate Samples B and C sequentially, i.e., add 1 mL to B and then 1 mL to C.) Record the initial pH and the pH at ca. 1.0 mL intervals until a pH of 9.0 is reached. Also, observe any color changes that occur during the titration to determine

when the phenolphthalein endpoint is reached. Sample A is intended to help you remember the original color of the apple juice. Sample B (without phenolphthalein) does not need to be followed with the pH meter but is to be titrated along with the other beakers to aid in observing color changes.

22.3.9 Data and Calculations

22.3.9.1 Soda

Using the volume of NaOH used, calculate the titratable acidity (TA) of each soda sample as percentage citric acid, and then calculate the mean TA of each type of sample (MW citric acid = 192.14; equivalent weight = 64.04) (*Note*: See Eq. 22.20 in Tyl, 2024)

Rep	Buret start	Buret end	Vol. NaOH titrant	Color fades? Cl		TA
Unboiled soda:						
1						
2						
					$\bar{X} =$	
					SD =	
Boiled soda:						
1						
2						
					$\bar{X} =$	
					SD =	

Sample calculation:

$$\%Acid = \frac{(\text{mL base titrant}) \times (N \text{ of base in mol}/\text{L}) \times (\text{Eq.wt. of acid})}{(\text{sample volume in mL}) \times 10}$$

N NaOH = 0.1019 N
mL base = 7 mL
Eq. wt. citric acid = 64.04
Vol. sample = 20 mL

$$\%Acid = (7\,\text{mL} \times 0.1019 \times 64.04)/(20\,\text{mL} \times 10)$$
$$= 0.276\% \text{ citric acid}$$

22.3.9.2 Apple Juice

Sample B

Color change during titration:
Color at end of titration:

Sample C (titrate to >pH 9.0)

| mL NaOH pH | 1 | 2 | 3 | 4 | 5 | 6 | 7 | 8 | 9 | 10 |
| mL NaOH pH | 11 | 12 | 13 | 14 | 15 | 16 | 17 | 18 | 19 | 20 |

Plot pH versus mL of 0.1 N NaOH (but use the normality of your own NaOH solution) (pH on the y-axis) for the sample that contained phenolphthalein (Sample C). Interpolate to find the volume of titrant at pH 8.2 (the phenolphthalein endpoint).

Calculate the titratable acidity of the apple juice as percentage malic acid (MW malic acid = 134.09; equivalent weight = 67.04).

22.3.10 Questions

1. Soda samples. (a) Did any color changes occur in either the boiled or the unboiled sample within several minutes of the phenolphthalein endpoint being reached? (b) How did boiling the sample affect the determination of titratable acidity? (c) Explain the differences in color changes and titratable acidity between the two samples.

2. What caused the color changes in the apple juice titrated without any phenolphthalein present? (Hint: Consider the pigments in apples.) How would you recommend determining the endpoint in the titration of tomato juice?

3. You are determining the titratable acidity of a large number of samples. You ran out of freshly boiled dd H_2O with an Ascarite® trap on the water container, so you switch to using tap-distilled H_2O. Would this likely affect your results? Explain.

4. The electrode of your pH meter has a slow response time and seems to need cleaning, since it is heavily used for a variety of solutions high in proteins, lipids, and minerals. You would ideally check the electrode instructions for specific recommendations on cleaning, but the instructions were thrown away. (As the new lab supervisor, you have since started a policy of filing all instrument/equipment instructions.) What solutions would you use to try to clean the electrode?

RESOURCE MATERIALS

AOAC International (2023) Official methods of analysis, 22nd edn. (On-line). AOAC International, Rockville, MD

Tyl C (2024) pH and titratable acidity. Ch. 22. In: Ismail BP, Nielsen SS (eds) Nielsen's food analysis, 6th edn. Springer, New York

Fat Characterization

23

S. Suzanne Nielsen

Contents

S. S. Nielsen (✉)
Department of Food Science, Purdue University,
West Lafayette, IN, USA
e-mail: nielsens@purdue.edu

© The Author(s), under exclusive license to Springer Nature Switzerland AG 2024
B. P. Ismail, S. S. Nielsen (eds.), *Nielsen's Food Analysis Laboratory Manual*, Food Science Text Series,
https://doi.org/10.1007/978-3-031-44970-3_23

23.1 INTRODUCTION

23.1.1 Background

Lipids in food are subjected to many chemical reactions during processing and storage. While some of these reactions are desirable, others are undesirable; so, efforts are made to minimize the reactions and their effects. The laboratory deals with characterization of fats and oils with respect to composition, structure, and reactivity.

23.1.2 Reading Assignment

Pike OA (2024) Fat characterization. Ch. 23. In: Ismail BP, Nielsen SS (eds) Nielsen's Food Analysis, 6th edn. Springer, New York.

23.1.3 Overall Objective

The overall objective of this laboratory is to determine aspects of the composition, structure, and reactivity of fats and oils by various methods.

23.2 SAPONIFICATION VALUE

23.2.1 Objective

Determine the saponification number of fats and oils.

23.2.2 Principle of Method

Saponification is the process of treating a neutral fat with alkali, breaking it down to glycerol and fatty acids. The saponification value (or number) is defined as the amount of alkali needed to saponify a given quantity of fat or oil, expressed as mg potassium hydroxide to saponify 1 g sample. Excess alcoholic potassium hydroxide is added to the sample, the solution is heated to saponify the fat, the unreacted potassium hydroxide is back-titrated with standardized hydrochloric acid using a phenolphthalein indicator, and the calculated amount of reacted potassium hydroxide is used to determine the saponification value.

23.2.3 Chemicals

	CAS no.	Hazards
Ethanol	64-17-5	Highly flammable
Hydrochloric acid (HCl)	7647-01-0	Corrosive
Phenolphthalein	77-09-8	Irritant
Potassium hydroxide (KOH)	1310-58-3	Corrosive

23.2.4 Reagents

(**It is recommended that these solutions be prepared by the laboratory assistant before class.)

- Alcoholic potassium hydroxide, ca. 0.7 N**

- Dissolve 40 g KOH, low in carbonate, in 1 L of distilled ethanol, keeping temperature below 15.5 °C while the alkali is being dissolved. The solution should be clear.
- Hydrochloric acid, ca. 0.5 *N*, accurately standardized**
- Prepare ca. 0.5 *N* HCl. Determine the exact normality using solution of standard base.
- Phenolphthalein indicator solution**
  1%, in 95% ethanol

23.2.5 Hazards, Precautions, and Waste Disposal

Use hydrochloric acid in a fume hood. Otherwise, adhere to normal laboratory safety procedures. Wear safety glasses at all times. Wastes likely may be put down the drain using a water rinse, but follow good laboratory practices outlined by environmental health and safety protocols at your institution.

23.2.6 Supplies

- Air (reflux) condenser (650 mm long, minimum)
- Beaker, 250 mL (to melt fat)
- Buchner funnel (to fit side-arm flask)
- Boiling beads
- Burets, 50 mL (2)
- Fat and/or oil samples
- Filter paper (to fit Buchner funnel, to filter oil and melted fat)
- Flasks, 250–300 mL, to fit condenser (4)
- Graduated cylinder, 50 mL (1)
- Mechanical pipettor, 1000 μL, with plastic tips (or 1-mL volumetric pipette)
- Side-arm flask

23.2.7 Equipment

- Analytical balance
- Hot plate or water bath (with variable heat control)

23.2.8 Procedure

(Instructions are given for analysis in duplicate.)

1. Melt any solid samples. Filter melted fat sample and oil sample through filter paper to remove impurities.
2. Weigh accurately ca. 5 g melted fat or oil into each of two 250–300 flasks that will connect to a condenser. Record the weight of the sample. Prepare sample in duplicate.
3. Add accurately (using a 50-mL graduated cylinder) 50 mL of alcoholic KOH into the flask.

4. Prepare duplicate blank samples with just 50 mL of alcoholic KOH in a 250–300 mL flask.
5. Add several boiling beads to the flasks with fat or oil sample.
6. Connect the flasks with the sample to a condenser. Boil gently but steadily on a hot plate (or water bath) until the sample is clear and homogenous, indicating complete saponification (requires ca. 30–60 min). (*Note*: The fumes should condense as low as possible in the condenser; otherwise, a fire hazard will be created.)
7. Allow samples to cool somewhat. Wash down the inside of the condenser with a little deionized distilled (dd) water. Disconnect the flask from the condenser. Allow the samples to cool to room temperature.
8. Add 1 mL phenolphthalein to samples and titrate with 0.5 *N* HCl (from a buret) until the pink color just disappears. Record the volume of titrant used.
9. Repeat Steps 5–8 with sample blanks. Reflux the blanks for the same time period as used for the sample.

23.2.9 Data and Calculations

Sample	Weight (g)	Titrant volume (mL)	Saponification value
1			
2			
			$\bar{X} =$
			SD =

Oil/fat sample type tested:
Blank titration (mL):
Blank 1 =
Blank 2 =
$\bar{X} =$
Calculate the saponification number (or value) of each sample as follows:

$$\text{Saponification value} = \frac{(B - S) \times N \times 56.1}{W}$$

where:
Saponification value = mg KOH per g of sample
B = volume of titrant (mL) for blank
S = volume of titrant (mL) for sample
N = normality of HCl (mmol/mL)
56.1 = molecular weight (MW) of KOH (mg/mmol)
W = sample mass (g)

23.2.10 Questions

1. What is meant by unsaponifiable matter in lipid samples? Give an example of such a type of compound.
2. What does a high versus a low saponification value tell you about the nature of a sample?

23.3 IODINE VALUE

23.3.1 Objective

Determine the iodine value of fats and oils.

23.3.2 Principle of Method

The iodine value (or number) is a measure of the degree of unsaturation, defined as the grams of iodine absorbed per 100-g sample. In the assay, a measured quantity of fat or oil dissolved in solvent is reacted with a measured excess amount of iodine or some other halogen, which reacts with the carbon-carbon double bonds. After a solution of potassium iodide is added to reduce excess ICl to free iodine, the liberated iodine is titrated with a standardized solution of sodium thiosulfate using a starch indicator. The calculated amount of iodine reacted with the double bonds is used to calculate the iodine value.

23.3.3 Chemicals

	CAS no.	Hazards
Acetic acid (glacial)	64-19-7	Corrosive
Carbon tetrachloride (CCl$_4$)	56-23-5	Toxic, dangerous for the environment
Chloroform	67-66-3	Harmful
Hydrochloric acid (HCl)	7647-01-0	Corrosive
Iodine	7553-56-2	Harmful, dangerous for the environment
Potassium dichromate (K$_2$Cr$_2$O$_7$)	7789-00-6	Toxic, dangerous for the environment
Potassium iodide (KI)	7681-11-0	
Sodium thiosulfate	7772-98-7	
Soluble starch	9005-25-8	

23.3.4 Reagents

(**It is recommended that these solutions be prepared by the laboratory assistant before class.)

- Potassium iodide solution, 15%**
  Dissolve 150 g KI in dd water and dilute to 1 liter.
- Sodium thiosulfate, 0.1 N standardized solution (AOAC Method 942.27)**
  Dissolve ca. 25 g sodium thiosulfate in 1 L dd water. Boil gently for 5 min. Transfer while hot to a storage bottle (make sure the bottle has been well cleaned and is heat resistant). Store the solution in a dark, cool place. Use the following procedure to standardize the sodium thiosulfate

solution: Accurately weigh 0.20–0.23 g potassium dichromate (K$_2$Cr$_2$O$_7$) (previously dried for 2 h at 100 °C) into a glass-stoppered flask. Dissolve 2 g potassium iodide (KI) in 80 mL chlorine-free water. Add this water to the potassium dichromate. To this solution, add, with swirling, 20 mL ca. 1 M HCl and immediately place in the dark for 10 min. Titrate a known volume of this solution with the sodium thiosulfate solution, adding starch solution after most of the iodine has been consumed.

- Starch indicator solution, 1% (prepare fresh daily)**
  Mix ca. 1 g soluble starch with enough cold dd water to make a thin paste. Add 100 mL boiling dd water. Boil ca. 1 min while stirring.
- Wijs iodine solution**
  Dissolve 10 g ICl$_3$ in 300 mL CCl$_4$ and 700 mL glacial acetic acid. Standardize this solution against 0.1 N sodium thiosulfate (25 mL of Wijs solution should consume 3.4–3.7 mEq of thiosulfate). Then, add enough iodine to the solution such that 25 mL of the solution will require at least 1.5 times the milliequivalency of the original titration. Place the solution in an amber bottle. Store in the dark at less than 30 °C.

23.3.5 Hazards, Precautions, and Waste Disposal

Carbon tetrachloride and potassium chromate are toxic and must be handled with caution. Use acetic acid and hydrochloric acid in a fume hood. Otherwise, adhere to normal laboratory safety procedures. Wear safety glasses at all times. Carbon tetrachloride, chloroform, iodine, and potassium chromate must be handled as hazardous wastes. Other wastes likely may be put down the drain using a water rinse, but follow good laboratory practices outlined by environmental health and safety protocols at your institution.

23.3.6 Supplies

- Beakers, 250 mL (one to melt fat; one to boil water) (2)
- Buchner funnel (to fit side-arm flask)
- Buret, 10 or 25 mL
- Fat and/or oil samples
- Filter paper (to fit Buchner funnel, to filter melted fat and oil)
- Flasks, 500 mL, glass stoppered (4)
- Graduated cylinder, 25 mL
- Graduated cylinder, 100 mL
- Mechanical pipettor, 1000 μL, with plastic tips (or 1 mL volumetric pipette)
- Side-arm flask
- Volumetric pipette, 10 mL
- Volumetric pipette, 20 mL

23.3.7 Equipment

- Analytical balance
- Hot plate

23.3.8 Procedure (Instructions Are Given for Analysis in Duplicate)

1. Melt any samples that are solid at room temperature by heating to a maximum of 15 °C above the melting point. Filter melted fat sample and oil sample through filter paper to remove impurities.
2. Weigh accurately 0.25 to 0.3 g sample (amount used depends on the expected iodine number) into each of two dry 500-mL glass-stoppered flasks. Add 10 mL chloroform to dissolve the fat or oil.
3. Prepare two blanks by adding only 10 mL chloroform to 500 mL glass-stoppered flasks.
4. Pipette 25 mL Wijs iodine solution into the flasks. (The amount of iodine must be 50–60% in excess of that absorbed by the fat.)
5. Let flasks stand for 30 min in the dark with occasional shaking.
6. After incubation in the dark, add 20 mL of the 15% potassium iodide solution to each flask. Shake thoroughly. Add 100 mL freshly boiled and cooled water, washing down any free iodine on the stopper.
7. Titrate the iodine in the flasks with standard sodium thiosulfate (0.1 N), adding it gradually with constant and vigorous shaking until the yellow color almost disappears. Then add 1–2 mL of starch indicator and continue the titration until the blue color entirely disappears. Toward the end of the titration, stopper the flask and shake violently so that any iodine remaining in the chloroform can be taken up by the potassium iodide solution. Record the volume of titrant used.

23.3.9 Data and Calculations

Sample	Weight (g)	Titrant volume (mL)	Iodine value
1			
2			
			$\bar{X} =$
			SD =

Oil/fat sample type tested:
Blank titration (mL):
Blank 1 =
Blank 2 =
$\bar{X} =$
Calculate the iodine value of each sample as follows:

$$\text{Iodine value} = \frac{(B - S) \times N \times 126.9}{W \times 1000} \times 100$$

where:
Iodine value = g iodine absorbed per 100 g of sample
B = volume of titrant (mL) for blank
S = volume of titrant (mL) for sample
N = normality of $Na_2S_2O_3$ (mol/1000 mL)
126.9 = MW of iodine (g/mol)
W = sample mass (g)

23.3.10 Questions

1. In the iodine value determination, why is the blank volume higher than that of the sample?
2. What does a high versus a low iodine value tell you about the nature of the sample?

23.4 FREE FATTY ACID VALUE

23.4.1 Objective

Determine the free fatty acid (FFA) value of fats and oils.

23.4.2 Principle of Method

Free fatty acid value, or acid value, reflects the amount of fatty acids hydrolyzed from triacylglycerols. Free fatty acid is the percentage by weight of a specific fatty acid. Acid value is defined as the milligrams of potassium hydroxide needed to neutralize the free acids present in 1 g of fat or oil. A liquid fat sample combined with neutralized 95% ethanol is titrated with standardized sodium hydroxide to a phenolphthalein endpoint. The volume and normality of the sodium hydroxide are used, along with the weight of the sample, to calculate the free fatty acid value.

23.4.3 Chemicals

	CAS no.	Hazards
Ethanol	64-17-5	Highly flammable
Phenolphthalein	77-09-8	Irritant
Sodium hydroxide (NaOH)	1310-73-2	Corrosive

23.4.4 Reagents

(**It is recommended that these solutions be prepared by the laboratory assistant before class.)

- Ethanol, neutralized**

- Neutralize 95% ethanol to a permanent pink color with alkali and phenolphthalein.
- Phenolphthalein indicator**
- In a 100 mL volumetric flask, dissolve 1 g phenolphthalein in 50 mL 95% ethanol. Dilute to volume with dd water.
- Sodium hydroxide, 0.1 N, standardized**

- Use commercial product, or prepare as described in other laboratory manual chapters (Chaps. 2 and 21).

23.4.5 Hazards, Precautions, and Waste Disposal

Adhere to normal laboratory safety procedures. Wear safety glasses at all times. Wastes likely may be put down the drain using a water rinse, but follow good laboratory practices outlined by environmental health and safety protocols at your institution.

23.4.6 Supplies

- Beaker, 250 mL (to melt fat)
- Buchner funnel (to fit side-arm flask)
- Buret, 10 mL
- Erlenmeyer flasks, 250 mL (4)
- Fat and/or oil samples
- Filter paper (to fit Buchner funnel; to filter melted fat and oil)
- Graduated cylinder, 100 mL
- Mechanical pipettor, 1000 µL, with plastic tips (or 1 mL volumetric pipette)
- Side-arm flask

23.4.7 Equipment

- Analytical balance
- Hot plate

23.4.8 Procedure

(Instructions are given for analysis in triplicate.)

1. Melt any samples that are solid at room temperature by heating to a maximum of 15 °C above the melting point. Filter melted fat sample and oil sample through filter paper to remove impurities.

2. As a preliminary test, accurately weigh ca. 5 g melted fat or oil into a 250 mL Erlenmeyer flask.
3. Add ca. 100 mL neutralized ethanol and 2 mL phenolphthalein indicator.
4. Shake to dissolve the mixture completely. Titrate with standard base (ca. 0.1 N NaOH), shaking vigorously until the endpoint is reached. This is indicated by a slight pink color that persists for 30 s. Record the volume of titrant used. Use the information below to determine if the sample weight you have used is correct for the range of acid values under which your sample falls. This will determine the sample weight to be used for Step 5.

The *Official Methods and Recommended Practices of the AOCS* (AOCS, 2013) recommends the following sample weights for ranges of expected acid values:

FFA range (%)	Sample (g)	Alcohol (mL)	Strength of alkali
0.00–0.2	56.4 ± 0.2	50	0.1 N
0.2–1.0	28.2 ± 0.2	50	0.1 N
1.0–30.0	7.05 ± 0.05	75	0.25 N

5. Repeat Steps 1–3 more carefully in triplicate, recording each weight of the sample and the volume of the titrant.

23.4.9 Data and Calculations

Sample	Weight (g)	Titrant volume (mL)	FFA value
1			
2			
3			
			$\bar{X} =$
			SD =

Oil/fat sample type tested:
Calculate the FFA value of each sample as follows:

$$\%FFA\left(as\,oleic\right)=\frac{V \times N \times 282}{W \times 1000}\times 100$$

where:
% FFA = percent free fatty acid (g/100 g), expressed as oleic acid
V = volume of NaOH titrant (mL)
N = normality of NaOH titrant (mol/1000 mL)
282   = MW of oleic acid (g/mol)
W = sample mass (g)

23.4.10 Questions

1. What is a high FFA value indicative of relative to product history?
2. Why is the FFA content of frying oil important?

3. In a crude fat extract, FFA are naturally present, but they are removed during processing to enhance the stability of the fat. State and describe the processing step that removes the FFA naturally present.

23.5 PEROXIDE VALUE

23.5.1 Objective

Determine the peroxide value of fats and oils, as an indicator of oxidative rancidity.

23.5.2 Principle of Method

Peroxide value is defined as the milliequivalents of peroxide per kilogram of fat, as determined in a titration procedure to measure the amount of peroxide or hydroperoxide groups. To a known amount of fat or oil, excess potassium iodide is added, which reacts with the peroxides in the sample. The iodine liberated is titrated with standardized sodium thiosulfate using a starch indicator. The calculated amount of potassium iodide required to react with the peroxide present is used to determine the peroxide value.

23.5.3 Chemicals

	CAS no.	Hazards
Acetic acid (glacial)	64-19-7	Corrosive
Chloroform	67-66-3	Harmful
Hydrochloric acid (HCl)	7647-01-0	Corrosive
Potassium chromate ($K_2Cr_2O_7$)	7789-00-6	Toxic, dangerous for environment
Potassium iodide (KI)	7681-11-0	
Sodium thiosulfate	7772-98-7	
Soluble starch	9005-25-8	

23.5.4 Reagents

(**It is recommended that these solutions be prepared by the laboratory assistant before class.)

- Acetic acid-chloroform solution**

  Mix three volumes of concentrated acetic acid with two volumes of chloroform.
- Potassium iodide solution, saturated**

  Dissolve excess KI in freshly boiled dd water. Excess solid must remain. Store in the dark. Test before use by adding 0.5 mL acetic acid-chloroform solution, then add 2 drops 1% starch indicator solution. If the solution turns blue, requiring >1 drop 0.1 N thiosulfate solution to discharge color, prepare a fresh potassium iodide solution.

- Sodium thiosulfate, 0.1 N, standard solution (AOAC Method 942.27)**

  Dissolve ca. 25 g sodium thiosulfate pentahydrate in 1 L dd water. Boil gently for 5 min. Transfer while hot to a storage bottle (make sure the bottle has been well cleaned and is heat resistant). Store the solution in a dark, cool place. Use the following procedure to standardize the sodium thiosulfate solution: Accurately weigh 0.20–0.23 g potassium chromate ($K_2Cr_2O_7$) (previously dried for 2 h at 100 °C) into a glass-stoppered flask. Dissolve 2 g potassium iodide (KI) in 80 mL chlorine-free water. Add this water to the potassium chromate. To this solution, add, with swirling, 20 mL ca. 1 M HCl, and immediately place in the dark for 10 min. Titrate a known volume of this solution with the sodium thiosulfate solution, adding starch solution after most of the iodine has been consumed.
- Starch indicator solution, 1% (prepare fresh daily)**

  Mix ca. 1 g soluble starch with enough cold dd water to make a thin paste. Add 100 mL boiling dd water. Boil ca. 1 min while stirring.

23.5.5 Hazards, Precautions, and Waste Disposal

Potassium chromate is toxic and must be handled with caution. Use hydrochloric acid in a fume hood. Otherwise, adhere to normal laboratory safety procedures. Wear gloves and safety glasses at all times. Chloroform and potassium chromate must be handled as hazardous wastes. Other wastes likely may be put down the drain using a water rinse, but follow good laboratory practices outlined by environmental health and safety protocols at your institution.

23.5.6 Supplies

- Beaker, 250 mL (to melt fat)
- Buchner funnel (to fit side-arm flask)
- Buret, 25 mL or 50 mL
- Erlenmeyer flasks, 250 mL, glass stoppered (4)
- Fat and/or oil samples
- Filter paper (to fit Buchner funnel, to filter melted fat and oil)
- Graduated cylinders, 50 mL (2)
- Mechanical pipettor, 1000 µL, with plastic tips (or 1 mL volumetric pipette)
- Side-arm flask

23.5.7 Equipment

- Analytical balance
- Hot plate

23.5.8 Procedure

(Instructions are given for analysis in duplicate.)

1. Melt any samples that are solid at room temperature by heating to a maximum of 15 °C above the melting point. Filter melted fat sample and oil sample through filter paper to remove impurities.
2. Accurately weigh ca. 5 g fat or oil (to the nearest 0.001 g) into each of two 250-mL glass-stoppered Erlenmeyer flasks.
3. Add 30 mL acetic acid-chloroform solution and swirl to dissolve.
4. Add 0.5 mL saturated KI solution. Let stand with occasional shaking for 1 min. Add 30 mL dd water.
5. Slowly titrate samples with 0.1 N sodium thiosulfate solution (you might need lower concentration depending on the nature of the sample, your TA will guide you), with vigorous shaking until the yellow color is almost gone.
6. Add ca. 0.5 mL 1% starch solution, and continue titration, shaking vigorously to release all iodine from the chloroform layer, until the blue color just disappears. Record the volume of titrant used. (If <0.5 mL of the sodium thiosulfate solution is used, repeat determination.)
7. Prepare (omitting only the oil) and titrate a blank sample. Record the volume of titrant used.

23.5.9 Data and Calculations

Sample	Weight (g)	Titrant volume (mL)	Peroxide value
1			
2			
			$\bar{X}=$
			SD =

Oil/fat sample type tested:
Blank titration (mL):
Blank 1 =
Blank 2 =
$\bar{X}=$
Calculate the peroxide value of each sample as follows:

$$\text{Peroxide value} = \frac{(S-B) \times N}{W} \times 1000$$

where:
Peroxide value = mEq peroxide per kg of sample
S = volume of titrant (mL) for sample
B = volume of titrant (mL) for blank
N = normality of $Na_2S_2O_3$ solution (mEq/mL)
1000 = conversion of units (g/kg)
W = sample mass (g)

23.5.10 Questions

1. What are some cautions in using peroxide value to estimate the amount of autoxidation in foods?
2. The peroxide value method was developed for fat or oil samples. What must be done to a food sample before measuring its peroxide value using this method?

23.6 THIN-LAYER CHROMATOGRAPHY SEPARATION OF SIMPLE LIPIDS

23.6.1 Objective

Separate and identify the lipids in some common foods using thin-layer chromatography (TLC).

23.6.2 Principle of Method

Like all types of chromatography, TLC is a separation technique that allows for the distribution of compounds between a mobile phase and a stationary phase. Most classes of lipids can be separated from each other by adsorption chromatography on thin layers. In TLC, a thin layer of stationary phase is bound to an inert support (i.e., glass plate, plastic, or aluminum sheet). The sample and standards are applied as spots near one end of the plate. For ascending chromatography, the plate is placed in a developing chamber, with the end of the plate nearest the spots being placed in the mobile phase at the bottom of the chamber. The mobile phase migrates up the plate by capillary action, carrying and separating the sample components. The separated bands can be visualized or detected and compared to the separation of standard compounds.

23.6.3 Chemicals

	CAS no.	Hazards
Acetic acid	64-19-7	Corrosive
Diethyl ether	60-29-7	Harmful, extremely
Flammable		
Hexane	110-54-3	Harmful, highly flammable, dangerous for the environment
Sulfuric acid	7664-93-9	Corrosive

23.6.4 Reagents

- Chloroform:methanol, 2:1, v/v
- Mobile phase
  Hexane:diethyl ether:acetic acid, 78:20:2
- Standards
  Triacylglycerol, fatty acid, cholesteryl ester, and cholesterol
- Sulfuric acid solution
  Concentrated H_2SO_4, in 50% aqueous solution

23.6.5 Hazards, Precautions, and Waste Disposal

Use acetic acid and sulfuric acid in a fume hood. Diethyl ether is extremely flammable, is hygroscopic, and may form explosive peroxides. Otherwise, adhere to normal laboratory safety procedures. Wear safety glasses at all times. Diethyl ether and hexane must be handled as hazardous wastes. Other wastes likely may be put down the drain using a water rinse, but follow good laboratory practices outlined by environmental health and safety protocols at your institution.

23.6.6 Supplies

- Capillary tubes (or syringes) (to apply samples to plates)
- Developing tank, with lid
- Filter paper, Whatman No. 1 (to line developing tank)
- Oil/fat food samples (e.g., hamburger, safflower oil) (prepare at a concentration of 20 µg/mL in 2: 1 v/v chloroform-methanol solution)
- Pencil
- Thin-layer chromatography plates: Silica Gel 60, 0.25 mm thick coating on glass backing, 20 × 20 cm (EM Science)

23.6.7 Equipment

- Air blower (e.g., blow hair dryer)
- Oven

23.6.8 Procedure

23.6.8.1 Preparation of Silica Gel Plates

1. Place plates in oven at 110 °C for 15 min, then cool to ambient temperature (5 min).
2. With a pencil, draw a line to mark the origin, 2.5 cm from the bottom of the plate.
3. Make marks with a pencil to divide the plate into 10 "lanes" of equal width.

4. Use capillary tubes or syringes to apply approximately 10 µl of each standard and sample to a separate lane (use the middle eight lanes). The application should be done as a streak across the center of the lane origin. This is best accomplished with four spots of 2.5 µl each.
5. Below the origin line, write the identity of the sample/standard in each lane.
6. Allow spots to dry. You may accelerate drying by using a low-temperature air blower.
7. Write your name in the top right corner of the plate.

23.6.8.2 Development of Plates

1. Line the developing tank with Whatman no. 1 or similar filter paper.
2. Pour the mobile phase gently over the filter paper until the depth of solvent in the tank is approximately 0.5 cm. About 200 mL is required.
3. Place the lid on the tank and allow 15 min for the atmosphere in the tank to become saturated with solvent vapor.
4. Place the spotted TLC plate in the developing tank and allow it to develop until the solvent front reaches a point about 2 cm from the top of the plate.
5. Remove the plate from the tank and *immediately* mark the position of the solvent front. Evaporate the solvent in the fume hood.

23.6.8.3 Visualization of Lipids

1. In a well-ventilated fume hood, spray lightly with 50% aqueous H_2SO_4. Allow to dry.
2. Heat plate for 5–10 min at 100–120 °C. Remove from oven, cool, and inspect. Handle the plate with caution as the surface still contains sulfuric acid.
3. Mark all visible spots at their center, and note the color of the spots.

23.6.9 Data and Calculations

For the spots of each of the standards and the samples, report the distance from the origin for the spot. Also for each spot, calculate the R_f value, as the distance from the origin to the spot divided by the distance from the origin to the solvent front. Using the R_f value of the standards, identify as many of the spots (bands) in the samples as possible.

Standard	Distance from origin	R_f Value
Triacylglycerol		
Fatty acid		
Cholesteryl ester		
Cholesterol		

Sample spot number	Distance from origin	R_f Value	Identity

Oil/fat sample type tested:

23.6.10 Questions

1. Explain the chemical structure of an ester of cholesterol.
2. Besides the four fat constituents used as standards, what other fat constituents might be found using a TLC method such as this?

Acknowledgments The authors of this chapter thank Michael C. Qian (Department of Food Science and Technology, Oregon State University, Corvallis, OR, USA) and Oscar A. Pike (Department of Nutrition, Dietetics, and Food Science, Brigham Young University, Provo, UT, USA) for contributing material for this chapter when it was originally developed.

RESOURCE MATERIALS

AOCS (2020) Official methods and recommended practices of the AOCS, 7th edn, 2rd printing. American Oil Chemists' Society, Champaign, IL

Pike OA (2024) Fat characterization. Ch. 23. In: Ismail BP, Nielsen SS (eds) Nielsen's food analysis, 6th edn. Springer, New York

Proteins: Extraction, Quantitation, and Electrophoresis

24

S. Suzanne Nielsen and Denise M. Smith

Contents

24.1 INTRODUCTION

24.1.1 Background

Electrophoresis can be used to separate and visualize protein banding patterns. In sodium dodecyl sulfate-polyacrylamide gel electrophoresis (SDS-PAGE), proteins are dissociated into subunits and then separated based on size within a gel matrix by applying an electric field. It is usually necessary to apply a sample volume to the gel that contains a known amount of total

S. S. Nielsen (✉)
Department of Food Science, Purdue University,
West Lafayette, IN, USA
e-mail: nielsens@purdue.edu

D. M. Smith
School of Food Science, Washington State University,
Pullman, WA, USA
e-mail: denise.smith@wsu.edu

protein to allow comparisons between samples. While it is possible to use an official method (e.g., Kjeldahl, N combustion) to determine total protein, it is often convenient to use a rapid colorimetric method of protein analysis that requires only a small amount of sample. The bicinchoninic acid (BCA) assay method will be used for this purpose.

In this experiment, sarcoplasmic muscle proteins are extracted with a 0.15 M salt solution, the protein content of the extract is measured by the BCA colorimetric assay, and the proteins in the fish extracts are separated and visualized by SDS-PAGE. Visualization of the protein banding patterns makes it possible to distinguish among different types of fish since many fish have a characteristic protein subunit pattern. For example, one might use this technique as part of a scheme to detect economic adulteration when an inexpensive fish species is substituted for a more expensive fish in the wholesale or retail marketplace.

24.1.2 Reading Assignment

Baraem BP, Smith DM (2024) Protein separation and characterization. Ch. 24. In: Ismail BP, Nielsen SS (eds) Nielsen's food analysis, sixth edn. Springer, New York.

Dee DR, Chang SKC (2024) Protein analysis. Ch. 18. In: Ismail BP, Nielsen SS (eds) Nielsen's food analysis, sixth edn. Springer, New York.

24.1.3 Objective

Extract proteins from the muscles of freshwater and saltwater fish, measure the protein content of the extracts, separate the proteins by electrophoresis, and then compare the different protein banding patterns that result based on subunit size and relative quantity.

24.1.4 Principle of Method

Sarcoplasmic proteins can be extracted from fish muscle with 0.15 M salt. Protein content of the extract can be determined by the colorimetric BCA assay method. In this assay, protein reduces cupric ions to cuprous ions under alkaline conditions. The amount of cuprous ions formed is proportional to the amount of protein present. The cuprous ions react with the BCA reagent to form a purple color that can be quantified spectrophotometrically and related to the protein content by comparison to a standard curve. Proteins in the extract can be separated by SDS-PAGE. Proteins become negatively charged when bound to SDS, so they move through the gel matrix toward the anode (pole with positive charge) at a rate based on size alone. The molecular mass of a given protein subunit

can be estimated by comparing its electrophoretic mobility with proteins of known molecular weight using a standard curve. A linear relationship is obtained by plotting the logarithm of the molecular mass of standard proteins against their respective electrophoretic mobilities (R_f).

24.1.5 Notes

This experiment may be done over two laboratory sessions. The protein can be extracted, quantified, and prepared for electrophoresis in the first session. The prepared protein samples can be frozen until electrophoresis is completed in the second laboratory session. Alternatively, in a single laboratory session, one group of students could do the protein extraction and quantitation, while a second group of students prepares the electrophoresis gels. Also, different groups of students could be assigned different fish species. Multiple groups could run their samples on a single electrophoresis gel. The gels for electrophoresis can be purchased commercially (e.g., Bio-Rad, Mini-PROTEAN TGX precast gels, 12% resolving gel) or made as described below.

Some fish species work better than others for preparing the extracts and comparing differences in protein patterns. Catfish (freshwater) and tilapia (saltwater) work well as extracts and show some differences. Trout gives very thick extracts. Freshwater and saltwater salmon show few differences.

24.1.6 Chemicals

	CAS No.	Hazards
Sample extraction:		
Sodium chloride (NaCl)	7647-14-5	Irritant
Sodium phosphate, monobasic (NaH$_2$PO$_4$. H$_2$O)	7558-80-7	Irritant
Protein determination (BCA method):		
Bicinchoninic acid	979-88-4	Irritant, toxic
Bovine serum albumin (BSA)	9048-46-8	
Copper sulfate (CuSO$_4$)	7758-98-7	Irritant
Sodium bicarbonate (NaHCO$_3$)	144-55-8	
Sodium carbonate (Na$_2$CO$_3$)	497-19-8	Irritant
Sodium hydroxide (NaOH)	1310-73-2	Corrosive
Sodium tartrate	868-18-8	
Electrophoresis:		
Acetic acid (CH$_3$COOH)	64-19-7	Corrosive
Acrylamide	79-06-1	Toxic
Ammonium persulfate (APS)	7727-54-0	Harmful, oxidizing
Bis-acrylamide	110-26-9	Harmful
Bromophenol blue	115-39-9	
Butanol	71-36-3	Harmful

	CAS No.	Hazards
Coomassie blue R-250	6104-59-2	
Ethylenediaminetetraacetic acid, disodium salt (Na₂EDTA·2H₂O)	60-00-4	Irritant
Glycerol (C₃H₈O₃)	56-81-5	
Glycine	56-40-6	
Hydrochloric acid (HCl)	7647-01-0	Corrosive
β-Mercaptoethanol	60-24-2	Toxic
Methanol (CH₃OH)	67-56-1	Highly flammable
Protein molecular weight standards (e.g., bio-rad 161-0374, Precision plus protein dual Color standards, 10–250 KD)		
Sodium dodecyl sulfate (SDS, dodecyl sulfate, sodium salt)	151-21-3	Harmful
N, N, N′, N′-Tetramethylethylenediamine (TEMED)	110-18-9	Highly flammable, corrosive
Tris base	77-86-1	

24.2 REAGENTS

(** It is recommended that these solutions be prepared by a laboratory assistant before class.)

24.2.1 Sample Extraction

- Extraction buffer, 300 mL/fish species **.
  Buffer of 0.15 M sodium chloride, 0.05 M sodium phosphate, pH 7.0 (Students asked to show the calculations for this buffer later in Questions.)

24.2.2 Protein Determination (BCA Method)

Commercial BCA test kits can be purchased, such as the Pierce BCA Protein Assay Kit (cat. No. 23225, Rockford, IL), which includes procedures to complete either test tube (2 mL working reagent/tube) or microplate assays (200 μL working reagent/well).

This kit includes:

- Bovine serum albumin (BSA) standard ampules, 2 mg/mL in 0.9% saline, and 0.05% sodium azide as a preservative
- BCA Reagent A: Contains sodium carbonate, sodium bicarbonate, BCA detection reagent, and sodium tartrate in 0.1 M sodium hydroxide
- Reagent B: 4% cupric sulfate solution

24.2.3 Electrophoresis

(*Note:* Running buffers (Tris/glycine/SDS), sample buffers (Laemmli sample buffer), and precast gels can be purchased commercially or made in the laboratory. β-Mercaptoethanol may need to be added to a commercial sample preparation buffer.)

- Acrylamide: bis-acrylamide solution**
  29.2 g acrylamide and 2.4 g methylene bis-acrylamide, with dd water to 100 mL
- Ammonium persulfate (APS), 7.5%, in dd water, 1 mL, prepared fresh daily**
- Bromophenol blue, 0.05%
- Coomassie Brilliant Blue stain solution**
  Purchase premixed or prepare: 454 mL dd water, 454 mL methanol, 92 mL acetic acid, and 1.5 g Coomassie Brilliant Blue R-250 (*Note*: Bio-Rad offers Coomassie G-250 stains that do not require traditional destaining with methanol or acetic acid. A water wash stops the staining process.)
- Destain solution**
  850 mL dd water, 75 mL methanol, 75 mL acetic acid
- EDTA, disodium salt, 0.2 M, 50 mL **.
- Glycerol, 37% (use directly).
- Electrophoresis sample preparation buffer **.
  Purchase premixed or prepare: 1 mL of 0.5 M Tris (pH 6.8), 0.8 mL glycerol, 1.6 mL 10% SDS, 0.4 mL β-mercaptoethanol, and 0.5 mL 0.05% (wt/vol) bromophenol blue, diluted to 8 mL with dd water.
- Sodium dodecyl sulfate, 10% solution in dd water, 10 mL**.
- TEMED (use directly).
- Tray (running) buffer, 25 mM Tris, 192 mM glycine, 0.1% SDS, pH 8.3**.
  Purchase premixed or prepare.
- Tris buffer, 1.5 M, pH 8.8, 50 mL [separating (resolving) gel buffer]**.
  Purchase premixed or prepare
- Tris buffer, 0.5 M, pH 6.8, 50 mL (stacking gel buffer)**.
  Purchase premixed or prepare

Gel preparation: Use the formula that follows and the instructions in the procedure to prepare two 8.4 × 5.0 cm SDS-PAGE slab gels, 15% acrylamide, 0.75 mm thick. Actual formulation will depend on the size of slab gels.

Reagent	Separating gel 15% gel	Stacking gel 4.5% gel
Acrylamide:bis-acrylamide	2.4 mL	0.72 mL
10% SDS	80 μL	80 μL
1.5 M Tris, pH 8.8	2.0 mL	–

Reagent	Separating gel 15% gel	Stacking gel 4.5% gel
0.5 M Tris, pH 6.8	–	2.0 mL
dd water	3.6 mL	5.3 mL
37% glycerol	0.15 mL	–
10% APS[a]	40 μL	40 μL
TEMED	10 μL	10 μL

[a]APS is added to separating and stacking gels after all other reagents are combined, solution is degassed, and each gel is ready to be poured

24.2.4 Hazards, Precautions, and Waste Disposal

Acrylamide monomers may cause cancer and are very toxic in contact with skin and if swallowed. β-Mercaptoethanol is harmful if swallowed, toxic in contact with skin, and irritating to eyes. Adhere to normal laboratory safety procedures. Wear gloves and safety glasses at all times. Acrylamide and β-mercaptoethanol wastes must be disposed of as hazardous wastes. Gloves and pipette tips in contact with acrylamide and β-mercaptoethanol also should be handled as hazardous wastes. Other waste likely can be washed down the drain with a water rinse, but follow good laboratory practices outlined by environmental health and safety protocols at your institution.

24.3 SUPPLIES

24.3.1 Sample Extraction

- Beaker, 250 mL
- Centrifuge tubes, 50 mL
- Cutting board
- Erlenmeyer flask, 125 mL
- Graduated cylinder, 50 mL
- Filter paper, Whatman No. 1
- Fish, freshwater (e.g., catfish) and saltwater species (e.g., tilapia)
- Funnel
- Knife
- Pasteur pipettes and bulbs
- Test tube with cap
- Weighing boat

24.3.2 Protein Determination (BCA Method)

- Beaker, 50 mL
- Graduated cylinder, 25 mL
- Mechanical, adjustable volume pipettor, 1000 μL, with plastic tips
- Test tubes

24.3.3 Electrophoresis

- Beaker, 250 mL (for boiling samples)
- Erlenmeyer flasks, 2 L (for stain and destain solutions) (2)
- Glass boiling beads (for boiling samples)
- Graduated cylinder, 100 mL
- Graduated cylinder, 500 mL
- Hamilton syringe (to load samples on gels)
- Mechanical, adjustable volume pipettors, 1000 μL, 100 μL, and 20 μL, with plastic tips
- Pasteur pipettes, with bulbs
- Rubber stopper (to fit 25-mL sidearm flasks)
- Sidearm flasks, 25 mL (2)
- Test tubes or culture tubes, small size, with caps
- Tubing (to attach to vacuum system to degas gel solution)
- Weighing paper/boats

24.4 EQUIPMENT

- Analytical balance
- Aspirator system (for degassing solutions)
- Blender
- Centrifuge
- Electrophoresis unit
- pH meter
- Power supply
- Spectrophotometer
- Top loading balance
- Vortex mixer
- Water bath

24.5 PROCEDURE

(Single sample extracted)

24.5.1 Sample Preparation

1. Coarsely cut up about 100 g fish muscle (representative sample) with a knife. Accurately weigh out 90 g on a top loading balance.
2. Blend one part fish with three parts extraction buffer (90 g fish and 270 mL extraction buffer) for 1.0 min in a blender. (*Note*: Smaller amounts of fish and buffer, but in the same 1:3 ratio, can be used for a small blender.)
3. Pour 30 mL of the muscle homogenate into a 50-mL centrifuge tube. Label tube with tape. Balance your tube against a classmate's sample. Use a spatula or Pasteur pipette to adjust tubes to an equal weight.

4. Centrifuge the samples at 2000 x g for 15 min at room temperature. Collect the supernatant.
5. Filter a portion of the supernatant by setting a small funnel in a test tube. Place a piece of Whatman No. 1 filter paper in the funnel, and moisten it with the extraction buffer. Filter the supernatant from the centrifuged sample. Collect about 10 mL of filtrate in a test tube. Cap the tube.
6. Determine protein content of filtrate using the BCA method and prepare sample for electrophoresis (see below).

24.5.2 BCA Protein Assay

(Instructions are given for duplicate analysis of each concentration of standard and sample.)

1. Prepare the Working Reagent for the BCA assay by combining Pierce Reagent A with Pierce Reagent B, 50:1 (v/v), A:B. For example, use 50-mL Reagent A and 1.0-mL Reagent B to prepare 51.0-mL Working Reagent, which is enough for the BSA standard curve and testing the extract from one type of fish. (*Note*: This volume is adequate for assaying duplicates of five standard samples and two dilutions of each of two types of fish.)
2. Prepare the following dilutions of the supernatant (filtrate from Procedure, Sample Preparation, Step 5): dilutions of 1:5, 1:10, and 1:20 in *extraction buffer*. Mix well.
3. In test tubes, prepare duplicates of each reaction mixture of diluted extracts and BSA standards (using 2 mg BSA/mL solution) as indicated in the table that follows:
4. Mix each reaction mixture with a vortex mixer, and then incubate in a water bath at 37 °C for 30 min.
5. Read the absorbance of each tube at 562 nm using a spectrophotometer.
6. Use the data from the BSA samples to create a standard curve of absorbance at 562 nm versus µg protein/tube. Determine the equation of the line for the standard curve. Calculate the protein concentration (µg/mL) of the extract from each fish species using the equation of the line from the BSA standard curve and the absorbance value for a dilution of the fish extract that had an absorbance near the middle point on the standard curve. Remember to correct for dilution used. (*Note*: Do not use the diluted samples for electrophoresis. Use the original extract prepared as described below.)

Tube identity	dd water	BSA std.	Fish extract (µL)	Working reagent (mL)
Blank	100	0	–	2.0
Std. 1	80	20	–	2.0
Std. 2	60	40	–	2.0
Std. 3	40	60	–	2.0
Std. 4	20	80	–	2.0
Std. 5	0	100	–	2.0
Sample 1:5	50	–	50	2.0
Sample 1:10	50	–	50	2.0
Sample 1:20	50	–	50	2.0

24.5.3 Electrophoresis

1. Assemble the electrophoresis unit according to the manufacturer's instructions.
2. Use the table that follows the list of electrophoresis reagents to combine appropriate amounts of all reagents for the *separating* gel, except APS, in a side-arm flask. Degas the solution and then add APS. Proceed immediately to pour the solution between the plates to create the separating (resolving) gel. Pour the gel to a height approximately 1 cm below the bottom of the sample well comb. Immediately, add a layer of butanol across the top of the separating gel, adding it carefully so as not to disturb the upper surface of the separating gel. This butanol layer will prevent a film from forming and help obtain an even surface. Allow the separating gel to polymerize for 30 min, and then remove the butanol layer just before the stacking gel is ready to be poured.
3. Use the table that follows the list of electrophoresis reagents to combine appropriate amounts of all reagents for the *stacking* gel, except APS, in a sidearm flask. Degas the solution for 15 min (as per the manufacturer's instructions) and then add APS. Proceed immediately to pour the solution between the plates to create the stacking gel. Immediately, place the well comb between the plates and into the stacking gel. Allow the stacking gel to polymerize for 30 min before removing the well comb. Before loading the samples into the wells, wash the wells twice with dd water.
4. Mix the fish extract samples well (filtrate from Sect. 24.5.1, Step 5). For each sample, combine 0.1-mL sample with 0.9 mL electrophoresis sample buffer in a screw cap culture tube. Apply cap.
5. Heat capped tubes for 3 min in boiling water.
6. Apply 10 and 20 µg protein of each fish extract to wells of the stacking gel using a syringe. Calculate the volume to apply based on the protein content of the extract and the dilution used when preparing the extract in electrophoresis sample buffer.

7. Apply 10 µl of molecular weight standards to one sample well.
8. Follow the manufacturer's instructions to assemble and run the electrophoresis unit. When the line of bromophenol blue tracking dye has reached the bottom of the separating gel, shut off the power supply. Disassemble the electrophoresis unit, and carefully remove the separating gel from between the plates. Place the gel in a flat dish with the Coomassie Brilliant Blue stain solution. Allow the gel to stain for at least 30 min. (If possible, place the dish with the gel on a gentle shaker during staining and destaining.) Pour off the stain solution and then destain the gel for at least 2 h using the destain solution with at least two changes of the solution.
9. Measure the migration distance (cm) from the top of the gel to the center of the protein band for the molecular weight standards and for each of the major protein bands in the fish extract samples. Also measure the migration distance of the bromophenol blue tracking dye from the top of the gel.
10. Observe and record the relative intensity of the major protein bands for each fish extract.

24.6   DATA AND CALCULATIONS

24.6.1 Protein Determination

Tube identity	Absorbance	µg Protein/tube	µg/mL sample
Std. 1, 20 µL BSA			
Std. 1, 20 µL BSA			
Std. 2, 40 µL BSA			
Std. 2, 40 µL BSA			
Std. 3, 60 µL BSA			
Std. 3, 60 µL BSA			
Std. 4, 80 µL BSA			
Std. 4, 80 µL BSA			
Std. 5, 100 µL BSA			
Std. 5, 100 µL BSA			
Sample 1:5			
Sample 1:5			
			$\bar{X} =$
Sample 1:10			
Sample 1:10			
			$\bar{X} =$
Sample 1:20			
Sample 1:20			
			$\bar{X} =$

Sample calculation for fish extract protein concentration:

For fish extract diluted 1:20 and 50 µL analyzed with absorbance of 0.677:

Equation of the line: $y = 0.0108x + 0.0022$

If $y = 0.677$, $x = 62.48$

$C_i = C_f(V_2/V_1)(V_4/V_3)$

(See Ch. 3 in this laboratory manual; C_i = initial concentration; C_f = final concentration)

$$C_i = \left(62.48\,\mu g\,\text{protein}\,/\,\text{tube}\right) \times \left(20\,\text{mL}\,/\,1\text{mL}\right) \times \left(\text{tube}\,/\,50\,\text{uL}\right)$$
$$= 24.99\,\mu g\,\text{protein}\,/\,\text{uL fish extract}$$

Sample calculation to determine the volume of prepared extract to apply 20 ug protein to each sample well:

How many µL are needed to get 20 µg protein? Remember the electrophoresis sample buffer dilution is 1:10:

24.99 µg protein/µL × Z µL × 1 mL/10 mL = 20 µg

Z = 8.00 uL

24.6.2 Electrophoresis

1. Calculate the relative mobility of three major protein bands and all the molecular weight standards. To determine the relative mobility (R_f) of a protein band, divide its migration distance from the top of the gel to the center of the protein band by the migration distance of the bromphenol blue tracking dye from the top of the gel:

$$R_f = \frac{\text{distance of protein migration}}{\text{distance of tracking dye migration}}$$

2. Prepare a standard curve by plotting relative mobility (x-axis) versus log molecular weight of standards (y-axis).
3. Using the standard curve, estimate the molecular weight of the major protein subunits in the freshwater and saltwater fish extracts.

Sample identity	Distance of protein migration	Distance of tracking dye migration	Relative mobility	Molecular weight
Molecular Weight Standards				
1				
2				
3				
4				
5				
Fish species:				
Freshwater				
Saltwater				

24.7   QUESTIONS

1. Describe how you would prepare 1 L of the buffer used to extract the fish muscle proteins (0.15 M sodium chloride, 0.05 M sodium phosphate, pH 7.0). Show all calculations.

2. Discuss the differences between the fish species, regarding the presence or absence of major protein bands identified by the molecular mass and the relative amounts of these proteins.

RESOURCE MATERIALS

Baraem BP, Smith DM (2024) Protein separation and characterization. Ch. 24. In: Ismail BP, Nielsen SS (eds) Nielsen's food analysis, 6th edn. Springer, New York

Bio-Rad (2023) A Guide to Polyacrylamide Gel Electrophoresis and Detection. Bulletin 6040 Rev B. Bio-Rad Laboratories https://www.bio-rad.com/webroot/web/pdf/lsr/literature/Bulletin_6040.pdf Accessed June 15, 2023

Dee DR and Chang SKC (2024) Protein analysis. Ch. 18. In: Ismail BP, Nielsen SS (eds) Nielsen's food analysis, 6th edn. Springer, New York

Etienne M et al. (2000) Identification of fish species after cooking by SDS-PAGE and urea IEF: a collaborative study. J Agr Food Chem 48:2653-2658

Etienne M et al. (2001) Species identification of formed fishery products and high pressure-treated fish by electrophoresis: a collaborative study. Food Chem 72:105-112

Laemmli UK (1970) Cleavage of structural proteins during the assembly of the head of bacteriophage T4. Nature 227:680-685

Olsen BJ and Markwell J (2007) Assays for the determination of protein concentration. Unit 3.4 Basic Protocol 3. BCA Assay. Current Protocols in Protein Science John Wiley and Sons, New York

Piñeiro C et al (1999) Development of a sodium dodecyl sulfate-polyacrylamide gel electrophoresis reference method for the analysis and identification of fish species in raw and heat-processed samples: a collaborative study. Electrophoresis 20:1425-1432

ThermoFisher Scientific (2020) User guide: Pierce™ BCA protein assay kit. Pierce Biotechnology, Waltham, MA https://assets.thermofisher.com/TFS-Assets/LSG/manuals/MAN0011430_Pierce_BCA_Protein_Asy_UG.pdf Accessed 6-15-2023

Glucose Determination by Enzyme Analysis

25

B. Pam Ismail and Robert E. Ward

Contents

25.1 INTRODUCTION

25.1.1 Background

Enzyme analysis is used for many purposes in food science and technology. Enzyme activity is used to indicate adequate processing, assess enzyme preparations, and measure constituents of foods that are enzyme substrates. In this experiment, the glucose content of corn syrup solids is determined using the enzymes glucose oxidase and peroxidase. Glucose oxidase catalyzes the oxidation of glucose to form hydrogen peroxide (H_2O_2), which then reacts with a dye in the presence of peroxidase to give a stable-colored product.

As described, this experiment uses individual, commercially available reagents, but enzyme test kits that include all the reagents to quantitate glucose in food also are available as a package [e.g., Sigma Glucose (HK) Assay Kit]. Enzyme test kits also are available to quantitate various other components of foods. Companies that sell enzyme test kits usually provide detailed instructions for the use of these kits, including information about the following: (1) principle of the assay, (2) contents of the test kit, (3) preparation of solutions, (4) stability of solutions, (5) procedure to follow, (6) calculations, and (7) further instructions regarding dilutions and recommendations for specific food samples.

25.1.2 Reading Assignment

Huber KC, BeMiller JN (2024) Carbohydrate analysis, Ch. 19. In: Ismail BP, Nielsen SS (eds) Nielsen's Food Analysis, 6th edn. Springer, New York.

B. P. Ismail (✉)
Department of Food Science and Nutrition, University of Minnesota, St. Paul, MN, USA
e-mail: bismailm@umn.edu

R. E. Ward
Department of Nutrition, Dietetics and Food Sciences, Utah State University, Logan, UT, USA
e-mail: robert.ward@usu.edu

© The Author(s), under exclusive license to Springer Nature Switzerland AG 2024
B. P. Ismail, S. S. Nielsen (eds.), *Nielsen's Food Analysis Laboratory Manual*, Food Science Text Series,
https://doi.org/10.1007/978-3-031-44970-3_25

Reyes-De-Coreuera JI (2024) Application of enzymes in food analysis. Ch. 26. In: Ismail BP, Nielsen SS (eds) Nielsen's Food Analysis, 6th edn. Springer, New York.

25.1.3 Objective

Determine the glucose content of food products using the enzymes, glucose oxidase and peroxidase.

25.1.4 Principle of Method

Glucose is oxidized by glucose oxidase to form hydrogen peroxide, which then reacts with a dye in the presence of peroxidase to give a stable-colored product that can be quantitated spectrophotometrically (coupled reaction).

25.1.5 Chemicals

	CAS no.	Hazards
Acetic acid (Sigma A6283)	64-19-7	Corrosive
o-Dianisidine. 2HCl (Sigma D3252)	20325-40-0	Tumor causing,
		Carcinogenic
D-Glucose (Sigma G8720)	50-99-7	
Glucose oxidase (Sigma G6641)	9001-37-0	
Horseradish peroxidase (Sigma P6782)	9003-99-0	
Sodium acetate (Sigma S2889)	127-09-3	
Sulfuric acid (Aldrich 320,501)	7664-93-9	Corrosive

25.1.6 Reagents

(**It is recommended that these solutions be prepared by the laboratory assistant before class.)

- Acetate buffer**, 0.1 M, pH 5.5
    Dissolve 8 g sodium acetate in ca. 800 mL water in a 1-L beaker. Adjust pH to 5.5 using 1 M HCl. Dilute to 1 L in a volumetric flask.
- Glucose test solution**
    In a 100-mL volumetric flask, dissolve 20 mg glucose oxidase (~ 300 to 1000 units), 40 mg horseradish peroxidase, and 40 mg o-dianisidine·2HCl in the 0.1 M acetate buffer. Dilute to volume with the acetate buffer and filter as necessary.
- Glucose standard solution, 1 mg/mL
    Use commercial D-glucose solution (e.g., Sigma)
- Sulfuric acid, diluted** (1 part H_2SO_4 + 3 parts water)
    In a 500-mL beaker in the hood, add 150 mL water, and then add 50 mL H_2SO_4. This will generate a lot of heat.

25.1.7 Hazards, Precautions, and Waste Disposal

Concentrated sulfuric acid is extremely corrosive; avoid contact with skin and clothes and breathing vapors. Acetic acid is corrosive and flammable. Wear safety glasses at all times and corrosive-resistant gloves. Otherwise, adhere to normal laboratory safety procedures. The o-dianisidine·2HCl must be disposed of as hazardous waste. Other waste likely may be put down the drain using a water rinse, but follow good laboratory practices outlined by environmental health and safety protocols at your institution.

25.1.8 Supplies

- Beaker, 1 L
- Corn syrup solids (or high fructose corn syrup), 0.5 g
- Spatulas (5)
- Test tubes, 18 × 150 mm, heavy walled to keep from floating in water bath (14)
- Test tube rack
- Volumetric flasks, 100 mL (2)
- Volumetric flask, 250 mL
- Volumetric pipette, 10 mL
- Volumetric flasks, 1 L (2)
- Weighing paper

25.1.9 Equipment

- Analytical balance
- Mechanical, adjustable volume pipettors, 200 µL, 1000 µL, and 5000 µL, with tips
- pH meter
- Spectrophotometer
- Water bath, 30 °C

25.2 PROCEDURE

(Instructions are given for analysis in duplicate.)

1. Prepare dilutions for standard curve. Use the adjustable pipettors to deliver aliquots of glucose standard solution (1 mg/mL) and deionized distilled (dd) water as indicated in the table below into clean test tubes. These dilutions will be used to create a standard curve of 0–0.2 mg glucose/mL.

	mg glucose/mL				
	0	0.05	0.10	0.15	0.20
mL glucose std. solution	0	0.150	0.300	0.450	0.600
mL dd water	3.000	2.850	2.700	2.550	2.400

2. Prepare sample solution and dilutions. Accurately weigh ca. 0.50 g corn syrup solids and dilute with water to volume in a 250-mL volumetric flask (Sample A). Using volumetric pipettes and flasks, dilute 10.00 mL of Sample A to 100 mL with water (Sample B). These sample dilutions will let you determine glucose concentrations in samples containing 1–100% glucose.
3. Add 1.000 mL of water to each of the 14 test tubes. In duplicate, add 1.000 mL of the individual standard and sample dilutions to the test tubes.
4. Put all tubes in the water bath at 30 °C for 5 min. Add 1.000 mL glucose test solution to each tube at 30 s intervals.
5. After exactly 30 min, stop the reactions by adding 10 mL of the diluted H_2SO_4. Cool to room temp.
6. Zero spectrophotometer with water in the reference position using a double-beam spectrophotometer. Take two readings (repeated measures, msmt) using separate aliquots from each tube.

25.3 DATA AND CALCULATIONS

Weight original sample: _____ g
   Absorbance of standard solutions:

Tube	Msmt	(mg glucose/mL)				
		0	0.05	0.10	0.15	0.20
1	1					
	2					
2	1					
	2					
Average						
Absorbance:						

   Absorbance of samples:

Tube	Msmt	Sample A	Sample B
1	1		
	2		
2	1		
	2		
Average			
Absorbance:			

   Calculation of glucose concentration in sample:

1. Plot absorbance of standards on the y-axis versus mg glucose/mL on the x-axis.
2. Calculate the concentration of glucose for the sample dilution A or B that had an absorbance within the working range of the standard curve:

$$\left(Absorbance - y - intercept\right)/slope = mg\,glucose\,/\,mL$$

3. Calculate the glucose concentration in the original sample, as a percentage.

   Example calculations:

   Original sample of 0.512 g

   Average measured absorbance sample dilution B: 0.200

   Calculation from standard curve:
   0.200–0.003/(2.98 mL/mg glucose) = 0.066 mg glucose/mL B)
   C_{sample} = (0.066 mg glucose/mL B) × (100 mL B/10 mL A) × (250 mL A/512 mg sample)

   = 0.323 mg glucose/mg sample
   = 0.323 g glucose per g sample
   = 32.3% glucose

25.4 QUESTIONS

1. Explain why this experiment is said to involve a coupled reaction. Write in words the equations for the reactions. What conditions must be in place to ensure accurate results for such a coupled reaction?
2. How do the results obtained compare to specifications for the commercial product analyzed?

Acknowledgments The authors of this chapter thank Charles E. Carpenter (Department of Nutrition, Dietetics and Food Sciences, Utah State University, Logan UT, USA) for contributing material for the chapter when it was originally developed.

Resource Materials

Huber KC, BeMiller JN (2024) Carbohydrate analysis, Ch. 19. In: Ismail BP, Nielsen SS (eds) Nielsen's food analysis, 6th edn. Springer, New York
Reyes-De-Coreuera JI (2024) Application of enzymes in food analysis. Ch. 26. In: Ismail BP, Nielsen SS (eds) Nielsen's food analysis, 6th edn. Springer, New York

Gliadin Detection by Immunoassay

26

Qinchun Rao

Contents

26.1 INTRODUCTION

26.1.1 Background

Immunoassays are very sensitive and efficient tests that are commonly used to identify a specific protein. Examples of applications in the food industry include the identification of proteins expressed in genetically modified foods, allergens, or proteins associated with a disease, including celiac disease. This genetic disease is associated with about 1% of the world's population and more than two million Americans. These individuals react immunologically to wheat proteins. Consequently, their own immune systems attack and damage their intestines. This disease can be managed if gluten is avoided in foods. Gluten consists of alcohol-insoluble glutenins and alcohol-soluble prolamins. Gluten is mainly found in wheat, oat, barley, rye, and other grain flours and related starch derivatives. Rice and corn are two common grains that do not contain significant gluten and are well tolerated by those with celiac disease. Wheat protein makes up 7–15% of a wheat grain. Prolamins in wheat are called gliadin. About 40% of wheat proteins are various forms of gliadin protein.

The immune system of animals can respond to many foreign substances by the development of specific antibodies. Antibodies bind strongly to and assist in the removal of a foreign substance in the body. Animals make antibodies against many different "antigens," defined as foreign substances that will elicit a specific immune response in the host. These include foreign proteins, peptides, carbohydrates, nucleic acids, lipids, and other naturally occurring or synthetic compounds.

Immunoassays are tests that take advantage of the remarkably specific and strong binding of antibodies to antigens. Immunoassays can be used to determine the presence and

Q. Rao (✉)
Department of Health, Nutrition, and Food Sciences,
Florida State University, Tallahassee, FL, USA
e-mail: qrao@fsu.edu

B. P. Ismail, S. S. Nielsen (eds.), *Nielsen's Food Analysis Laboratory Manual*, Food Science Text Series,
https://doi.org/10.1007/978-3-031-44970-3_26

quantity of either antibody or antigen. Antibodies that identify a specific protein (antigen) can be developed by immunizing a laboratory animal with this protein, similar to human vaccination against a specific disease. These antigen-specific antibodies can be used in vitro testing to identify the antigen in a food (e.g., detection of gliadin in food products) through the appropriate use of a label, such as an enzyme or fluorescent molecules linked covalently to either the antibody or a reference antigen. This type of immunoassay concept also can be used to determine the presence of specific antibodies in human blood. For example, by analyzing the presence of gliadin-specific antibodies in an individual's blood, one can determine if the individual has celiac disease.

26.1.2 Reading Assignment

Rao Q (2024) Immunoassays. Ch.27. In: Ismail BP, Nielsen SS (eds) Food analysis, 6th edn. Springer, New York

26.1.3 Objective

Determine the presence of gliadin in various food products using a rabbit anti-gliadin antibody horseradish peroxidase conjugate in a dot blot immunoassay.

26.1.4 Principle of Method

A simple dot blot immunoassay will be used in this lab to detect gliadin in food samples. Dot blot assays use nitrocellulose (NC) paper as a solid phase carrier. Initially, gliadin in food samples is isolated by differential centrifugation, in which most of the non-gliadin proteins are washed away with water and sodium chloride (NaCl) solution, then gliadin in the pellet is extracted with a detergent solution. A drop of this food sample extract or standard antigen (gliadin) is applied to the NC paper, which adheres nonspecifically. The remaining binding sites on the NC paper are then "blocked" using a protein unrelated to gliadin, such as bovine serum albumin (BSA), to minimize the nonspecific binding. The bound gliadin in the spots can then be detected with an antigen-specific antibody-enzyme conjugate. Theoretically, this antibody probe will bind only to gliadin bound to the NC paper. Next, this NC paper is washed free of unbound antibody-enzyme conjugate and then placed in a substrate solution in which an enzymatically catalyzed precipitation reaction can occur. Brown-colored "dots" indicate the presence of the gliadin-specific antibody and hence the antigen (gliadin). The color intensity indicates the amount of antigen present in the food sample. The stronger the color, the greater the amount of the antigen (gliadin) in the food sample extract.

26.1.5 Chemicals

	CAS no.	Hazards
Bovine serum albumin (BSA)	9048-46-8	
Chicken egg albumin (CEA)	9006-59-1	
3,3′-Diaminobenzidine tetrahydrochloride (DAB)	7411-49-6	
Gliadin standard protein	9007-90-3	
Hydrogen peroxide, 30% (H_2O_2)	7722-84-1	Oxidizing, corrosive
Rabbit anti-gliadin immunoglobin conjugated to horseradish peroxidase (RAGIg-HRP, Sigma, A1052)		
Sodium chloride (NaCl)	7647-14-5	Irritant
Sodium dodecyl sulfate (SDS)	151-21-3	Harmful
Sodium phosphate, monobasic ($NaH_2PO_4 \cdot H_2O$)	7558-80-7	Irritant
Tris(hydroxymethyl)aminomethane (TRIS)	77-86-1	Irritant
Tween-20 detergent	9005-64-5	

26.1.6 Reagents

(**It is recommended that these solutions be prepared by the laboratory assistant before class.)

- Blocking solution**
  3% (g/mL) BSA in PBST; 5–10 mL per student
- DAB substrate**
  60 mg DAB dissolved in 100 mL 50 mM TRIS (pH 7.6), then filtered through filter paper Whatman #1. (*Note*: DAB may not completely dissolve in this buffer if it is the free base form instead of the acid form. Just filter out the undissolved DAB, and it will still work well.) Just 5 min prior to use, add 100 μL 30% H_2O_2; 5–10 mL per student
- Gliadin antibody probe**
  1:500 (mL/mL) diluted RAGIg-HRP using 0.5% (g/mL) BSA in PBST; 5–10 mL per student
- Gliadin extraction detergent**
  1% (g/mL) SDS in water; 10 mL per student
- Gliadin standard protein, 4000 μg/mL, in 1% SDS**
  One vial of 150 μL per student
- Negative control sample**
  3% CEA (or other non-gliadin protein) in PBST; 100 μL per student.
- Phosphate-buffered saline (PBS)**
  0.05 *M* sodium phosphate, 0.9% (g/mL) NaCl, pH 7.2
- Phosphate-buffered saline + Tween 20 (PBST)**
  0.05 *M* sodium phosphate, 0.9% (g/mL) NaCl, 0.05% (mL/mL) Tween 20, pH 7.2; 250 mL per student

26.1.7 Hazards, Precautions, and Waste Disposal

Adhere to normal laboratory safety procedures. Wear gloves and safety glasses at all times. Handle the DAB substrate with care. Wipe up spills and wash hands thoroughly. The DAB, SDS, and hydrogen peroxide wastes should be disposed of as hazardous wastes. Other wastes likely may be put down the drain using a water rinse, but follow good laboratory practices outlined by environmental health and safety protocols at your institution.

26.1.8 Supplies

- 1–10 µL, 10–100 µL, and 200–1000 µL positive displacement pipettors
- Disposable tips for pipettors
- Filter paper, Whatman #1
- Food samples (e.g., flour, crackers, cookies, starch, pharmaceuticals, etc.)
- Funnels, tapered glass
- Mechanical, adjustable volume pipettors, for 2 µL, 100 µL, and 1000 µL ranges, with plastic tips (glass capillary pipettes can be substituted for 2 µL pipettors)
- Microcentrifuge tubes, 1.5 mL, 2 per sample processed
- Nitrocellulose paper (Bio-Rad 162-0145) cut into 1.7 cm × 2.3 cm rectangular strips (NC strips)
- Petri dishes, 3.5 cm
- Test tubes, 13 × 100 mm, six per student
- Test tube rack, one per student
- Tissue paper
- Tweezers, one set per student
- Wash bottles, one per two students for PBST
- Wash bottles, one per two students for distilled water

26.1.9 Equipment

- Mechanical platform shaker
- Microcentrifuge
- pH meter
- Vortex mixer

26.2 PROCEDURE

26.2.1 Sample Preparation

(*Note*: Sample preparation by the students may take place on a separate day prior to the immunoassay. In this initial sample preparation lab, the principles of differential centrifuga-tion, with respect to the Osborne protein classification system, may be studied. Sample preparation and immunoassay may not be reasonable to achieve in 1 day. If only 1 day can be allocated for this lab, the samples can be prepared ahead of time for the students by the laboratory assistant, and this lab will demonstrate concepts and techniques of a simple immunoassay.)

1. Weigh accurately (record the mass) about 0.1 g of flour, starch, or ground processed food, and add to a 1.5-mL microcentrifuge tube. Add 1.0 mL distilled water and vortex for 2 min. Centrifuge at $800 \times g$ for 5 min. Discard the supernatant (albumins). Repeat.
2. Add 1.0 mL of 1.5 M NaCl to the pellet from Step 1 and resuspend it by vortexing for 2 min. If the pellet is not resuspending, dislodge it with a spatula. Centrifuge at $800 \times g$ for 5 min. Discard the supernatant (globulins). Repeat.
3. Add 1.0 mL of 1% SDS detergent to the pellet from Step 2 and resuspend it to extract the gliadin. Vortex for 2 min. Centrifuge at $800 \times g$ for 5 min. Carefully pipette off most of the supernatant, and transfer to another 1.5-mL microcentrifuge tube. Discard the pellet.

26.2.2 Standard Gliadin

The standard pure gliadin is dissolved at a concentration of 4000 µg/mL in 1% SDS. To provide for a series of standards to compare unknown samples, dilute the standard serially by a factor of 10 in 13 × 100 mm test tubes to make 400 µg/mL, 40 µg/mL, and 4 µg/mL standards in 1% SDS. Use 100 µL of the highest standard transferred to 900 µL of 1% SDS detergent for the first tenfold dilution. Repeat this procedure serially to produce the last two standards. As each standard is made, mix it well on a vortex mixer.

26.2.3 Nitrocellulose Dot Blot Immunoassay

(*Note*: The nitrocellulose (NC) strips should only be handled with tweezers to prevent the binding of proteins and other compounds from your fingers. Hold the NC strips with the tips of the tweezers on the corners of the strips to avoid damaging or interfering with the spotted surface.)

1. Mark two NC strips with a pencil into six equal boxes each (see drawing below).
2. Pipette 2 µL each of the sample, standard or negative controls onto the NC strip. Lay the NC strips flat onto the tissue paper.

(a) On NC strip A, pipette 2 μL of four SDS-extracted food samples.

(b) On NC strip B, pipette 2 μL of four gliadin standards.

(c) On the remaining two squares (5 and 6) on strips A and B, add 2 μL of 1% SDS and 2 μL of 3% CEA (negative control), respectively.

(d) Let the spots air-dry on the strips.

Below are diagrams of the NC strips marked off in boxes and numbered by pencil. The circles are not penciled in; instead, they represent where the 2-μL sample or standard spots will be applied.

3. Place both NC strips into a petri dish containing 5 mL of the blocking solution (3% BSA in PBST), and incubate for 20 min on the mechanical shaker at room temperature so that the NC strips move around slightly in the solution.

4. Rinse the strips well with PBST using a wash bottle over a sink, holding the strips by the tips of their corners with tweezers.

5. Place the NC strips into a petri dish containing about 5 mL of 1:500 diluted RAGIg-HRP conjugate, and incubate for 60 min on the mechanical shaker.

6. Wash the NC strips with PBST using a wash bottle and then incubate them for 5 min in a clean petri dish half full with PBST. Rinse again well with PBST, and rinse one last time with distilled water.

7. Add the NC strips to a petri dish containing the 5 mL of DAB/H$_2$O$_2$ substrate, and watch for the development of a brown stain. (*Note*: Handle the substrate with care. Wipe up spills, wash hands thoroughly, and wear gloves. Although there is no specific evidence that DAB is a carcinogenic compound, it should be treated as if it were.) Stop the reaction in 10–15 min, or when the background nitrocellulose color is becoming noticeably brown, by rinsing each strip in distilled water.

8. Let the NC strips air-dry on the tissue paper.

A

(1) (2) (3)

B

(1) (2) (3)

(4) (5) (6)

(4) (5) (6)

A series (food sample):	B series (gliadin standards):
1 = food sample 1 @ 1× dilution	1 = gliadin standard @ 4000 μg/mL
2 = food sample 2 @ 1× dilution	2 = gliadin standard @ 400 μg/mL
3 = food sample 3 @ 1× dilution	3 = gliadin standard @ 40 μg/mL
4 = food sample 4 @ 1× dilution	4 = gliadin standard @ 4 μg/mL

A series (food sample):	B series (gliadin standards):
5 = 1% SDS	5 = 1% SDS
6 = 3% CEA (negative control)	6 = 3% CEA (negative control)

26.3 DATA AND CALCULATIONS

Make any observations you feel are pertinent to this laboratory. Attach the developed NC strips to your lab report with transparent tape.

Describe the results based on observations of the degree of brown-colored stain in standards and samples relative to negative controls. You can use a simple quantitative rating system like +++, ++, +, +/−, − to describe and report the relative color intensity of the immunoreactions (*Note*: the brown dot images will fade in several days).

Make very crude approximations of the quantity of gliadin in each food sample relative to (more or less than) the standard gliadin dots. Comment on this crude estimate relative to the food product's gluten status (i.e., gluten-free or not; *Codex Alimentarius* (https://www.fao.org/fao-who-codexalimentarius/en/) defines less than 20 mg of gluten/kg in total, based on the food as sold or distributed to the consumer to be classified as "gluten-free").

Tabulate your results in a manner that is easy to interpret.

To make the gluten status estimation, you must know the values of both the concentration of the food sample (mg food/mL extraction solution) extracted and the concentration of the gliadin standards (mg gliadin/mL extraction solution) to which you are making a comparison.

Example calculation:

If the food sample has a concentration of 100 mg/mL and reacts equivalently to a 4 μg/mL gliadin standard, it can be estimated that 4 μg gliadin is in 100 mg of the food sample. Because both are applied at equal volumes, the two concentrations can be related fractionally (i.e., 4 μg gliadin/100 mg or 40 mg gliadin/kg food sample). The gliadin content of gluten is generally taken as 50%. Since this sample has a gliadin concentration higher than the limit set by *Codex Alimentarius* (10 mg gliadin/kg food or 20 mg gluten/kg food), the food tested cannot be considered "gluten-free".

26.4 QUESTIONS

1. Draw a set of symbolic pictures representing the stages of the dot blot assay used in this laboratory, including the major active molecular substances being employed (i.e.,

nitrocellulose paper, antigen, blocking reagent, antibody-enzyme conjugate, substrate, and product).

2. Why should you block unbound sites on nitrocellulose strips with 3% BSA in a special blocking step after applying samples to the membrane?

3. Why is a spot of 1% SDS (gliadin extraction detergent) used in the dot blot?

4. Why is a spot of 3% CEA (negative control) used in the dot blot?

5. Describe the basic role of horseradish peroxidase (i.e., why is it labeled to the antibody). What roles do DAB and H_2O_2 play in the development of the colored dot reaction in this immunoassay? Do not describe the actual chemical reaction mechanisms, but rather explain why a color reaction can ultimately infer an antigen (i.e., gliadin) is present on the nitrocellulose paper.

Acknowledgments The author thanks Yun-Hwa Peggy Hsieh (Department of Nutrition and Integrative Physiology, Florida State University, Tallahassee, FL, USA), who updated the chapter for previous editions of the book.

RESOURCE MATERIALS

Rao Q (2024) Immunoassays. Ch.27. In: Ismail BP, Nielsen SS (eds) Food analysis, 6th edn. Springer, New York

Choung RS, Larson SA, Khaleghi S, Rubio-Tapia A, Ovsyannikova IG, King KS, Larson JJ, Lahr BD, Poland GA, Camilleri MJ, Murray JA (2017a) Prevalence and morbidity of undiagnosed celiac disease from a community-based study. Gastroenterology 152:830–839

Choung RS, Unalp-Arida A, Ruhl CE, Brantner TL, Everhart JE, Murray, JA (2017b) Less hidden celiac disease but increased gluten avoidance without a diagnosis in the united states: Findings from the national health and nutrition examination surveys from 2009 to 2014. Mayo Clin Proc 92:30–38

Codex Alimentarius Commission (2008) Codex standard for foods for special dietary use for persons intolerant to gluten. CXS 118–1979; Food and Agriculture Organization of the United Nations: Rome, Italy. World Health Organization: Geneva, Switzerland

Lacorn M, Dubois T, Weiss T, Zimmermann L, Schinabeck TM, Loos-Theisen S, Scherf K (2022) Determination of gliadin as a measure of gluten in food by R5 sandwich ELISA RIDASCREEN® gliadin matrix extension: Collaborative study 2012.01. J AOAC Int 105:442–455

Mendez E, Vela C, Immer U, Janssen FW (2005) Report of a collaborative trial to investigate the performance of the R5 enzyme linked immunoassay to determine gliadin in gluten-free food. Eur J Gastroenterol Hepatol 17:1053–1063

Rheological Measurements of Food Products

27

Helen S. Joyner

Contents

27.1 INTRODUCTION

27.1.1 Background

Whether working in product development, quality control, or process design and scale-up, rheology plays an integral role in the manufacturing of high-quality food products. Rheology is a science based on fundamental physical relationships concerned with how all materials respond to applied forces or deformations. Flow behaviors and fracture are two such responses to force or deformation.

Determination and control of the flow properties of fluid foods are critical for optimizing processing conditions and obtaining the desired sensory qualities for the consumer. Understanding fluid flow behaviors is necessary to properly size equipment used to transport fluid food products. For example, the equipment used to pump a dough mixture would be very different from that used for milk. Additionally,

H. S. Joyner (✉)
Premier Nutrition, Emeryville, CA, USA

the amount of time the food needs to flow through a heat exchanger, and therefore the amount of heating or "thermal dose" received, is directly related to its flow properties.

Fracture properties are a key rheological behavior for solid (self-supporting) food products. Understanding how foods deform and fracture is important for understanding how they behave during eating. Deformation and fracture behaviors also can be used to determine what type of packaging and conveying is needed for solid foods to reach the consumer without breaking. For example, the air in potato chip bags is present to protect the chips against breakage during transport.

The rheological properties of a material are a function of composition, temperature, and other processing conditions, such as the speed a fluid travels through a pipe. Identifying how these parameters influence rheological behaviors may be accomplished by instrumental testing.

27.1.2 Reading Assignment

Joyner, HS (2024) Rheological principles for food analysis. Ch 29. In: Ismail BP, Nielsen SS (eds) Food analysis, 6th edn. Springer, New York

Singh RP, Heldman DR (2001) Introduction to food engineering, 3rd edn. Academic Press, San Diego, CA, pp 69–78, 144–157

27.1.3 Overall Objective

The overall objective of this laboratory is to determine commonly measured rheological behaviors of several fluid and solid food products using rheological instruments widely used throughout the food industry.

27.2 VISCOSITY MEASUREMENT BY BROOKFIELD VISCOMETER

27.2.1 Objectives

1. Explain the basic principles of fluid rheology.
2. Gain experience measuring fluid viscosity using a Brookfield viscometer.
3. Describe the effects of temperature and (shear) speed on viscosity.

27.2.2 Supplies

- Beakers, 250 mL (6)
- Dish soap

- French salad dressing
- Honey (make sure there are no additional ingredients aside from honey)
- Thermometer or thermocouple with digital reader

27.2.3 Equipment

- Brookfield rotational viscometer (AMETEK Brookfield, Middleborough, MA), model LV and spindle #3
- Refrigerator

27.2.4 Procedure

1. You will be testing honey and salad dressing samples in duplicate. Honey will be tested at just room temperature, and salad dressing will be tested at both room and refrigerated temperatures. Fill two 250-mL beakers with 200 mL honey and the four remaining 250-mL beakers each with 200 mL salad dressing. Label the beakers appropriately and cover them with plastic wrap until ready to test. Place two of the beakers of salad dressing in a refrigerator for at least 2 h prior to analysis. The remaining beakers shall be allowed to equilibrate to room temperature.
2. Prior to evaluating the samples, make sure the viscometer is level. Use the leveling bubble and circle on the viscometer to check. Adjust the feet on the viscometer if needed until the bubble is in the center of the circle.
3. On the data sheet provided, record the viscometer model number and spindle number, product information (type, brand, etc.), and the sample temperature. Because rheological properties are strongly dependent on temperature, sample temperatures must be measured and recorded prior to performing measurements.
4. Immerse the spindle into the test fluid (i.e., honey, salad dressing) up to the notch cut in the shaft. The viscometer motor should be off.
5. Zero the digital viscometers before turning on the motor.
6. Set the motor at the lowest speed revolutions per minute (rpm) setting. Once the digital display shows a stable value, record the percentage of full-scale torque reading. Increase the rpm setting to the next speed and again record the percentage of full-scale torque reading. Repeat this procedure until the maximum rpm setting has been reached or 100% (but not higher) of the full-scale torque reading is obtained. *Do not increase the speed once 100% of the full-scale torque has been reached.*

7. Stop the motor, then slowly raise the spindle from the sample. Remove the spindle and clean with soap and water, then dry. Be sure not to use any abrasive scrubbers or soaps on the spindle, since these can scratch the spindle and result in measurement errors.

8. Repeat Steps 2–7 to test all samples.

9. Once all the data have been collected for the salad dressing and honey at room temperature, remove the salad dressing sample from the refrigerator and repeat Steps 2–7. Remove each beaker just before testing and be sure to record the sample temperature just before performing the measurements.

10. Data from the duplicate samples collected under identical conditions may be pooled to generate an average reading.

27.2.5 Data and Calculations

Date:
Product information:
Viscometer make and model:
Spindle number:

Sample	Sample temp. (°C)	Spindle speed (rpm)	% Reading	Factor	Viscosity (mPa-s)

1. Sketch the experimental apparatus and label the major parts.

2. Calculate the viscosity of the test fluids at each rpm using the spindle factors. A factor exists for each spindle-speed combination (Table 27.1). For every dial reading (percentage full-scale torque), multiply the display value by the corresponding factor to calculate the viscosity with units of mPa-s.

3. Plot viscosity versus rpm for each fluid on a single graph.

Table 27.1 Factors for Brookfield model LV (spindle #3)

Speed (rpm)	Factor
0.3	4000
0.6	2000
1.5	800
3	400
6	200
12	100
30	40
60	20

4. Label the plots with the type of fluid (e.g., Newtonian, pseudoplastic, Herschel-Bulkley) based on the response of viscosity to speed (rpm). *Keep in mind that speed is proportional to shear rate. In other words, as the speed is doubled, the shear rate is doubled.*

27.3 VISCOSITY MEASUREMENT BY BOSTWICK CONSISTOMETER

27.3.1 Objectives

1. Explain the basic principles of fluid rheology.
2. Gain experience measuring fluid viscosity using a Bostwick consistometer.
3. Describe the effects of temperature and (shear) speed on viscosity.

27.3.2 Supplies

- Beakers, 250 mL (6)
- Dish soap
- French salad dressing
- Honey (make sure there are no additional ingredients aside from honey)
- Thermometer or thermocouple with digital reader
- Stopwatch

27.3.3 Equipment

- Bostwick consistometer (CSC Scientific Company, Inc., Fairfax, VA)
- Refrigerator

27.3.4 Procedure

1. You will be testing honey and salad dressing samples in duplicate. Honey will be tested at just room temperature, and salad dressing will be tested at both room and refrigerated temperatures. Fill two 250-mL beakers with 200 mL honey and the four remaining 250-mL beakers each with 200 mL salad dressing. Label the beakers appropriately and cover them with plastic wrap until ready to test. Place two of the beakers of salad dressing in a refrigerator for at least 2 h prior to analysis. The remaining beakers shall be allowed to equilibrate to room temperature.

2. Prior to evaluating the samples, make sure the consistometer is at the correct angle. Level the consistometer by adjusting the leveling screws under the consistometer until the leveling bubble is in the center of the black circle.
3. On the data sheet provided, record the consistometer make and model, product information (type, brand, etc.), and the sample temperature. Because rheological properties are strongly dependent on temperature, sample temperatures must be measured and recorded prior to performing measurements.
4. Close the product gate and hold it down while pulling the lever arm as far up as it can go.
5. Pour the test fluid (i.e., honey, salad dressing) in the product reservoir until it reaches the top of the product gate.
6. Push the lever arm down to raise the product gate; start the stopwatch as you push the lever arm down. Record the consistency or the distance the fluid traveled 30 s after the gate is raised. The numbers on the slope are distances in cm.
7. Clean the consistometer with soap and water, then dry. Be sure the consistometer is completely dry before using it again.
8. Repeat Steps 2–7 to test all samples.
9. Once all the data have been collected for the salad dressing and honey at room temperature, remove the salad dressing sample from the refrigerator, and repeat Steps 2–7. Remove each beaker just before testing and be sure to record the sample temperature just before performing the measurements.
10. Data from duplicate samples collected under identical conditions may be pooled to generate an average reading.

27.3.5 Data and Calculations

Date:
Product information:
Consistometer make and model:

Sample	Sample temp. (°C)	Consistency (cm)

1. Sketch the experimental apparatus and label the major parts.

2. Compare the consistency of the test fluids with the consistometer to the viscosities measured with the Brookfield viscometer. Determine which rpm gives you the best match between the two data sets.

27.4 LARGE-STRAIN COMPRESSION MEASUREMENT

27.4.1 Objectives

1. Compare the large-strain compression behaviors of different cheeses.
2. Gain experience performing large-strain compression tests using a TA.XT Texture Analyzer.
3. Describe the effects of temperature on fracture behaviors.

27.4.2 Chemicals

	CAS No.	Hazards
Ethanol (95 or 100%)	64-17-5	Highly flammable
Mineral oil	8012-95-1	

27.4.3 Reagents

- Ethanol, 70%. 100 mL
  Use 95% or 100% ethanol and deionized distilled water to prepare 100 mL of a 70% ethanol solution

27.4.4 Supplies

- 1-inch diameter cork borer
- 2 kg weight
- Calipers
- Dairy-free (e.g., Daiya) medium cheddar cheese (two 1 lb blocks; the cheese must be in block form)
- Disposable pipets
- Kimwipes
- Kraft medium cheddar cheese (two 1 lb blocks; the cheese must be in block form)
- Mineral oil

27.4.5 Equipment

- TA.XT Texture Analyzer (Texture Technologies, Hamilton, MA), with a 5 kN load cell and a 2-inch diameter cylindrical aluminum probe
- Refrigerator

27.4.6 Procedure

1. Place one of each of the two types of cheese blocks in a refrigerator at least 2 h prior to analysis. The remaining two blocks should be allowed to equilibrate to room temperature.
2. A one-cycle testing program should be set up using the following parameters:
   (a) Return to Start method from the Test Library
   (b) 50 mm/min pretest, test, and post-test speed
   (c) Compression to 50% strain
   (d) Peak force and total area post-test macro
3. Prior to evaluating the samples, make sure the aluminum probe is attached to the TA.XT and the force and height are calibrated. The force should be calibrated using a 2 kg weight using the Calibrate Force option in the TA.XT software (Exponent). The height should be calibrated using the Calibrate Height option in Exponent with the probe attached.
4. On the data sheet provided, record the TA.XT model, product information (type, brand, etc.), and the sample temperature. Because rheological properties are strongly dependent on temperature, sample temperatures must be measured and recorded prior to performing measurements.
5. Unwrap one of the room temperature cheese blocks and use the 1-inch diameter cork borer to cut a cylinder of cheese from the block. Inspect the cylinder and discard if there are cracks or if the sides of the cylinder are not straight. Mineral oil can be used to coat the inside of the cork borer so that the sample can be removed more easily.
6. Measure the height and diameter of the cylinder and record the measurements on the data sheet provided.
7. Apply a thin layer of mineral oil to the flat ends of the cylinder and place it on the testing platform. (The oil needs to be applied to the ends so that they do not stick to the test surfaces during compressions.) This can be done by putting a few drops of mineral oil on the testing platform with a disposable pipet and dabbing each end into the oil.
8. Lower the aluminum plate so that it is just above the top of the cylinder but does not touch it. Enter the sample name and testing date into the software, then start the test.
9. After the test has completed, record the peak force (maximum compressive force), area under the curve (compressive work), and maximum compressive distance. The first two values can be found on the Results tab. The third value can be found in the raw data, which is available in the View menu. Find the maximum distance traveled in the data.
10. Discard the sample and clean the testing surfaces with a 70% ethanol solution, then dry with Kimwipes. Be sure the surfaces are completely dry before testing an additional sample.
11. Repeat Steps 4–10 to test all samples. It is recommended to test at least three cylinders of each cheese type.
12. Once all the data have been collected for the cheeses at room temperature, remove the cheese blocks from the refrigerator, and repeat Steps 4–10. Return the cheese blocks to the refrigerator when not preparing samples to keep them cold. Be sure to record the sample temperature before performing any measurements and work quickly to prevent the samples from becoming too warm during testing.
13. Data from samples collected under identical conditions may be pooled to generate an average reading.

27.4.7 Data and Calculations

Date:
Product information:
TA.XT model:
TA.XT probe used:

Sample	Sample temp. (°C)	Sample diameter (mm)	Sample diameter (m)	Sample height (mm)	Sample height (m)	Maximum Compressive Force (N)	Compressive Work (N.s)	Maximum Compressive Distance (mm)

1. Sketch the experimental apparatus and label the major parts.
2. For each sample, convert the maximum compressive force to stress using the following formula:

$$\tau = \frac{F}{A}$$

   where τ is stress (Pa), F is force (N), and A is the cross-sectional area of the cheese cylinder (m).
3. For each sample, determine the Cauchy strain using the following formula:

$$\varepsilon_c = \frac{L}{L_0}$$

   where ε_c is the Cauchy strain (unitless), L is the length of the cheese cylinder at maximum compression (m), and L_0 is the initial length of the cylinder (m).

   Hint: L is NOT the maximum compressive distance, but that distance can be used to determine L.
4. For each sample, determine the Hencky (true) strain using the following formula:

$$\varepsilon_h = \ln\left(1 + \frac{L}{L_0}\right)$$

   where ε_h is the Hencky strain (unitless), L is the length of the cheese cylinder at maximum compression (m), and L_0 is the initial length of the cylinder (m).

   Hint: L is NOT the maximum compressive distance, but that distance can be used to determine L.

27.5   QUESTIONS

1. Differentiate a Newtonian fluid from a non-Newtonian fluid. Were your materials Newtonian or non-Newtonian? Explain your choice.
2. Describe the importance of viscosity and flow properties in food processing, quality control, and consumer satisfaction. How might food composition impact its viscosity or flow properties? What ingredient(s) in the salad dressing may have caused deviations from Newtonian behavior? What effect does temperature have on the viscosity of fluid foods?
3. For samples at similar temperatures and identical speeds, was the viscosity of honey ever less than the viscosity of salad dressing? Is this behavior representative of the sample flow behavior at all speeds?
4. Why is it important to test samples at more than 1 speed?
5. Compare the viscosities determined from the two different viscosity measurement methods. Are they similar? Do they give you the same information about the fluids?

If there are differences in the viscosity results, what might be responsible for the differences?
6. Which viscosity measurement method gives you the most information about the test fluids? How could you adjust the other measurement method to give you more information about the test fluid?
7. Consistency is not the same as viscosity. Considering that different fluids have different flow behaviors, do you think it would be possible to develop an equation to convert consistency to viscosity? Why or why not?
8. A fluid food product was designed to have a certain viscosity profile. When the product was being developed, the viscosity was measured at several different speeds on a Brookfield viscometer. Now the product is in full-scale production in two locations. The quality control team in Location 1 uses a different model of Brookfield viscometer equipped with a different size spindle to check the viscosity of each batch of product. The quality control team at Location 2 uses a Bostwick consistometer to check the viscosity of each batch of product. What issues might occur if there is a problem with product viscosity at one of the production facilities or if the two production facilities want to compare product viscosity data?
9. Why might some companies choose to measure viscosity or consistency with certain instruments? In your answer, consider product type and formulation, manufacturing process, and other factors about the company in your answer
10. What is fracture? Describe two different types of fracture that can be seen in food products. Provide examples for your discussion.
11. Was the maximum compressive force the same as the fracture force for the cheese samples? Why or why not?
12. For cheese samples at the same temperature, which sample required greater stress to fracture? Which sample required greater strain to fracture? How might the differences in composition of the cheeses have caused these results?
13. What differences did you observe between the Hencky and Cauchy strains for the cheese samples? Which strain is more appropriate for large-strain compressive testing? Why?
14. What effect did temperature have on the stress and strain values you calculated for each cheese? What temperature-related changes in these foods could have contributed to these results?
15. Why is it a good idea to convert force and deformation to stress and strain before comparing rheological data? What might happen if this is not done?

16. Compression data from a TAXT or similar instrument is often used to estimate texture attributes of food products. Discuss the advantages and disadvantages of using instrumental compressive testing, including Texture Profile Analysis, to predict texture attributes such as firmness, stickiness, compressibility, fracturability, and cohesiveness.

Acknowledgments This laboratory was adapted from a chapter originally created by:

Dr Christopher R. Daubert, College of Agriculture, Food and Natural Resources, Columbia, MO, USA, and Dr Brian E. Farkas, McCormick & Company, Hunt Valley, MD, USA.

RESOURCE MATERIALS

Joyner, HS (2024) Rheological principles for food analysis. Ch 29. In: Ismail BP, Nielsen SS (eds) Nielsen's food analysis, 6th edn. Springer, New York

Singh RP, Heldman DR (2001) Introduction to food engineering, 3rd edn. Academic Press, San Diego, CA, pp. 69–78, 144–157

Color Measurements of a Solid and Calculation of Color Specifications from Spectral Data

28

M. Monica Giusti, Danielle M. Voss,
and Gonzalo Miyagusuku-Cruzado

Contents

28.1 INTRODUCTION

28.1.1 Background

Food color is arguably one of the most important determinants of acceptability and is, therefore, an important specification for food products. The development of compact and easy to use colorimeters and spectrocolorimeters (spectrophotometers designed for color measurements) as well as applications and attachments designed for use with mobile devices has made the quantitative measurement of color a routine part of research, crop harvesting, product development, and quality assurance.

There are several widely employed systems of color specification: notably the Commission Internationale de l'Eclairage (CIE) tristimulus, the CIELAB, and the Munsell color system. Color is a three-dimensional characteristic, and to describe any color three attributes are necessary regardless of the color system. The CIE tristimulus system uses mathematical coordinates (XYZ) to represent the amount of red, green, and blue primaries required by a "standard observer" to give a color match. These XYZ values are then used to determine chromaticity coordinates x and y. This system, however, does not correlate well to human color perception. The CIELAB system was later developed to produce a more linear, uniform color space. CIELAB values L^* (lightness), a^* (red-green axis), and b^* (yellow-blue axis) are calculated from the CIE XYZ tristimulus values and provide visually linear color specification. These values can be converted into coordinates to express color based on lightness (L^*), chroma (C^*_{ab}), and hue angle (h_{ab}) which is expressed in degrees. Alternatively, the Munsell system relies on matching colors to standard colored chips. Value, hue, and chroma are employed to express lightness, "color," and saturation, respectively. Available software, often incorporated into modern instruments, enables the investigator to report data in the different notations.

There are different instruments designed specifically to measure color and provide quantitative measure of each color dimension. Most instruments (colorimeters and spectrocolorimeters) allow us to determine how the color of an object will appear under different types of light and express the color

M. M. Giusti (✉) · D. M. Voss · G. Miyagusuku-Cruzado
Department of Food Science and Technology, The Ohio State University, Columbus, OH, USA
e-mail: giusti.6@osu.edu

© The Author(s), under exclusive license to Springer Nature Switzerland AG 2024
B. P. Ismail, S. S. Nielsen (eds.), *Nielsen's Food Analysis Laboratory Manual*, Food Science Text Series,
https://doi.org/10.1007/978-3-031-44970-3_28

using different color scales. The color of a solid or opaque sample is determined using the spectra of reflected or scattered light. Instruments measuring reflectance can account for the influence of texture on the color (i.e., shiny, matte) by choosing to include or exclude the specular component of the light. For liquid or transparent samples, the instrument uses transmittance and calculates color from the object's light transmittance spectra. Different reading methods can include total transmittance, regular transmittance, and haze measurements to account for the presence of suspended or opaque particles. If a color instrument is not available, it is possible to calculate color parameters of a liquid from absorbance spectral data obtained with any spectrophotometer. Yet, the geometry of the instruments designed for color measurements is typically optimized for color analyses.

When reporting color data, it is important to specify the conditions used for measuring color including the instrument, the way the sample was presented to the instrument, and the specific parameters chosen for the measurement. Understanding the different color specification systems and the means of interconversion aids the food scientist in selecting appropriate means of reporting and comparing color measurements.

28.1.2 Reading Assignment

Giusti MM, Gordillo B, Gonzalez-Miret Martin ML, (2024) Color analysis. Ch. 31. In: Ismail BP, Nielsen SS (eds) Nielsen's food analysis, 6th edn. Springer, New York.

28.1.3 Objectives

1. Learn how to measure the color of a solid material and color differences.
2. Gain understanding on how to use absorbance spectral data to calculate the following:
   (a) % Transmission data
   (b) Tristimulus color values X, Y, and Z
   (c) CIELAB color values
3. Become familiar with readily available software related to color:
   (a) Adobe Color to create visual representations of color data
   (b) ColorBySpectra to interconvert between color specification systems

28.1.4 Materials

1. Colorimeter or spectrocolorimeter capable of measuring the color of a solid material (reflectance instrument) like handheld or benchtop instruments

2. Two red apples with noticeable differences between each other in the color of the skin (such as Red Delicious and Honeycrisp)
3. Calculator
4. Absorbance spectrum (a spectrum of cranberry juice is provided in Table 28.3)
5. Appropriate software for color interconversion and color representation
   Examples:
   (a) Adobe Color (produced by Adobe Inc., San Jose, CA) is a free, online software which creates color swatches based on CIELAB data, helpful for visualizing color coordinate data. https://color.adobe.com/create/color-wheel.
   (b) ColorBySpectra (developed by Farr and Giusti (2017) at The Ohio State University, Columbus, OH) is a downloadable software that converts absorbance data into color coordinates including XYZ, HunterLab (L, a, b), CIELAB (L^*, a^*, b^*, and L^*, C^*_{ab}, h_{ab}) with multiple illuminant choices available. A free version for academic use is available for download from https://u.osu.edu/giustilab/innovations/. We recommend downloading the software ahead of time as you may need some IT help to install it on your computer.

28.2   PROCEDURE

28.2.1 Color Measurement and Comparison of Solid Samples

1. Record the name and model of the instrument that you are using. This must always be included when reporting color. Additionally, note if the instrument measures reflectance with the specular component included or excluded.
2. Select the parameters to use for the color measurement and record beneath Table 28.1. We suggest you choose reflectance specular excluded (vs. included), 10° observer angle (vs. 2°), illuminant D65 (simulating sunlight). Set the scale to CIELAB.
3. Calibrate the colorimeter according to the manufacturer instructions.
4. Place an apple in direct contact with the instrument's aperture and press the read button or trigger.
5. Record the CIELAB color values (L^*, a^*, b^*, C^*_{ab}, h_{ab}) in Table 28.1.
6. Rotate the apple and take a second reading. Repeat the step for a third reading. You should have three sets of L^*, a^*, and b^* color coordinates for one apple.
7. Follow the same procedure (repeat steps 2–5) with the second apple.

8. Report your results listing the following: mean L^*, a^*, b^*, C^*_{ab}, h_{ab} values, measurement parameters used, and instrument and manufacturer.
9. Use the following formula to quantify the color difference (ΔE^*_{ab}) between the apples and record it in Table 28.1:

$$\Delta E^*_{ab} = \left[\left(L^*_1 - L^*_2 \right)^2 + \left(a^*_1 - a^*_2 \right)^2 + \left(b^*_1 - b^*_2 \right)^2 \right]^{1/2}$$

Table 28.1 Color measurements of apples

		L^*	a^*	b^*	C^*_{ab}	h_{ab}
Apple 1	Reading 1					
	Reading 2					
	Reading 3					
	Average					
Apple 2	Reading 1					
	Reading 2					
	Reading 3					
	Average					

Illuminant used:
Observer angle used:
Reflectance specular component included or excluded?
Color difference (ΔE^*_{ab}):

Optional Activity: Evaluating the Influence of the Illuminant on Color

1. On the same instrument from Sect. 28.2.1, select a different illuminant. Common illuminants besides D65 are F2 and A (simulating fluorescent and tungsten lights, respectively).
2. Measure the color of an apple chosen as described in Sect. 28.2.1, recording data in Table 28.2. Note that some instruments will convert color data from one illuminant to another using previously taken measurements and mathematical calculations.
3. Compare and contrast the color coordinates obtained under the different illuminants used. Note that you could calculate the color differences (ΔE^*_{ab}) produced by changing illuminant using the equation presented before.
4. Enter the mean CIELAB L^*, a^*, b^* color coordinates from the apple measurements into AdobeColor software (https://color.adobe.com/create/color-wheel). On the color wheel tab, select "Custom" as the Color Harmony Rule and LAB as the Color Mode. Type the mean L^* a^* b^* coordinates, rounded to the nearest whole number, into the spaces below each color swatch. Discuss the similarities and differences between the color swatch and the observed color.

Table 28.2 Color measurements of apples under different illuminant (optional activity)

		L^*	a^*	b^*	C^*_{ab}	h_{ab}
Apple 1 (Changed illuminant)	Reading 1					
	Reading 2					
	Reading 3					
	Average					
Apple 2 (Changed illuminant)	Reading 1					
	Reading 2					
	Reading 3					
	Average					

Illuminant used:
Observer angle used:
Reflectance specular component included or excluded?
Color difference (ΔE^*_{ab}):

28.2.2 Determining Color of a Liquid from an Absorbance Spectra

An absorption spectra of cranberry juice is provided in Table 28.3 to use for the following procedure.

1. From the spectral data in Table 28.3, determine the $\lambda_{vis\text{-}max}$ (wavelength of maximum absorption in the visible range) for the cranberry juice and record it in Table 28.4.
2. Convert all absorbance values to %transmittance $(\%T)$ using the equation below:

$$\%\text{Transmittance} = 10^{(2-\text{Absorbance})}$$

3. Multiply $\%T$ by $E\bar{x}$, $E\bar{y}$, and $E\bar{z}$ (Table 28.3). These factors $(E\bar{x}, E\bar{y}, \text{and } E\bar{z})$ correspond to the 1964 CIE standard supplemental observer curves for x, y, and z with illuminant D_{65} and 10° observer angle for spectra collected in 10 nm measurement step.
4. Sum the values of $\%T \times E\bar{x}$, $\%T \times E\bar{y}$, and $\%T \times E\bar{z}$ and use these sums and the equations listed below to obtain X, Y, and Z values.

$$X = \left[\Sigma \left(\%T \times E\bar{x} \right) \right] / \left[\Sigma \left(E\bar{y} \right) \right]$$
$$Y = \left[\Sigma \left(\%T \times E\bar{y} \right) \right] / \left[\Sigma \left(E\bar{y} \right) \right]$$
$$Z = \left[\Sigma \left(\%T \times E\bar{z} \right) \right] / \left[\Sigma \left(E\bar{y} \right) \right]$$

The sums of each are divided by the sum of $E\bar{y}$ (which is 100.00 under these measurement conditions). By doing this, the three values are normalized to $Y = 100$, which is "perfect" white. Objects are specified relative to luminosity of perfect white rather than the absolute level of light.

Table 28.3 Absorbance data for cranberry juice[a] and calculation of %Transmission (%T) and XYZ values by the weighted ordinance method[b]

λ_{nm}	Absorbance	%T	$E\overline{x}$	$E\overline{x} \cdot \%T$	$E\overline{y}$	$E\overline{y} \cdot \%T$	$E\overline{z}$	$E\overline{z} \cdot \%T$
360	1.88		0.000		0.000		0.000	
370	1.61		0.000		0.000		−0.001	
380	1.29		0.001		0.000		0.004	
390	0.95		0.005		0.000		0.020	
400	0.73		0.097		0.010		0.436	
410	0.58		0.616		0.064		2.808	
420	0.53		1.660		0.171		7.868	
430	0.53		2.377		0.283		11.703	
440	0.55		3.512		0.549		17.958	
450	0.57		3.789		0.888		20.358	
460	0.61		3.103		1.277		17.861	
470	0.68		1.937		1.817		13.085	
480	0.78		0.747		2.545		7.510	
490	0.88		0.110		3.164		3.743	
500	0.99		0.007		4.309		2.003	
510	1.08		0.314		5.631		1.004	
520	1.09		1.027		6.896		0.529	
530	1.02		2.174		8.136		0.271	
540	0.91		3.380		8.684		0.116	
550	0.73		4.735		8.903		0.030	
560	0.56		6.081		8.614		−0.003	
570	0.41		7.310		7.950		0.001	
580	0.30		8.393		7.164		0.000	
590	0.22		8.603		5.945		0.000	
600	0.16		8.771		5.110		0.000	
610	0.13		7.996		4.067		0.000	
620	0.10		6.476		2.990		0.000	
630	0.09		4.635		2.020		0.000	
640	0.08		3.074		1.275		0.000	
650	0.07		1.814		0.724		0.000	
660	0.07		1.031		0.407		0.000	
670	0.07		0.557		0.218		0.000	
680	0.06		0.261		0.102		0.000	
690	0.06		0.114		0.044		0.000	
700	0.06		0.057		0.022		0.000	
710	0.06		0.028		0.011		0.000	
720	0.06		0.011		0.004		0.000	
730	0.07		0.006		0.002		0.000	
740	0.07		0.003		0.001		0.000	
750	0.07		0.001		0.000		0.000	
760	0.07		0.000		0.000		0.000	
SUM			94.81		100.00		107.30	

[a]1 cm pathlength; Molecular Devices Spectra Max M2 plate reader spectrophotometer
[b]1964 CIE color matching functions for 10° standard supplemental observer, illuminant D_{65}, and 10° measurement step. $E\overline{x}$, $E\overline{y}$, and $E\overline{z}$ values were obtained from ASTM International E308–22

Table 28.4 Color characteristics worksheet for cranberry juice

	Cranberry juice color characteristics
$\lambda_{\text{vis-max}}$	
X	
Y	
Z	
L*	
a*	
b*	
C*$_{\text{ab}}$	
h$_{\text{ab}}$	

5. Calculate the CIELAB color coordinates using the X, Y, and Z values using the following equations:

$$L^* = \left[116 \times \left(Y/100\right)^{1/3}\right] -$$
$$16a^* = 500 \times \left[\left(X/94.81\right)^{1/3} - \left(Y/100\right)^{1/3}\right]$$
$$b^* = 200 \times \left[\left(Y/100\right)^{1/3} - \left(Z/107.30\right)^{1/3}\right]$$
$$\text{Chroma}\left(C^*_{\text{ab}}\right) = \left(a^{*2} + b^{*2}\right)^{1/2}$$
$$\text{hue angle}\left(h_{\text{ab}}\right) = \text{arctangent}\left(b^*/a^*\right)$$

The values of 94.81, 100, and 107.30 correspond to the white spot value of X_n, Y_n, and Z_n, respectively, for D65 illuminate, 10° observer angle, and 10 nm step. If these measurement conditions change, these constants must also be updated.

Optional Activity: Collection of Absorbance Spectra of Other Samples

1. To determine color values for a different liquid sample, collect the absorption spectra of the sample using a spectrophotometer. Place your sample in a cuvette or a microplate and collect the spectra from at least 400 to 700 nm (although a wider range from 360 to 780 nm is ideal), measuring every 10, 5, 2, or 1 nm. Ensure absorbance values remain within the linear range for your spectrophotometer (commonly below Abs. 1.2). If needed, dilute your sample to reach the linear range.

2. Calculate the color data using the equations in Sect. 28.2.2 or using ColorBySpectra software. To use the software, save the spectral data of your sample as a CSV file. Open ColorBySpectra on your computer and upload the saved CSV file of absorbance data, choose the illuminant, and process the data. The converted color data will be saved as a separate file.

28.3   QUESTIONS

1. For a single apple measured in Sect. 28.2.1, were there variations in the L*, a*, and b* values for the three replicate readings? If so, what challenges does this present when trying to represent an object's color? How well did the color swatches produced in Sect. 28.2.1 match the color of that apple?

2. How much did the CIELAB color coordinates (L*, a*, b*, C*$_{\text{ab}}$, and h$_{\text{ab}}$) differ between the two apples measured and what does this say about their colors? Describe the color using visual descriptors (i.e., darker, more red) and use the coordinates and ΔE^*_{ab} to support your statements.

3. Did you expect the cranberry juice to have a $\lambda_{\text{vis-max}}$ of 520 nm? How could you explain the fact that a solution that absorbs light at a wavelength corresponding to green light expresses red color?

4. Which has the greater hue angle, a banana with coordinates L* = 70.39, a* = −1.87, b* = 54.06 or L* = 61.96, a* = −10.70, b* = 43.62? Which banana has greater chroma? Based on the color data, can you know which banana would be considered riper? Note that for h$_{\text{ab}}$ calculations, if a* is a negative number, you will need to add 180° to your answer.

Acknowledgments The authors of this chapter thank Ronald E. Wrolstad and Daniel E. Smith, (Department of Food Science & Technology, Oregon State University, Corvallis, OR, USA) for originally developing this chapter.

RESOURCE MATERIALS

ASTM International (2022) ASTM E308–22 standard practice for computing the colors of objects by using the CIE system. ASTM International, West Conshohocken, Pennsylvania.

Berns RS (2019) Billmeyer and Saltzman's principles of color technology, 4th edn. Wiley, New York.

CIE (2004). CIE15:2004–CIE technical report colorimetry, 3rd edn. ISBN: 3901906339.

Farr, JE, Giusti, MM. (2017) ColorBySpectra: An application to automate conversion of spectral data to color spaces. Software developed at the Giusti Phytochemicals Laboratory, Food Science and Technology Department, The Ohio State University. Retrieved from: https://u.osu.edu/giustilab/innovations/

Giusti MM, Gordillo B, Gonzalez-Miret Martin ML (2024) Color analysis. Ch. 31. In: Ismail BP, Nielsen SS (eds) Nielsen's food analysis, 6th edn. Springer, New York.

Extraneous Matter Examination

29

S. Suzanne Nielsen

29

Contents

29

S. S. Nielsen (✉)
Department of Food Science, Purdue University,
West Lafayette, IN, USA
e-mail: nielsens@purdue.edu

© The Author(s), under exclusive license to Springer Nature Switzerland AG 2024
B. P. Ismail, S. S. Nielsen (eds.), *Nielsen's Food Analysis Laboratory Manual*, Food Science Text Series,
https://doi.org/10.1007/978-3-031-44970-3_29

29.1 INTRODUCTION

29.1.1 Background

Extraneous materials are any foreign substances in foods that are associated with objectionable conditions or practices in production, storage, or distribution of foods. Extraneous materials include: (a) filth or objectionable matter contributed by animal contamination (rodent, insect, or bird matter) or unsanitary conditions, (b) decomposed material or decayed tissues due to parasitic or nonparasitic causes, and (c) miscellaneous matter (sand, soil, glass, rust, or other foreign substances). Bacterial contamination is excluded from these substances.

Filth is classified according to its extractability. Light filth is oleophilic and lighter than water (separated from product by floating it in an oil-aqueous mixture). Insect fragments, rodent hairs, and feather barbules are examples of light filth. Heavy filth is heavier than water and separated from the product by sedimentation based on different densities of filth, food particles, and immersion liquids ($CHCl_3$, CCl_4, etc.). Examples of heavy filth are sand, soil, and nutshell fragments. Sieved filth involves particles separated from the product by the use of selected mesh sizes. Whole insects, stones, sticks, and bolts are examples of sieved filth.

Various methods of isolation of extraneous matter from various food commodities can be found in the *Official Methods of Analysis* of the AOAC International and in the *Approved Methods of Analysis* of the AACC International. Presented here are a few procedures for some food commodities, with descriptions based on AOAC methods, but the quantities reduced to half.

29.1.2 Reading Assignment

Dogan H, Subramanyam B (2024) Extraneous matter. Ch 34. In: Ismail BP, Nielsen SS (eds) Nielsen's food analysis, 6th edn. Springer, New York.

29.1.3 Notes

Regulatory examination of samples by the Food and Drug Administration (FDA) would be based on replicate samples using official methods, including the specified sample size. However, for instructional purposes, the costs associated with adequate commercial 1-L Wildman trap flasks, reagents, and food samples specified in official methods may be prohibitive. Procedures given below are based on AOAC methods, but all quantities are reduced to half, and a 500-mL Wildman trap flask (vs. 1-L trap flask) is specified in most procedures. Commercially available 1-L trap flasks with the standard stopper rod would ideally be used (with all quantities in the procedures doubled). However, 500-mL trap flasks can be made for use in this experiment. To do this, drill a hole through a rubber stopper of a size just too large for a 500-mL Erlenmeyer flask. Thread a heavy string through the hole in the stopper, and knot both ends of the string. Coat the sides of the rubber stopper with glycerin and *carefully* force it (with larger end of stopper pointed up) through the top of the flask. Note that the string could be a trap for contaminants such as rodent hair and insect fragments.

For the parts of this laboratory exercise that require filter paper, S&S #8 (Schleicher & Schuell, Inc., Keene, NH) is recommended. It meets the specifications set forth in the AOAC Method 945.75 Extraneous Materials (Foreign Matter) in Products, Isolation Techniques Part B(i), which suggests using "smooth, high wet strength, rapid acting filter paper ruled with oil-, alcohol-, and water-proof lines 5 mm apart." The S&S #8 ruled filter paper is 9 cm in diameter and fits well into the *top* of standard 9-cm *plastic* petri dishes. The *bottom* of the plastic petri dish can be used as a protective cover over the sample filter paper in the top of the petri dish. The top of the plastic petri dish provides a 9 cm flat surface (as opposed to glass petri dishes) for examining the filter paper, making it easier to view the plate without having to continuously refocus the microscope. The 5-mm ruled lines provide a guide for systematically examining and enumerating contaminants on the filter paper at 30× magnification. To obtain a moist surface on which contaminants can be manipulated and observed, apply a small amount of glycerin: 60% alcohol (1:1) solution to the top of the petri dish before transferring the filter paper from the Buchner funnel. Using both overhead and substage lighting with the microscope will assist in identifying contaminants.

29.1.4 Objective

The objective of this laboratory is to utilize techniques to isolate the extraneous matter from various foods: cottage cheese, jam, infant food, potato chips, and citrus juice.

29.1.5 Principle of Methods

Extraneous materials can be separated from food products by particle size, sedimentation, and affinity for oleophilic solutions. Once isolated, extraneous materials can be examined microscopically.

29.2 EXTRANEOUS MATTER IN SOFT CHEESE

29.2.1 Chemicals

	CAS no.	Hazards
Phosphoric acid (H_3PO_4)	7664-38-2	Corrosive

29.2.2 Reagents

- Phosphoric acid solution, 400–500 mL
    Combine 1 part phosphoric acid with 40 parts deionized distilled (dd) water (vol/vol).

29.2.3 Hazards, Precautions, and Waste Disposal

Adhere to normal laboratory safety procedures. Wear safety glasses at all times. Waste likely may be put down the drain using a water rinse, but follow good laboratory practices outlined by environmental health and safety protocols at your institution.

29.2.4 Supplies

- Beaker, 1 L (for phosphoric acid solution)
- Beaker, 600 mL (to boil water)
- Buchner funnel
- Cottage cheese, 115 g
- Filter paper
- Heavy gloves
- Pipette, 10 mL (to prepare phosphoric acid solution)
- Pipette bulb or pump
- Spoon
- Sidearm flask, 500 mL or 1 L
- Stirring rod

- Tap water, ca. 500 mL (boiling)
- Tweezers
- Volumetric flask, 500 mL (to prepare phosphoric acid solution)
- Weighing boat

29.2.5 Equipment

- Hot plate
- Microscope
- Top loading balance
- Water aspirator system

29.2.6 Procedure

(Based on AOAC Method 960.49, Filth in Dairy Products.)

1. Weigh out 115 g cottage cheese and add it to 400–500 mL boiling phosphoric acid solution (1 + 40 mixture) in a 1-L beaker, stirring with a glass stirring rod continuously to disperse the cottage cheese.
2. Filter the mixture through filter paper in a Buchner funnel, using a vacuum created by a water aspirator. Do not let the mixture accumulate on the paper, and continually wash filter with a stream of hot water to prevent clogging. Make sure the cheese mixture is hot as it is filtered. When filtration is impeded, add hot water or phosphoric acid solution (1 + 40 mixture) until the paper clears. (May also use dilute (1–5%) alkali or hot alcohol to aid in filtration.) Resume addition of sample and water until sample is filtered.
3. Examine filter paper microscopically.

29.3 EXTRANEOUS MATTER IN JAM

29.3.1 Chemicals

	CAS no.	Hazards
Heptane (12.5 mL)	142-82-5	Harmful, highly flammable, dangerous for the environment
Hydrochloric acid, concentrated (HCl) (5 mL)	7647-01-0	Corrosive

29.3.2 Hazards, Precautions, and Waste Disposal

Heptane is an extremely flammable liquid; avoid open flames, breathing vapors, and contact with skin. Otherwise, adhere to normal laboratory safety procedures. Wear safety glasses at all times. Dispose of heptane waste as hazardous waste. Other waste may be put down the drain using a water rinse.

29.3.3 Supplies

- Beakers, 250 mL (for weighing jam and heating water) (2)
- Buchner funnel
- Filter paper
- Glass stirring rod
- Graduated cylinder, 100 mL
- Graduated pipette, 10 mL (for heptane)
- Ice water bath (to cool mixture to room temperature)
- Jam, 50 g
- Pipette bulb or pump
- Sidearm flask, 500 mL or 1 L
- Spoon
- Thermometer
- Tweezers
- Volumetric pipette, 5 mL (for conc. HCl)
- Waste jar (for heptane)
- Water, dd, 100 mL (heated to 50 °C)
- Wildman trap flask, 500 mL

29.3.4 Equipment

- Hot plate
- Microscope
- Top loading balance
- Water aspirator system

29.3.5 Procedure

(Based on AOAC Method 950.89, Filth in Jam and Jelly)

1. Empty contents of jam jar into beaker and mix thoroughly with glass stirring rod.
2. Weigh 50 g of jam into a beaker, add ca. 80 mL dd water at 50 °C, transfer to a 500-mL trap flask (use the other ca. 20 mL dd water to help make transfer), add 5 mL conc. HCl, and boil for 5 min.
3. Cool to room temperature (with an ice water bath).
4. Add 12.5 mL heptane and stir thoroughly.
5. Add dd water to a level so heptane is just above rubber stopper when in the "trap" position.
6. Trap off the heptane, and filter the heptane through filter paper in a Buchner funnel using vacuum created by a water aspirator.
7. Examine filter paper microscopically.

29.4 EXTRANEOUS MATTER IN INFANT FOOD

29.4.1 Chemicals

	CAS no.	Hazards
Light mineral oil (10 mL)	8012-95-1	

29.4.2 Hazards, Precautions, and Waste Disposal

Adhere to normal laboratory safety procedures. Wear safety glasses at all times. Waste may be put down the drain using water rinse.

29.4.3 Supplies

- Baby food, ~113 g (1 jar)
- Buchner funnel
- Filter paper
- Glass stirring rod
- Graduated cylinder, 10 or 25 mL
- Pipette bulb or pump
- Sidearm flask, 500 mL or 1 L
- Spoon
- Tweezers
- Volumetric pipette, 10 mL
- Water, deaerated, 500 mL
- Wildman trap flask, 500 mL

29.4.4 Equipment

- Microscope
- Water aspirator system

29.4.5 Procedure

(Based on AOAC Method 970.73, Filth in Pureed Infant Food, A. Light Filth.)

1. Transfer 113 g (one jar) of baby food to a 500-mL trap flask.
2. Add 10 mL of light mineral oil and mix thoroughly.
3. Fill the trap flask with deaerated water (can use dd water) at room temperature.
4. Let stand 30 min, stirring 4–6 times during this period.
5. Trap off mineral oil in a layer above the rubber stopper then filter the mineral oil through filter paper in a Buchner funnel using vacuum created by a water aspirator.
6. Examine filter paper microscopically.

29.5 EXTRANEOUS MATTER IN POTATO CHIPS

29.5.1 Chemicals

	CAS no.	Hazards
Ethanol, 95%	64-17-5	Highly flammable
Heptane (9 mL)	142-82-5	Harmful, highly flammable, dangerous to environment
Petroleum ether (200 mL)	8032-32-4	Harmful, highly flammable, dangerous to environment

29.5.2 Reagents

- Ethanol, 60%, 1 L
    Use 95% ethanol to prepare 1 L of 60% ethanol; dilute 632 mL of 95% ethanol with water to 1 L.

29.5.3 Hazards, Precautions, and Waste Disposal

Petroleum ether, heptane, and ethanol are fire hazards; avoid open flames, breathing vapors, and contact with skin. Otherwise, adhere to normal laboratory safety procedures. Wear safety glasses at all times. Heptane and petroleum ether wastes must be disposed of as hazardous wastes. Other waste may be put down the drain using a water rinse.

29.5.4 Supplies

- Beaker, 400 mL
- Buchner funnel
- Filter paper
- Glass stirring rod
- Graduated cylinder, 1 L (to measure 95% ethanol)
- Ice water bath
- Potato chips, 25 g
- Sidearm flask, 500 mL or 1 L
- Spatula
- Wildman trap flask, 500 mL
- Tweezers
- Volumetric flask, 1 L (to prepare 60% ethanol)
- Waste jars (for heptane and petroleum ether)

29.5.5 Equipment

- Hot plate
- Microscope
- Top loading balance
- Water aspirator system

29.5.6 Procedure

(Based on AOAC Method 955.44, Filth in Potato Chips.)

1. Weigh 25 g of potato chips into a 400-mL beaker.
2. With a spatula or glass stirring rod, crush chips into small pieces.
3. In a hood, add petroleum ether to cover the chips. Let stand 5 min. Decant petroleum ether from the chips through filter paper. Again add petroleum ether to the chips, let stand 5 min and decant through filter paper. Let petroleum ether evaporate from chips in hood.
4. Transfer chips to a 500-mL trap flask, add 125 mL 60% ethanol, and boil for 30 min. Mark initial level of ethanol on flask. During boiling and at the end of boiling, replace ethanol lost by evaporation as a result of boiling.
5. Cool in ice water bath.
6. Add 9 mL heptane, mix, and let stand for 5 min.
7. Add enough 60% ethanol to the flask so that only the heptane layer is above the rubber stopper. Let stand to allow heptane layer to form at the top, trap off the heptane layer, and filter it through filter paper in a Buchner funnel.
8. Add 9 mL more heptane to solution. Mix and then let stand until heptane layer rises to the top. Trap off the heptane layer, and filter it through filter paper (i.e., new piece of filter paper, not piece used in Parts 3 and 4) in a Buchner funnel.
9. Examine the filter paper microscopically.

29.6 EXTRANEOUS MATTER IN CITRUS JUICE

29.6.1 Supplies

- Beaker, 250 mL
- Buchner funnel
- Cheesecloth
- Citrus juice, 125 mL
- Graduated cylinder, 500 mL or 1 L
- Sidearm flask, 250 mL
- Tweezers

29.6.2 Equipment

- Microscope
- Water aspirator system

29.6.3 Procedure

(Based on AOAC Method 970.72, Filth in Citrus and Pineapple Juice (Canned), Method A. Fly Eggs and Maggots)

1. Filter 125 mL of juice through a Buchner funnel fitted with a double layer of cheesecloth. Filter with a vacuum created by a water aspirator. Pour the juice slowly to avoid accumulation of excess pulp on the cheesecloth.

2. Examine material on cheesecloth microscopically for fly eggs and maggots.

29.7   QUESTIONS

1. Summarize the results for each type of food analyzed for extraneous materials.

2. Why are contaminants such as insect fragments found in food, and when does the Pure Food and Drug Act prohibit adulteration?

RESOURCE MATERIALS

AOAC International (2023) Official methods of analysis, 22th edn. (On-line). AOAC International, Rockville, MD

Dogan H, Subramanyam B (2024) Extraneous matter. Ch 34. In: Ismail BP, Nielsen SS (eds) Nielsen's food analysis, 6th edn. Springer, New York

Food Forensics

Jinping Dong

Contents

30.1 INTRODUCTION

30.1.1 Background

Food product safety and quality concerns can be caused by unexpected situations that inevitably happen during food manufacturing (e.g., occasional fault from the operators, physical wear and tear from the equipment and tools, failure in environmental controls, and sudden change in operating conditions). Typical changes that result include introduction of chemical contaminations and foreign materials, change of color and odor, viscosity loss, flocking, sedimentation, watering off, and filter plugging. Food forensics aims to identify the root cause of these issues and find a solution to correct or prevent the recurrence of the problem.

Many forensics cases require fast response from the analytical team because of the interruption to the production or potential harm to the consumers. Scientists need to decide quickly what tools and methods to use based on the nature of the problem. Qualitative information very often is sufficient to give directions to the operations team.

In this exercise, you are asked to help with an urgent problem from the sucrose production plant of your company. There are several foreign materials in the powdered sucrose product they produce. (Details of the situation are explained in Sect. 30.2.2 below.) After conducting your analysis to track down the root cause, you will summarize your findings and make suggestions to the plant manager.

30.1.2 Reading Assignment

Aimutis WR, Mortenson MA, Dong J (2024) Food forensic investigation. Ch. 35. In: Ismail BP, Nielsen SS (ed) Nielsen's food analysis, 6th edn. Springer, New York.

J. Dong (✉)
Core Research & Development, Cargill Research and Development Center, Cargill, Inc., Plymouth, MN, USA
e-mail: jinping_dong@cargill.com

Dogan H, Subramanian, B (2024) Analysis for extraneous matter. Ch. 34. In: Ismail BP, Nielsen SS (ed) Nielsen's food analysis, 6th edn. Springer, New York.

Dong J (2024) Food microstructure techniques. Ch. 32. In: Ismail BP, Nielsen SS (ed) Nielsen's food analysis, 6th edn. Springer, New York.

Nielsen SS (2024) Extraneous matter examination. Ch. 29. In: Ismail BP, Nielsen SS (ed) Nielsen's food analysis laboratory manual, 4th edn. Springer, New York.

Rodriguez-Saona L, Ayvaz H (2024) Infrared and raman spectroscopy. Ch. 8. In: Ismail BP, Nielsen SS (ed) Nielsen's food analysis, 6th edn. Springer, New York.

30.2 ASSESSING FOREIGN MATERIALS IN A POWDERED FOOD PRODUCT

30.2.1 Objective

The objective of this laboratory is to evaluate a powdered sucrose product that has potential safety issues. The main goals are to identify the foreign materials and assess the color change.

30.2.2 Details of Situation and Nature of Samples

As stated in Sect. 30.1.1, a sucrose production plant in your company has asked for urgent help because there are foreign materials found in the final product they produce. The inline x-ray detector sounded an alarm and the quality control check failed in the color test. The plant was shut down and the production will not resume until the root cause can be identified.

Information provided by the production plant included the following:

- A technician noticed some cracks on the clear plastic shield along the conveyor belt.
- A process engineer noticed an error in spray-drying equipment and is currently working to figure out what went wrong.
- Some plastic foreign materials were visible in the product.
- The alarm from the x-ray detector suggested possible metal contamination.
- The quality control team found some damage to the plastic bags used for packaging.

Several samples were provided by the production plant for analysis:

- Degraded/colored sucrose particle powder
- Foreign materials manually picked out of the sucrose product: two plastic films, one plastic fragment, and four metal pieces

- Sucrose particle powder (from normal production)

Given the complexity of the situation, before the lab exercise, the teaching assistant will prepare the materials to be tested by students (Teaching Assistant: Please refer to Sect. 30.2.4 for sample preparation. Students will need to determine which samples are standard quality and which are defective.)

30.2.3 Notes on Methods of Analysis

1. Quantitative information such as the amount of foreign material is not needed from the analysis.
2. There is no official method of analysis for food forensics.
3. Determination of the right tools is the first step for the investigation. Observation of the sample under a light microscope helps with that decision. (See Food Analysis Textbook Chap. 32, Sect. 32.2.2)
4. The first techniques to start with for chemical identification are spectroscopy methods, such as infrared (IR) and Raman, and elemental analysis with inductively coupled plasma (ICP) or energy dispersive spectroscopy (EDS).

30.2.4 Samples to Be Obtained and Prepared by Teaching Assistant

1. Foreign materials (for each student or a group of students):
   Plastic film 1: Saran wrap (10×10 cm^2)
   Plastic film 2: Ziploc bag (10×10 cm^2)
   Plastic fragment: plastic petri dish (5×5 cm^2)
   Metal piece 1: aluminum foil (10×10 cm^2)
   Metal piece 2: a penny
   Metal piece 3: a paper clip
   Metal piece 4: a staple
2. Degraded/colored sucrose sample (for each student or a group of students):
   Use fire or open flame to heat up or burn ~0.5 g of sucrose on an aluminum weighting pan (or aluminum foil) until the color of the crystals turns brown/dark. Avoid completely burning of the material. After the crystals cool down, use tweezers to pick out some particles with yellow, brown, and black colors.

30.2.5 Instrumentation

- Stereo light microscope (1×–10× magnification)
- Fourier transform infrared spectrometer with the attenuated total reflection accessory (FTIR-ATR)

- Optional: Scanning electron microscope with energy dispersive spectroscopy (SEM-EDS)

30.2.6 Hazards, Precautions, and Waste Disposal

Adhere to normal laboratory safety procedures. Wear safety glasses at all times. Waste likely may be put down the drain using a water rinse but follow good laboratory practices outlined by environmental health and safety protocols at your institution.

30.2.7 Supplies

- Acetone
- Aluminum weighing pan (use aluminum foil as replacement)
- Bunsen burner or an open flame
- Deionized (DI) or distilled water
- Degraded/colored sucrose particles (five to ten pieces) (per student or a group of students)
- Foreign materials: two plastic films, one plastic fragment, and four metal pieces (per student or a group of students)
- Glass vials (10 mL)
- Magnet (if no magnet is available, use a magnetic stir bar) (per student or a group of students)
- Gloves
- Microscope glass slide (per student or a group of students)
- Scissors
- Sucrose particles (reference), 100 g
- Tweezers

30.2.8 Procedure

1. Chemical identification of the plastics
   (a) Use scissors to cut a roughly 1×1 cm^2 piece from the provided plastic samples.
      Examine the three plastic samples under the light microscope (take notes on their appearance and other characteristics such as thickness/geometry, stiffness), following the standard operating procedure of the stereo microscope
      (i) Turn on the illumination source of the microscope. Place the sample on a glass slide and move the sample under the objective.
      (ii) Adjust the focus knob until the sample surface can be clearly seen through the eye pieces.
      (iii) Switch to a higher magnification objective to observe details on the sample.

   (b) Use FTIR-ATR to identify the chemical composition, following the standard operating procedure of FTIR:
      (i) Log into the computer, start the control software, and set up collection parameters (spectral resolution 4 cm^{-1}, spectral range 650–4000 cm^{-1}, and scan number of 32).
      (ii) Clean the ATR crystal surface (using DI water and acetone) and collect a background spectrum.
      (iii) Place the sample on the ATR crystal surface, use the clamp to press the sample down, and collect the sample spectrum.
      (iv) Save the data in JPEG and TXT format.

2. Elemental identification of the metals
   (a) Observe the four metal samples under the microscope and take notes on their color and surface morphology.
   (b) Test the flexibility of the aluminum foil, paper clip, and staple by bending the object with fingers or tweezers. Make notes on the test results.
   (c) Use the magnet to test all four metals and make notes of their magnetism.
   (d) If a SEM-EDS instrument is available, determine the elemental composition of the four metal samples, following the standard operating procedures below:
      (i) Start the SEM control program. Vent the sample chamber and wait until the chamber vacuum is gone.
      (ii) Place the metal sample on the aluminum sample holder and mount it to the sample stage.
      (iii) Close the sample chamber and start the vacuum pump. Wait until the system is ready and turn on the electron beam.
      (iv) Choose "Back Scattered" imaging mode, adjust the magnification, and focus to find the surface of the metal sample.
      (v) Turn on the EDS detector and collect the EDS spectrum. Select the automatic elemental identification function to assign all peaks.
      (vi) Save the spectrum in .csv or .dat format.

3. Structural and chemical identification of the degraded/colored sucrose
   (a) Make observations on the colored particles under the light microscope and compare them to the reference sucrose crystals. Make notes on their surface morphology, structural characteristics, and color.
   (b) Place two to three pieces of the colored particles in a glass vial. Add about 2 mL of DI water. Gently shake the vial to observe the solubility of the particles.
   (c) Use FTIR-ATR to examine the chemical composition of the colored particles and compare to the reference sucrose. Compare the two spectra and make notes on the differences.

30.3 FORENSICS REPORT

You are asked to put together a report to summarize the analysis that you performed on the samples. The purpose of the report is to help the plant manager understand the results to then determine the nature of the problem and address the foreign material contamination as well as the color issue. The following is what you want to consider including in the report:

1. Identification of the plastics: Look up the literature (or internet) to find and report the IR spectra for polyethylene and polystyrene and compare the spectra of the three plastic samples to the search results.
2. Potential sources of plastics: Look up and report typical applications of polyethylene and polystyrene, especially in the food industry.
3. Compare and describe your findings for the four metal samples.
4. Make suggestions to the plant manager on the potential sources of these foreign materials (use the information from Sect. 30.2.2).
5. Compare the FTIR spectra between the colored and reference sucrose particles. Describe your microscopy findings. Suggest what may have gone wrong in the spray drying process.
6. Make recommendations for future avoidance of these issues.

30.4 QUESTIONS

1. If all the foreign materials that you analyzed have not been separated from the sucrose product, what are the best approaches to make the separation?
2. How does an x-ray detector work to find foreign materials in food products?
3. What happens (chemically and physically) when sucrose is heated/burnt?
4. Briefly describe how FTIR works. Why can it not be used to analyze metals?

RESOURCE MATERIALS

Aimutis WR, Mortenson MA, Dong J (2024) Food forensic investigation. Ch. 35. In: Ismail BP, Nielsen SS (ed) Nielsen's food analysis, 6th edn. Springer, New York

Dogan H, Subramanian, B (2024) Analysis for extraneous matter. Ch. 34. In: Ismail BP, Nielsen SS (ed) Nielsen's food analysis, 6th edn. Springer, New York

Dong J (2024) Food microstructure techniques. Ch. 32. In: Ismail BP, Nielsen SS (ed) Nielsen's food analysis, 6th edn. Springer, New York

Ismail BP, Nielsen SS (2024) Extraneous matter examination. Ch. 29. In: Ismail BP, Nielsen SS (ed) Nielsen's food analysis laboratory manual, 4th edn. Springer, New York

Rodriguez-Saona L, Ayvaz H (2024) Infrared and Raman spectroscopy. Ch. 8. In: Ismail BP, Nielsen SS (ed) Nielsen's food analysis, 6th edn. Springer, New York

Part III

Answers to Practice Problems

Answers to Practice Problems in Chapter 2, Preparation of Reagents and Buffers

Catrin Tyl and B. Pam Ismail

1. (a) This problem can be solved by using Eq. (2.5): The molecular weight of NaH_2PO_4 is 120 g/mol. Make sure to use the same units throughout. Molecular weights are stated in mol/g, and the unit of concentration is in mol/L; thus, the 500 mL should also be converted into L:

$$m[g] = M\left[\frac{mol}{L}\right] \times v[L] \times MW\left[\frac{g}{mol}\right] \quad (2.5)$$

$$m[g] = 0.1\left[\frac{mol}{L}\right] \times 0.5[L] \times 120\left[\frac{g}{mol}\right] = 6\,g$$

   (b) Just like for Example A2, the only change in the calculation is the use of 156 instead of 120, thus:

$$m[g] = 0.1\left[\frac{mol}{L}\right] \times 0.5[L] \times 156\left[\frac{g}{mol}\right] = 7.8\,g$$

2. According to the definition of % wt/vol, as found in Table 2.1:

$$\%\frac{wt}{v} = \frac{weight\ solute[g] \times 100}{total\ volume[mL]}$$

$$weight\ of\ solute = \%\frac{wt}{v} \times \frac{1}{100} \times volume$$

$$= 10\left[\frac{g}{mL}\right] \times \frac{1}{100} \times 150[mL] = 15\,g$$

3. Determine NaOH's molarity in a 40% wt/vol solution: Due to NaOH's equivalence of 1, molarity equals normality. Use the definition of wt % to obtain the mass of NaOH in 1 L and Eq. (2.9) to calculate normality:

$$\%\frac{w}{v} = \frac{weight\ solute[g] \times 100}{total\ volume[mL]}$$

$$weight\ of\ solute[g] = \%\frac{w}{v} \times \frac{1}{100} \times volume$$

$$= 40\left[\frac{g}{mL}\right] \times \frac{1}{100} \times 1000[mL] = 400[g]$$

$$n[mol] = \frac{m[g]}{MW\left[\frac{g}{mol}\right]} = \frac{400}{40} = 10[mol](in\,1\,L)$$

   Thus, the molarity and the normality of this solution are 10.

4. The number of equivalents for H_2SO_4 is 2, because it can donate 2 H^+, and so the normality is two times the molarity. The mL of NaOH can be found through inserting into Eq. (2.19):

$$mL\ of\ NaOH = \frac{mL\ of\ sulfuric\ acid \times N\ of\ sulfuric\ acid}{N\ of\ NaOH}$$

$$mL\ of\ NaOH = \frac{200 \times 4}{10} = 80\,mL$$

5. For HCl, normality and molarity are equal, because 1 H^+ is released per molecule HCl. The problem can be solved like Example A3 by using Eq. (2.13) to calculate M of concentrated HCl, followed by Eq. (2.15):

$$M\left[\frac{mol}{L}\right] = \frac{d \times 1000\left[\frac{g}{L}\right]}{MW\left[\frac{g}{mol}\right]} \times \%\frac{wt}{wt} \quad (2.13)$$

C. Tyl (✉)
Department of Chemistry, Biotechnology and Food Science, Norwegian University of Life Sciences, As, Norway
e-mail: catrin.tyl@nmbu.no

B. P. Ismail
Department of Food Science and Nutrition, University of Minnesota, St. Paul, MN, USA
e-mail: bismailm@umn.edu

© The Author(s), under exclusive license to Springer Nature Switzerland AG 2024
B. P. Ismail, S. S. Nielsen (eds.), *Nielsen's Food Analysis Laboratory Manual*, Food Science Text Series,
https://doi.org/10.1007/978-3-031-44970-3_31

$$M\left[\frac{mol}{L}\right] = \frac{1.2 \times 1000\left[\frac{g}{L}\right]}{36.5\left[\frac{g}{mol}\right]} \times 0.37 = 12.16\left[\frac{mol}{L}\right]$$

$$M_1 \times v_1 = M_2 \times v_2 \qquad (2.15)$$

v of concentrated HCl[L]

$$= \frac{\text{vol of diluted HCl}[L] \times M \text{ of diluted HCl}\left[\frac{mol}{L}\right]}{M \text{ of concentrated HCl}\left[\frac{mol}{L}\right]}$$

$$v \text{ of concentrated HCl}[L] = \frac{0.25[L] \times 2\left[\frac{mol}{L}\right]}{12.16\left[\frac{mol}{L}\right]}$$

$$= 0.041 \text{ L or } 41 \text{ mL}$$

6. Just like for Examples A3 and A5, determine the molarity of the concentrated acetic acid with Eq. (2.13) (ignore %wt/wt):

$$M \text{ acetic acid} = \frac{1.05 \times 1000}{60.06}\left[\frac{g \times mol}{g \times L}\right] = 17.5\left[\frac{mol}{L}\right]$$

The desired amount is 0.04 mol; thus, take the amount calculated below with Eq. (2.16) and dilute to 1 L:

$$v[L] = \frac{n}{M}\left[\frac{mol \times L}{mol}\right] = \frac{0.04}{17.5} = 0.0023 \text{ L}$$

7. The weight of acetic acid in 1 L of solution can be found analogously to Problem 2:

$$\%\frac{wt}{v} = \frac{\text{weight solute}[g] \times 100}{\text{total volume}[mL]}$$

$$\text{weight of acetic acid} = \%\frac{wt}{v} \times \frac{1}{100} \times \text{volume}$$

$$= 1\left[\frac{g}{mL}\right] \times \frac{1}{100} \times 1000[mL] = 10 \text{ g}$$

The corresponding number of moles of 10 g can be found using Eq. (2.9):

$$n[mol] = \frac{m[g]}{MW\left[\frac{g}{mol}\right]} = \frac{10}{60.02} = 0.167[mol](\text{in } 1 \text{ L})$$

This already answers the question: A 1% acetic acid solution contains 0.167 moles/L, not 0.1 moles/L.

8. This problem can be solved analogously to Problem 14:

$$\%\frac{wt}{v} = \frac{\text{weight solute}[g] \times 100}{\text{total volume}[mL]}$$

$$\text{weight of sodium hydroxide} = \%\frac{wt}{v} \times \frac{1}{100} \times \text{volume}$$

$$= 10\left[\frac{g}{mL}\right] \times \frac{1}{100} \times 1000[mL]$$

$$= 100 \text{ g}$$

The corresponding number of moles of 100 g can be found using Eq. (2.9):

$$n[mol] = \frac{m[g]}{MW\left[\frac{g}{mol}\right]} = \frac{10}{40} = 0.25[mol](\text{in } 1 \text{ L})$$

This already answers the question: a 10% sodium hydroxide solution contains 0.25 moles/L, not 1 mol/L.

9. The normality of $K_2Cr_2O_7$ is six times the molarity. Use Eqs. (2.1) and (2.9) to calculate the molarity and then multiply with the number of equivalents to obtain normality:

$$\text{Molarity}(M)\left[\frac{mol}{L}\right] = \frac{\text{number of moles}(n)[mol]}{\text{vol}(v)[L]} \qquad (2.1)$$

$$n[mol] = \frac{m}{MW}[\times mol] \qquad (2.9)$$

$$M\left[\frac{mol}{L}\right] = \frac{\dfrac{0.2[g]}{294.187\left[\frac{g}{mol}\right]}}{0.1[L]} = 0.0068\left[\frac{mol}{L}\right]$$

$$N\left[\frac{\text{equivalents}}{L}\right] = M \times \text{number of equivalents}$$

$$= 0.0068 \times 6 = 0.04\left[\frac{\text{equivalents}}{L}\right]$$

10. The molecular weight of KHP is 204.22 g/mol, and since it only contains one unionized carboxyl group, its number of equivalents is 1 and its molarity equals its normality. According to Eq. (2.5), 100 mL would contain:

$$m[g] = M\left[\frac{mol}{L}\right] \times v[L] \times MW\left[\frac{g}{mol}\right] \qquad (2.5)$$

$$m[g] = 0.1 \times 0.1 \times 204.22 = 2.0422\,g$$

11. The first step is to find the desired amount of Ca in the 1000 mL. As listed in Table 2.1, ppm corresponds to:

$$ppm = \frac{mg\,solute}{kg\,solution}$$

Thus, $1000\,ppm = \dfrac{1000\,mg\,Ca}{kg\,standard\,solution}$.

For our example, the solution's density can be assumed to be 1, and thus, there would be 1000 mg Ca/L, or 1 g Ca/L. If 110.98 g of $CaCl_2$ contains 40.078 g Ca, then 1 g Ca is supplied by:

$$m[g]\,of\,CaCl_2 = \frac{1 \times 110.98}{40.078} = 2.7691\,g$$

12. First, the correct ratio of sodium acetate/acetic acid needs to be determined; similar to Example C3, $[A^-]$ is expressed through $[AH]$ to have only one unknown quantity in the equation and then inserted into Eq. (2.2):

$$\left[A^-\right] = 0.1 - \left[AH\right]$$

$$5.5 = 4.76 + \log\frac{0.1 - \left[AH\right]}{\left[AH\right]}$$

$$10^{(5.5-4.76)} = \frac{0.1 - \left[AH\right]}{\left[AH\right]}$$

$$5.5 \times \left[AH\right] = 0.1 - \left[AH\right]$$

$$\left[AH\right] = \frac{0.1}{(5.5+1)} = 0.0154 \quad \left[A^-\right] = 0.0846$$

Preparing the buffer by approach 1 (Chap. 2, Sect. 2.3) (1 L of 0.1 M solutions of sodium acetate and acetic acid):

$$m\,of\,sodium\,acetate\,[g] = MW \times M = 82 \times 0.1 = 8.2\,[g]$$

$$v\,of\,acetic\,acid\,[mL] = \frac{MW \times M}{d} = \frac{60.02 \times 0.1}{1.05} = 5.7\,[mL]$$

Mix them so that the resulting concentrations correspond to 0.0846 M of sodium acetate and 0.0154 M of acetic acid using Eq. (2.15):

$$M_1 \times v_1 = M_2 \times v_2 \qquad (2.15)$$

v of sodium acetate solution [mL]

$$= \frac{M\,of\,\left[A^-\right]\,in\,buffer \times v\,of\,buffer}{M\,of\,stock}$$

$$= \frac{0.0846 \times 0.25}{0.1} = 211.5\,[mL]$$

v of acetic acid solution [mL]

$$= \frac{M\,of\,\left[AH\right]\,in\,buffer \times v\,of\,buffer}{M\,of\,stock}$$

$$= \frac{0.0154 \times 0.25}{0.1} = 38.5\,[mL]$$

Preparing the buffer via approach 2 (Chap. 2, Sect. 2.3) (directly dissolve appropriate amounts of sodium acetate and acetic acid in 250 mL):

$$m\,of\,sodium\,acetate\,[g] = M\,of\,\left[A^-\right]\,in\,buffer\left[\frac{mol}{L}\right]$$
$$\times v\,of\,buffer\,[L] \times MW\left[\frac{g}{mol}\right] \quad (2.5)$$

$$m\,of\,sodium\,acetate\,[g] = 0.25\,[\cancel{L}] \times 0.0846\left[\frac{\cancel{mol}}{\cancel{L}}\right]$$
$$\times 82\left[\frac{g}{\cancel{mol}}\right] = 1.73\,[g]$$

$$v\,of\,acetic\,acid\,[mL] = \frac{v\,of\,buffer\,[L] \times M\,of\,\left[AH\right]\,in\,buffer\left[\frac{mol}{L}\right] \times MW\left[\frac{g}{mol}\right]}{\left[\frac{g}{mL}\right]}$$

$$v\,of\,acetic\,acid\,[mL] = \frac{0.25\,[\cancel{L}] \times 0.0154\left[\frac{\cancel{mol}}{\cancel{L}}\right] \times 60.02\left[\frac{\cancel{g}}{\cancel{mol}}\right]}{1.05\left[\frac{\cancel{g}}{mL}\right]} = 2.2\,[mL]$$

These would be dissolved in, for example, 200 mL, and then the pH adjusted and the volume brought up to 250 mL.

Preparing the buffer via approach 3 (Chap. 2, Sect. 2.3) (dissolve the appropriate amount of acetic acid to yield 250 mL of a 0.1 M solution in <250 mL, and then adjust the pH with NaOH of a high molarity, e.g., 6 or 10 M to pH 5.5):

v of acetic acid [mL]

$$= \frac{v\,of\,buffer\,[L] \times M\,of\,buffer\left[\frac{mol}{L}\right] \times MW\left[\frac{g}{mol}\right]}{\left[\frac{g}{mL}\right]}$$

v of acetic acid $[mL]$

$$= \frac{0.25[\text{L}] \times 0.1 \left[\frac{\text{mol}}{\text{L}}\right] \times 60.02 \left[\frac{\text{g}}{\text{mol}}\right]}{1.05 \left[\frac{\text{g}}{\text{mL}}\right]} = 1.43[mL]$$

13. The molarities of Na_2EDTA and $MgSO_4$ are found by rearranging Eq. (2.5):

$$m[g] = M \left[\frac{\text{mol}}{\text{L}}\right] \times v[L] \times MW \left[\frac{\text{g}}{\text{mol}}\right] \qquad (2.5)$$

$$M \left[\frac{\text{mol}}{\text{L}}\right] = \frac{m[g]}{v[L] \times MW \left[\frac{\text{g}}{\text{mol}}\right]}$$

$$M \left[\frac{\text{mol}}{\text{L}}\right] \text{ of } Na_2EDTA = \frac{1.179[\text{g}]}{0.25[L] \times 372.24 \left[\frac{\text{g}}{\text{mol}}\right]}$$

$$= 0.0127 \left[\frac{\text{mol}}{\text{L}}\right]$$

$$M \left[\frac{\text{mol}}{\text{L}}\right] \text{ of } MgSO_4 = \frac{0.78[\text{g}]}{0.25[L] \times 246.47 \left[\frac{\text{g}}{\text{mol}}\right]}$$

$$= 0.0127 \left[\frac{\text{mol}}{\text{L}}\right]$$

To calculate the buffer pH, determine the molarity of NH_4Cl using Eqs. (2.1) and (2.3) and the molarity of NH_3 with Eq. (2.27):

$$M \text{ of } NH_4Cl \text{ in buffer} \left[\frac{\text{mol}}{\text{L}}\right] = \frac{m[g]}{MW \left[\frac{\text{g}}{\text{mol}}\right] \times v[L]}$$

$$= \frac{16.9}{53.49 \times 0.25} = 1.26 \left[\frac{\text{mol}}{\text{L}}\right]$$

$$M \text{ of conc.} NH_3 \left[\frac{\text{mol}}{\text{L}}\right] = \frac{\left[\frac{\text{g}}{\text{mL}}\right] \times 1000}{MW \left[\frac{\text{g}}{\text{mol}}\right]} \times \%wt$$

$$= \frac{0.88 \times 1000}{17} \times 0.28 = 14.5 \left[\frac{\text{mol}}{\text{L}}\right]$$

$$M \text{ in buffer} \left[\frac{\text{mol}}{\text{L}}\right]$$

$$= \frac{M \text{ of stock solutions} \left[\frac{\text{mol}}{\text{L}}\right] \times v \text{ of stock solutions}[L]}{v \text{ of buffer}[L]}$$

$$\qquad (2.27)$$

$$M \text{ of } NH_3 \text{ in buffer} \left[\frac{\text{mol}}{\text{L}}\right]$$

$$= \frac{M \text{ of conc.} NH_3 \left[\frac{\text{mol}}{\text{L}}\right] \times v \text{ of conc.} NH_3 [L]}{v \text{ of buffer}[L]}$$

$$M \text{ of } NH_3 \text{ in buffer} \left[\frac{\text{mol}}{\text{L}}\right]$$

$$= \frac{14.5 \left[\frac{\text{mol}}{\text{L}}\right] \times 0.143[L]}{0.25[L]} = 8.29 \left[\frac{\text{mol}}{\text{L}}\right]$$

The normal form of the Henderson-Hasselbalch equation may be used after calculating the pK_a of NH_4^+. NH_4Cl acts as the acid; NH_4OH is the base (NH_4OH is just another way of writing NH_3 in water). Note that the actual pK_a and pK_b may be slightly different because of the added salts affecting the ionic strength.

$$pK_a \text{ of } NH_4^+ = 14 - pK_b = 14 - 4.74 = 9.26$$

$$pH = 9.26 + \log \frac{8.29}{1.26} = 9.26 + 0.82 = 10.08$$

14. (a) Use Eq. (2.5) to obtain the masses:

$$m[g] = M \left[\frac{\text{mol}}{\text{L}}\right] \times v[L] \times MW \left[\frac{\text{g}}{\text{mol}}\right] \qquad (2.5)$$

$$m[g] \text{ of } NaH_2PO_4.H_2O = 0.2 \times 0.5 \times 138 = 13.8[g]$$

$$m[g] \text{ of } Na_2HPO_4.7H_2O = 0.2 \times 0.5 \times 268 = 26.8[g]$$

(b) Use Eq. (2.38) to express $[A^-]$ through $[AH]$ to substitute into Eq. (2.25); find the pK_a in Table 2.2:

$$0.1 = [A^-] + [AH]$$

$$[A^-] = 0.1 - [AH]$$

$$6.2 = 6.71 + \log \frac{0.1 - AH}{AH}$$

$$-0.51 = \log \frac{0.1 - [AH]}{[AH]}$$

$$0.309 = \frac{0.1 - [AH]}{[AH]}$$

$$[AH] \times (0.309 + 1) = 0.1$$

$$[AH]\left[\frac{mol}{L}\right] = \frac{0.1}{1.219} = 0.0764\left[\frac{mol}{L}\right]\left[A^-\right]\left[\frac{mol}{L}\right]$$

$$= 1 - 0.0764 = 0.0236\left[\frac{mol}{L}\right]$$

The molarities of the stock solutions are 0.2 $\left[\frac{mol}{L}\right]$.
To find the volumes to mix, use Eq. (2.27):

$$M_1 \times v_1 = M_2 \times v_2 \qquad (2.27)$$

v of NaH_2PO_4 stock solution $[L]$
$$= \frac{M \text{ of } NaH_2PO_4 \text{ in buffer} \times v \text{ of buffer}}{M \text{ of stock solution}}$$

v of NaH_2PO_4 stock solution $[L]$
$$= \frac{0.0764 \times 0.2}{0.2} = 0.0764 [L] \text{ or } 76 \text{ mL}$$

v of Na_2HPO_4 stock solution $[L]$
$$= \frac{M \text{ of } Na_2HPO_4 \text{ in buffer} \times v \text{ of buffer}}{M \text{ of stock solution}}$$

$$v \text{ of } NaH_2PO_4 \text{ stock solution} [L] = \frac{0.0236 \times 0.2}{0.2}$$

$$= 0.024 [L] \text{ or } 24 \text{ mL}$$

(c) This problem is similar to Example C3. 1 mL of 6 M NaOH supplies (Eq. 2.2):

$$n \text{ of } NaOH [mol] = M \times v = 6 \times 0.001 = 0.006 [mol]$$

The amounts of NaH_2PO_4 and Na_2HPO_4 are found through Eq. (2.2) (use either the buffer molarity or values from the stock solutions calculated for Problem 12b):

$$n \text{ of } NaH_2PO_4 [mol] = M \times v$$

$$= 0.2 \times 0.076 = 0.015 [mol]$$

$$n \text{ of } NaH_2PO_4 [mol] = M \times v$$

$$= 0.2 \times 0.024 = 0.0048 [mol]$$

Addition of NaOH changes the ratio by increasing the amount of Na_2HPO_4 and decreasing NaH_2PO_4:

$$n \text{ of } NaH_2PO_4 \text{ after } HCl [mol] = 0.015 - 0.006$$

$$= 0.009 [mol]$$

$$n \text{ of } Na_2HPO_4 \text{ after } HCl [mol] = 0.0048 + 0.006$$

$$= 0.0108 [mol]$$

Substitute these values into Eq. (2.25) to find the new pH. (Note that you may insert molar ratios and you do not need to convert to concentrations, because the ratio would stay the same):

$$pH = 6.86 + \log\frac{0.0108}{0.009} = 6.94$$

15. To find the pH at 25 °C, the acid/base ratio needs to be substituted into Eq. (2.25):

$$pH = 8.06 + \log\frac{4}{1}$$

$$pH = 8.06 - 0.6 = 7.46$$

Calculate the pKₐ at 60 °C with Eq. (2.61), then insert into Eq. (2.25):

$$pK_a = 8.06 - \left[0.023 \times (60 - 25)\right] = 7.26$$

$$pH = 7.26 - 0.6 = 6.65$$

16. This problem is analogous to Example C3. Find the ratio of acid to base through Eqs. (2.25) and (2.38), then use Eq. (2.52) to find the volume of formic acid and Eq. (2.5) to find the mass of ammonium formate:

$$0.01 = \left[A^-\right] + \left[AH\right]$$

$$\left[A^-\right] = 0.01 - \left[AH\right]$$

$$3.5 = 3.75 + \log\frac{0.01 - \left[AH\right]}{\left[AH\right]}$$

$$-0.25 = \log\frac{0.01 - \left[AH\right]}{\left[AH\right]}$$

$$0.562 = \frac{0.01 - \left[AH\right]}{\left[AH\right]}$$

$$\left[AH\right] = \frac{0.01}{1.562} = 0.0064 \left[A^-\right] = 0.0036$$

$$v[mL] = \frac{M \times v \times MW}{d} \qquad (2.52)$$

$$v \text{ of formic acid} [\text{mL}] = \frac{M \times v \text{ of buffer} \times \text{MW}}{d}$$

$$m \text{ of ammonium formate} [\text{g}] = M \times v \times \text{MW of buffer}$$

$$= \frac{0.0064 \left[\frac{\cancel{\text{mol}}}{\cancel{\text{L}}} \right] \times 46 \left[\frac{\text{g}}{\cancel{\text{mol}}} \right] \times 1 [\cancel{\text{L}}]}{1.22 \left[\frac{\text{g}}{\text{mL}} \right]} = 0.24 [\text{mL}]$$

$$= 0.0036 \left[\frac{\cancel{\text{mol}}}{\cancel{\text{L}}} \right] \times 63.06 \left[\frac{\text{g}}{\cancel{\text{mol}}} \right] \times 1 [\cancel{\text{L}}] = 0.227 [\text{g}]$$

Andrew P. Neilson

1. A diagram of this scheme is shown *here*.

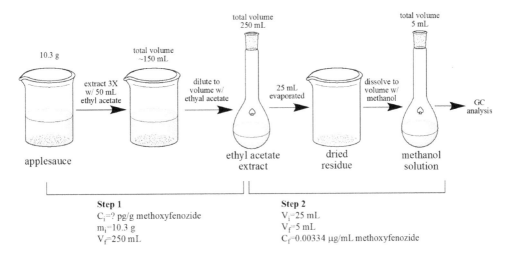

The calculations are performed as follows:

$$C_i = C_f \left(\frac{m \, or \, V_2}{m \, or \, V_1} \right) \left(\frac{m \, or \, V_4}{m \, or \, V_3} \right) \cdots \left(\frac{m \, or \, V_k}{m \, or \, V_{k-1}} \right)$$

$$C_i = \frac{0.00334 \, \mu g \, \text{methoxyfenozide}}{\text{mL methanol solution}} \left(\frac{250 \, \text{mL ethyl acetate extract}}{10.3 \, \text{g applesauce}} \right) \left(\frac{5 \, \text{mL methanol solution}}{25 \, \text{mL ethyl actetate extract}} \right)$$

$$= \frac{0.0162 \, \mu g \, \text{methoxyfenozide}}{\text{g applesauce}}$$

$$X = Cm = \left(\frac{0.0162 \, \mu g \, \text{methoxyfenozide}}{\text{g applesauce}} \right) (113 \, \text{g applesauce}) = 1.83 \, \mu g \, \text{methoxyfenozide}$$

A. P. Neilson (✉)
Department of Food, Bioprocessing and Nutrition Sciences, North Carolina State University, Raleigh, NC, USA
e-mail: aneilso@ncsu.edu

© The Author(s), under exclusive license to Springer Nature Switzerland AG 2024
B. P. Ismail, S. S. Nielsen (eds.), *Nielsen's Food Analysis Laboratory Manual*, Food Science Text Series,
https://doi.org/10.1007/978-3-031-44970-3_32

The concentration of methoxyfenozide in the applesauce is 0.0162 µg/g, and the total amount of methoxyfenozide in the entire applesauce cup is 1.83 µg.

2. A diagram of this scheme is shown *below* (note that the first dilution is "dilute to," while the last two dilutions are "dilute with").

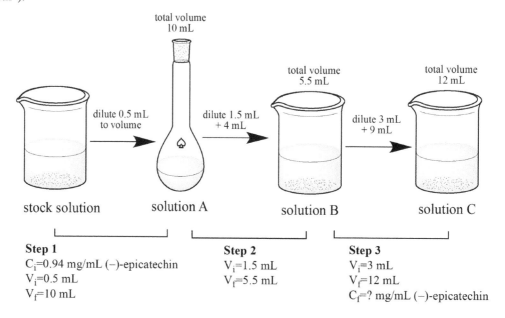

total volume
10 mL

total volume
5.5 mL

total volume
12 mL

dilute 0.5 mL
to volume

dilute 1.5 mL
+ 4 mL

dilute 3 mL
+ 9 mL

stock solution   solution A   solution B   solution C

Step 1
C_i=0.94 mg/mL (–)-epicatechin
V_i=0.5 mL
V_f=10 mL

Step 2
V_i=1.5 mL
V_f=5.5 mL

Step 3
V_i=3 mL
V_f=12 mL
C_f=? mg/mL (–)-epicatechin

The calculations are performed as follows:

$$C_f = C_i \left(\frac{m \text{ or } V_1}{m \text{ or } V_2} \right) \left(\frac{m \text{ or } V_3}{m \text{ or } V_4} \right) \cdots \left(\frac{m \text{ or } V_{k-1}}{m \text{ or } V_k} \right)$$

$$C_f = \frac{0.94 \text{ mg EC}}{\text{mL stock solution}} \left(\frac{0.5 \text{ mL stock solution}}{10 \text{ mL solution A}} \right) \left(\frac{1.5 \text{ mL solution A}}{5.5 \text{ mL solution B}} \right) \left(\frac{3 \text{ mL solution B}}{12 \text{ mL solution C}} \right) = \frac{0.00320 \text{ mg EC}}{\text{mL solution C}}$$

$$C_f = \frac{0.00320 \text{ mg EC}}{\text{mL solution C}} \left(\frac{1 \text{ g EC}}{1000 \text{ mg EC}} \right) \left(\frac{1 \text{ mol EC}}{290.26 \text{ g EC}} \right) \left(\frac{1,000,000 \text{ µmol EC}}{\text{mol EC}} \right) \left(\frac{1000 \text{ mL}}{1 \text{ L}} \right) = \frac{11.0 \text{ µmol EC}}{\text{L}} = 11.0 \, \mu M \text{ EC}$$

$$DF_\Sigma = \frac{C_f}{C_i} = \frac{\dfrac{0.00320 \text{ mg EC}}{\text{mL}}}{\dfrac{0.94 \text{ mg EC}}{\text{mL}}} = 0.00341$$

$$\text{dilution "fold" or "} X \text{"} = \frac{1}{\text{DF}} = \frac{1}{0.00341} = 293$$

The concentration of solution C is 0.00320 mg (–)-epicatechin/mL [or 11.0 µM (–)-epicatechin]. The dilution factor is 0.00341, a 293-fold (293X) dilution of the stock solution.

3. A diagram of this scheme is shown *below* (note that all of these are "dilute with," the standards are prepared in parallel, and 100 µL = 0.1 mL).

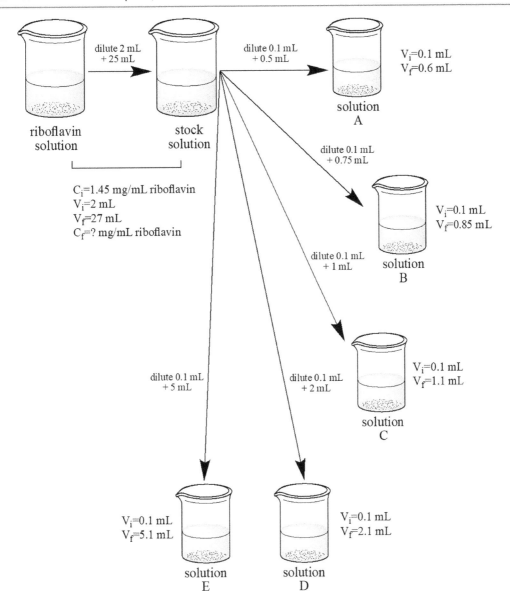

The calculations are performed as shown below. For the stock solution:

$$C_f = C_i\left(\frac{V_i}{V_f}\right) = \frac{1.45\,\text{mg riboflavin}}{\text{mL}}\left(\frac{2\,\text{mL}}{27\,\text{mL}}\right)$$

$$= \frac{0.107\,\text{mg riboflavin}}{\text{mL}}$$

For the standard solutions:

$$A: \quad C_f = C_i\left(\frac{V_i}{V_f}\right) = \frac{0.107\,\text{mg riboflavin}}{\text{mL}}$$

$$\frac{\left(\frac{0.1\,\text{mL}}{0.6\,\text{mL}}\right)}{}$$

$$= \frac{0.0179\,\text{mg riboflavin}}{\text{mL}}$$

$$B: \quad C_f = C_i\left(\frac{V_i}{V_f}\right) = \frac{0.107\,\text{mg riboflavin}}{\text{mL}}$$

$$\frac{\left(\frac{0.1\,\text{mL}}{0.85\,\text{mL}}\right)}{}$$

$$= \frac{0.0126\,\text{mg riboflavin}}{\text{mL}}$$

$$C: \quad C_f = C_i\left(\frac{V_i}{V_f}\right) = \frac{0.107\,\text{mg riboflavin}}{\text{mL}}$$

$$\frac{\left(\frac{0.1\,\text{mL}}{1.1\,\text{mL}}\right)}{}$$

$$= \frac{0.00976\,\text{mg riboflavin}}{\text{mL}}$$

$$D: \quad C_f = C_i\left(\frac{V_i}{V_f}\right) = \frac{0.107\,\text{mg riboflavin}}{\text{mL}}$$
$$\left(\frac{0.1\,\text{mL}}{2.1\,\text{mL}}\right)$$
$$= \frac{0.00511\,\text{mg riboflavin}}{\text{mL}}$$

$$E: \quad C_f = C_i\left(\frac{V_i}{V_f}\right) = \frac{0.107\,\text{mg riboflavin}}{\text{mL}}$$
$$\left(\frac{0.1\,\text{mL}}{5.1\,\text{mL}}\right)$$
$$= \frac{0.00211\,\text{mg riboflavin}}{\text{mL}}$$

The riboflavin concentration in the stock solution and the five standards is 0.107, 0.0179, 0.0126, 0.00976, 0.00511, and 0.00211 mg/mL, respectively.

4. First, 160 g/L = 160 mg/mL. Therefore, large dilutions are needed to get from 160 mg/mL to 0–0.5 mg/mL. From this point, there are several approaches that could be used. One approach would be to dilute the stock solution down to the most concentrated standard (0.5 mg/mL) and then do further dilutions from that concentration. To do this, we calculate the DF:

$$DF = \frac{C_f}{C_i} = \frac{0.5\,\text{mg}/\text{mL}}{160\,\text{mg}/\text{mL}} = \frac{1}{320} = 0.003125$$

From this DF, we can calculate the ratio of volumes needed. Recall that:

$$DF = \frac{V_i}{V_f} = \frac{1}{320}$$

Therefore, we need to find a ratio of volumes equal to 1/320 to do this dilution. You do not have a 320 mL volumetric flask to do a simple 1 mL into 320 mL total dilution. However, you could use a 250 mL volumetric flask for the final volume and calculate the starting volume of the stock solution needed:

$$\frac{V_i}{V_f} = \frac{1}{320}$$

$$V_i = \frac{V_f}{320} = \frac{250\,\text{mL}}{320} = 0.781\,\text{mL}$$

Therefore, the 0.5 mL standard is made by diluting 0.781 mL (using a 1 mL adjustable pipettor) to 250 mL final volume. You could use a variety of different dilutions to get the same final concentrations, as long as you do not use less than 0.2 mL as your starting volume (e.g., you could also dilute 0.313–100 mL and get the same concentration). From the 0.5 mg/mL standard, the other standards (0–0.3 mg/mL) can easily be made in 2 mL volumes by combining variable volumes of the 0.5 mg/mL standard with water to achieve a total volume of 2 mL. These volumes are calculated as follows:

$$C_iV_i = C_fV_f \quad \rightarrow \quad V_i = \frac{C_fV_f}{C_i}$$

For 0 mg/mL:

$$V_i = \frac{C_fV_f}{C_i} = \frac{(0\,\text{mg}/\text{mL})(2\,\text{mL})}{0.5\,\text{mg}/\text{mL}} = 0\,\text{mL}$$

For 0.1 mg/mL:

$$V_i = \frac{C_fV_f}{C_i} = \frac{\left(0.1\frac{\text{mg}}{\text{mL}}\right)(2\,\text{mL})}{0.5\frac{\text{mg}}{\text{mL}}} = 0.4\,\text{mL}$$

and so forth for each solution. The volume of water is the amount needed to bring the starting volume up to 2 mL:

For 0 mg/mL:

$$V_{water} = 2\,\text{mL} - 0\,\text{mL} = 2\,\text{mL}$$

For 0.1 mg/mL:

$$V_{water} = 2\,\text{mL} - 0.4\,\text{mL} = 1.6\,\text{mL}$$

and so forth. The dilutions from the diluted 0.5 mg/mL stock solution are shown in Table 32.1:

Table 32.1 Dilution example for a standard curve

Anthocyanins (mg/mL)	Diluted stock (mL)	Water (mL)	Total volume (mL)
0.5	2	0	2.0
0.3	1.2	0.8	2.0
0.2	0.8	1.2	2.0
0.1	0.4	1.6	2.0
0	0	2	2.0

A diagram of this scheme is shown *here*:

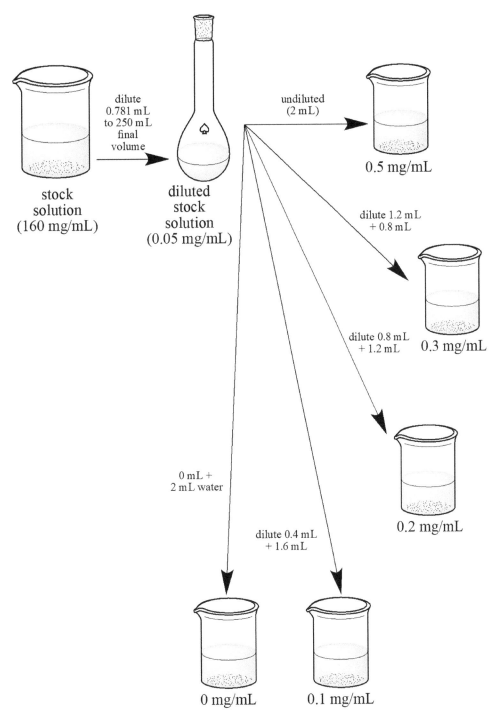

5. We know that the anthocyanin concentration of the juice is probably somewhere between 750 and 3000 μg/mL (0.75–3 mg/mL). However, it could literally be anywhere within this range. Samples with 0.75 and 3 mg/mL anthocyanins would require very different dilutions to get near the center of the standard curve (~0.25 mg/mL). How do we handle this? The solution is to design a dilution scheme with various dilutions of the same sample, which will likely yield at least 1 dilution in the acceptable range that can be used for quantification. To do this, we assume that the anthocyanin concentration in the juice could be on the extremes (~0.75 and ~3 mg/mL) and right in the middle of those extremes (~1.125 mg/mL). Then, we can calculate three different dilutions (assuming the three different

anthocyanin levels and diluting to the middle of the curve) so that we will have at least 1 useable sample regardless of the actual sample concentration. Assume that we want a final sample volume of 10 mL (we could pick any volume that corresponds to a volumetric flask we have available, but it would not make sense to make 50–1000 mL of diluted sample if we only need 2 mL for the analysis):

Assume 0.75 mg/mL:

$$V_i = \frac{C_f V_f}{C_i} = \frac{(0.25\,mg/mL)(10\,mL)}{0.75\,mg/mL} = 3.33\,mL$$

Assume 1.125 mg/mL:

$$V_i = \frac{C_f V_f}{C_i} = \frac{(0.25\,mg/mL)(10\,mL)}{1.125\,mg/mL} = 2.22\,mL$$

Assume 3 mg/mL:

$$V_i = \frac{C_f V_f}{C_i} = \frac{(0.25\,mg/mL)(10\,mL)}{3\,mg/mL} = 0.833\,mL$$

Therefore, we would make three dilutions using 3.33, 2.22, and 0.833 mL of the juice and dilute each to a final volume of 10 mL. We would then analyze the standard solutions and the three dilutions and use the diluted sample (and corresponding dilution factor) with an analytical response within the range of the standard curve to calculate the anthocyanin concentration in the juice.

6. This problem is set up as shown *below*. Note that all of the minerals from 2.8 mL milk are diluted to a total volume of 50 mL, so these steps can be simplified somewhat in the calculation as shown in the figure.

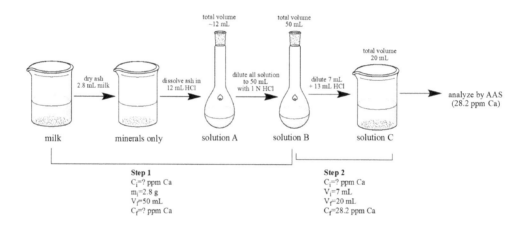

The problem can then be set up and solved as a multi-step dilution as follows:

$$C_i = C_f \left(\frac{m\,or\,V_2}{m\,or\,V_1} \right) \left(\frac{m\,or\,V_4}{m\,or\,V_3} \right) \cdots \left(\frac{m\,or\,V_k}{m\,or\,V_{k-1}} \right)$$

$$C_i = 28.2\,ppm\,Ca \left(\frac{50\,mL}{2.8\,mL} \right) \left(\frac{20\,mL}{7\,mL} \right) = 1440\,ppm\,Ca$$

7. First, you need to get the sample and standards into the same units. Convert the sample caffeine concentration to mM:

$$\frac{170\,mg\,caffeine}{400\,mL} \left(\frac{1000\,mL}{1L} \right) \left(\frac{1g}{1000\,mg} \right)$$
$$\left(\frac{1\,mol\,caffeine}{194.2\,caffeine} \right) \left(\frac{1 \times 10^6\,\mu mol\,caffeine}{1\,mol\,caffeine} \right)$$

$$= \frac{2190\,\mu mol\,caffeine}{L} = 2190\,\mu M\,caffeine$$

Then, determine the starting volume needed to dilute to 250 mL total volume to obtain a concentration at the center of the standard curve (50 μM caffeine):

$$C_i V_i = C_f V_f \rightarrow$$

$$V_i = \frac{C_f V_f}{C_i} = \frac{(50\,\mu M\,caffeine)(250\,mL)}{(2190\,\mu M\,caffeine)} = 5.71\,mL$$

Therefore, if 5.17 mL of the energy drink is diluted to 250 mL final volume, the caffeine concentration will be ~50 μM, which is right in the middle of your standard curve.

Acknowledgment The author thanks Sean F. O'Keefe (Department of Food Science and Technology, Virginia Polytechnic Institute and State University, Blacksburg, VA, USA) for his input on this chapter in the previous edition of the book.

Andrew P. Neilson

1. What is the mean of the observations (mg/cup)?

$$\bar{x} = \frac{\sum x_i}{n} = \frac{x_1 + x_2 + \ldots + x_{n-1} + x_n}{n} = \frac{1967.1\frac{mg}{cup}}{6}$$

$$= 327.85\frac{mg}{cup}; \text{round to } 328\,mg\,/\,cup$$

What is the standard deviation of the observations (mg/cup)? Since we have <30 observations, the correct formula for sample standard deviation is:

$$SD_{n-1} = \sqrt{\frac{\sum(x_i - \bar{x})^2}{n-1}} = \sqrt{\frac{\sum\left(x_i - 327.85\frac{mg}{cup}\right)^2}{6-1}}$$

$$= \sqrt{\frac{460.215\,mg^2\,/\,cup^2}{5}} = 9.593904\,mg\,/\,cup$$

Calculate the 96% confidence interval for the true population mean (use t-score, not Z-score, because n is small). We know that the formula for a CI is:

$$CI: \quad \bar{x} \pm t_{\frac{\alpha}{2}, df = n-1} \times \frac{SD}{\sqrt{n}}$$

We know the mean, SD, and n already, so we just need the t-score. First, calculate df:

$$df = n - 1 = 6 - 1 = 5$$

Next calculate C and then $\alpha/2$:

$$\text{for } 96\%\,CI, C = 0.96 \quad \rightarrow \quad \frac{\alpha}{2} = \frac{1-C}{2} = \frac{1-0.96}{2} = 0.02$$

Then, go to the t-table and find the t-score that corresponds to $df = 5$, $\alpha/2 = 0.02$

$$t_{\frac{\alpha}{2}, df = n-1} = t_{0.02, 5} = 2.757$$

Putting it all together:

$$CI: \quad \bar{x} \pm t_{\frac{\alpha}{2}, df = n-1} \times \frac{SD}{\sqrt{n}}$$

$$327.85\frac{mg}{cup} \pm 2.757 \times \frac{9.593904\frac{mg}{cup}}{\sqrt{6}}$$

$$\rightarrow \quad 327.85\,mg\,/\,cup \pm 10.7983\,mg\,/\,cup$$

What are the upper and lower limits of the 96% confidence interval for the population mean?
The upper limit of the CI is:

$$\bar{x} + t_{\frac{\alpha}{2}, df = n-1} \times \frac{SD}{\sqrt{n}} \quad \rightarrow \quad 327.85\frac{mg}{cup} + 10.7983\frac{mg}{cup}$$

$$= 338.648\,mg\,/\,cup$$

A. P. Neilson (✉)
Department of Food, Bioprocessing and Nutrition Sciences, North Carolina State University, Raleigh, NC, USA
e-mail: aneilso@ncsu.edu

© The Author(s), under exclusive license to Springer Nature Switzerland AG 2024
B. P. Ismail, S. S. Nielsen (eds.), *Nielsen's Food Analysis Laboratory Manual*, Food Science Text Series,
https://doi.org/10.1007/978-3-031-44970-3_33

The lower limit of the CI is:

$$\bar{x} - t_{\frac{\alpha}{2},df=n-1} \times \frac{SD}{\sqrt{n}} \rightarrow 327.85\frac{mg}{cup} - 10.7983\frac{mg}{cup}$$

$$= 317.052\,mg\,/\,cup$$

Use a t-test to determine if the sample mean provides strong enough evidence that the population is "out of spec" (i.e., the actual population mean $\neq 343$ mg cup). Since we have a mean "target" ($\mu = 343$ mg/cup) that we want to compare a sample against, we use a one-sample t-test:

$$t_{obs} = \frac{|\bar{x} - \mu|}{\frac{SD}{\sqrt{n}}}$$

Plugging in the mean, SD, n and μ values:

$$t_{obs} = \frac{|\bar{x} - \mu|}{\frac{SD}{\sqrt{n}}} = \frac{\left|327.85\frac{mg}{cup} - 343\frac{mg}{cup}\right|}{\frac{9.593904\frac{mg}{cup}}{\sqrt{6}}}$$

$$= \frac{|-15.15\,mg\,/\,cup|}{3.91669\,mg\,/\,cup} = 3.868$$

Based on t_{obs}, is there sufficient evidence that the population is "out of spec" (99% confidence)?

The decision rule is that we need to compare $t_{df,\,\alpha/2}$ vs. t_{obs}. Find $t_{df,\,\alpha/2}$.

For 99% confidence:

$$C = 0.99 \rightarrow \frac{\alpha}{2} = \frac{1-C}{2} = \frac{1-0.99}{2} = 0.005$$

$$df = n - 1 = 6 - 1 = 5$$

From the t-score table shown, $t_{5,\,0.005} = 4.032$. Next, compare $t_{df,\,\alpha/2}$ vs. t_{obs}:

> If $t_{obs} > t_{critical} \rightarrow$ there IS strong evidence that the true pop.mean $\neq \mu$

> If $t_{obs} < t_{critical} \rightarrow$ there IS NOT strong evidence that the true pop.mean $\neq \mu$

In this case, since t_{obs} (3.868) $< t_{critical}$ (4.032), there *IS NOT* sufficient evidence that the population is "out of spec" with 99% confidence.

2. You should use a two-sample t-test to determine if the means are statistically different. We need the mean and the SD_{n-1} for each population. Using your calculator or Excel, we get:

Lot A : $n = 5, \bar{x} = 86.98$ and SD $= 0.81670068$

Lot B : $n = 5, \bar{x} = 89.02$ and SD $= 0.192353841$

Next, calculate the pooled variance (s_p^2) from the two sample standard deviations:

$$pooled\ variance = s_p^2 = \frac{(n_1 - 1)SD_1^2 + (n_2 - 1)SD_2^2}{n_1 + n_2 - 2}$$

$$= \frac{(5-1)(0.81670068)^2 + (5-1)(0.192353841)^2}{5 + 5 - 2} = 0.352$$

Now, calculate t_{obs}:

$$t_{obs} = \frac{|\bar{x}_1 - \bar{x}_2|}{\sqrt{s_p^2\left(\frac{1}{n_1} + \frac{1}{n_2}\right)}} = \frac{|86.98 - 89.02|}{\sqrt{0.352 \times \left(\frac{1}{5} + \frac{1}{5}\right)}}$$

$$= \frac{|-2.04|}{\sqrt{0.1408}} = 5.43662$$

Based on the t-test, is there strong evidence that the sample means are statistically different? For a two-sample t-test, the decision rule is:

$$t_{obs} > t_{\frac{\alpha}{2},df=n_1+n_2-2} \rightarrow \text{strong evidence that}$$
$$\text{means are significantly different}$$

$$t_{obs} < t_{\frac{\alpha}{2},df=n_1+n_2-2} \rightarrow \text{insufficient evidence that}$$
$$\text{means are significantly different}$$

$$95\% confidence \rightarrow C = 0.95, \frac{\alpha}{2} = \frac{1-C}{2}$$

$$= \frac{1-0.95}{2} = \frac{0.05}{2} = 0.025$$

$$df = n_1 + n_2 - 2 = 5 = 5 - 2 = 8$$

Then, go to the t-table and find:

$$t_{critical}\left(t_{\frac{\alpha}{2},df=n_1+n_2-2}\right) : t_{0.025,8} = 2.306$$

Since t_{obs} (5.44) $> t_{critical}$ (2.31), there is strong evidence that the means differ (95% confidence).

Acknowledgment The author thanks Sean F. O'Keefe (Department of Food Science and Technology, Virginia Polytechnic Institute and State University, Blacksburg, VA, USA) for his input on this chapter in the previous edition of the book.

9 783031 449697